Engineering a Safer World

Engineering Systems

Editorial Board:

Flexibility in Engineering Design, by Richard de Neufville and Stefan Scholtes, 2011

Engineering a Safer World, by Nancy G. Leveson, 2011

Engineering Systems, by Olivier L. de Weck, Daniel Roos, and Christopher L. Magee, 2011

ENGINEERING A SAFER WORLD

Systems Thinking Applied to Safety

Nancy G. Leveson

The MIT Press
Cambridge, Massachusetts
London, England

For information about special quantity discounts, please email special_sales@mitpress.mit.edu

This book was set in Syntax and Times Roman by Toppan Best-set Premedia Limited. Printed and bound in the United States of America.

Library of Congress Cataloging-in-Publication Data

Leveson, Nancy.
Engineering a safer world : systems thinking applied to safety / Nancy G. Leveson.
   p.   cm. — (Engineering systems)
Includes bibliographical references and index.
ISBN 978-0-262-01662-9 (hardcover : alk. paper)
1. Industrial safety. 2. System safety. I. Title.
T55.L466   2012
620.8′6 — dc23
2011014046

10   9   8   7   6   5   4   3   2   1

We pretend that technology, our technology, is something of a life force, a will, and a thrust of its own, on which we can blame all, with which we can explain all, and in the end by means of which we can excuse ourselves.
—T. Cuyler Young, *Man in Nature*

To all the great engineers who taught me system safety engineering, particularly Grady Lee who believed in me. Also to those who created the early foundations for applying systems thinking to safety, including C. O. Miller and the other American aerospace engineers who created System Safety in the United States, as well as Jens Rasmussen's pioneering work in Europe.

Contents

Series Foreword

Engineering Systems is an emerging field that is at the intersection of engineering, management, and the social sciences. Designing complex technological systems requires not only traditional engineering skills but also knowledge of public policy issues and awareness of societal norms and preferences. In order to meet the challenges of rapid technological change and of scaling systems in size, scope, and complexity, Engineering Systems promotes the development of new approaches, frameworks, and theories to analyze, design, deploy, and manage these systems.

This new academic field seeks to expand the set of problems addressed by engineers, and draws on work in the following fields as well as others:

- Technology and Policy
- Systems Engineering
- System and Decision Analysis, Operations Research
- Engineering Management, Innovation, Entrepreneurship
- Manufacturing, Product Development, Industrial Engineering

The Engineering Systems Series will reflect the dynamism of this emerging field and is intended to provide a unique and effective venue for publication of textbooks and scholarly works that push forward research and education in Engineering Systems.

Series Editorial Board:

Joel Moses, Massachusetts Institute of Technology, Chair

Richard de Neufville, Massachusetts Institute of Technology

Manuel Heitor, Instituto Superior Técnico, Technical University of Lisbon

Granger Morgan, Carnegie Mellon University

Elisabeth Paté-Cornell, Stanford University

William Rouse, Georgia Institute of Technology

Preface

I began my adventure in system safety after completing graduate studies in computer science and joining the faculty of a computer science department. In the first week at my new job, I received a phone call from Marion Moon, a system safety engineer at what was then the Ground Systems Division of Hughes Aircraft Company. Apparently he had been passed between several faculty members, and I was his last hope. He told me about a new problem they were struggling with on a torpedo project, something he called "software safety." I told him I didn't know anything about it and that I worked in a completely unrelated field. I added that I was willing to look into the problem. That began what has been a thirty-year search for a solution and to the more general question of how to build safer systems.

Around the year 2000, I became very discouraged. Although many bright people had been working on the problem of safety for a long time, progress seemed to be stalled. Engineers were diligently performing safety analyses that did not seem to have much impact on accidents. The reason for the lack of progress, I decided, was that the technical foundations and assumptions on which traditional safety engineering efforts are based are inadequate for the complex systems we are building today.

The world of engineering has experienced a technological revolution, while the basic engineering techniques applied in safety and reliability engineering, such as fault tree analysis (FTA) and failure modes and effects analysis (FMEA), have changed very little. Few systems are built without digital components, which operate very differently than the purely analog systems they replace. At the same time, the complexity of our systems and the world in which they operate has also increased enormously. The old safety engineering techniques, which were based on a much simpler, analog world, are diminishing in their effectiveness as the cause of accidents changes.

For twenty years I watched engineers in industry struggling to apply the old techniques to new software-intensive systems—expending much energy and having little success. At the same time, engineers can no longer focus only on technical issues and ignore the social, managerial, and even political factors that impact safety if we are to significantly reduce losses. I decided to search for something new. This book describes the results of that search and the new model of accident causation and system safety techniques that resulted.

The solution, I believe, lies in creating approaches to safety based on modern systems thinking and systems theory. While these approaches may seem new or paradigm changing, they are rooted in system engineering ideas developed after World War II. They also build on the unique approach to engineering for safety, called System Safety, that was pioneered in the 1950s by aerospace engineers such as C. O. Miller, Jerome Lederer, and Willie Hammer, among others. This systems approach to safety was created originally to cope with the increased level of complexity in aerospace systems, particularly military aircraft and ballistic missile systems. Many of these ideas have been lost over the years or have been displaced by the influence of more mainstream engineering practices, particularly reliability engineering.

This book returns to these early ideas and updates them for today's technology. It also builds on the pioneering work in Europe of Jens Rasmussen and his followers in applying systems thinking to safety and human factors engineering.

Our experience to date is that the new approach described in this book is more effective, less expensive, and easier to use than current techniques. I hope you find it useful.

Relationship to *Safeware*

My first book, *Safeware*, presents a broad overview of what is known and practiced in System Safety today and provides a reference for understanding the state of the art. To avoid redundancy, information about basic concepts in safety engineering that appear in *Safeware* is not, in general, repeated. To make this book coherent in itself, however, there is some repetition, particularly on topics for which my understanding has advanced since writing *Safeware*.

Audience

This book is written for the sophisticated practitioner rather than the academic researcher or the general public. Therefore, although references are provided, an attempt is not made to cite or describe everything ever written on the topics or to provide a scholarly analysis of the state of research in this area. The goal is to provide engineers and others concerned about safety with some tools they can use when attempting to reduce accidents and make systems and sophisticated products safer.

It is also written for those who are not safety engineers and those who are not even engineers. The approach described can be applied to any complex, sociotechnical system such as health care and even finance. This book shows you how to "reengineer" your system to improve safety and better manage risk. If preventing potential losses in your field is important, then the answer to your problems may lie in this book.

Contents

The basic premise underlying this new approach to safety is that traditional models of causality need to be extended to handle today's engineered systems. The most common accident causality models assume that accidents are caused by component failure and that making system components highly reliable or planning for their failure will prevent accidents. While this assumption is true in the relatively simple electromechanical systems of the past, it is no longer true for the types of complex sociotechnical systems we are building today. A new, extended model of accident causation is needed to underlie more effective engineering approaches to improving safety and better managing risk.

The book is divided into three sections. The first part explains why a new approach is needed, including the limitations of traditional accident models, the goals for a new model, and the fundamental ideas in system theory upon which the new model is based. The second part presents the new, extended causality model. The final part shows how the new model can be used to create new techniques for system safety engineering, including accident investigation and analysis, hazard analysis, design for safety, operations, and management.

This book has been a long time in preparation because I wanted to try the new techniques myself on real systems to make sure they work and are effective. In order not to delay publication further, I will create exercises, more examples, and other teaching and learning aids and provide them for download from a website in the future.

Chapters 6–10, on system safety engineering and hazard analysis, are purposely written to be stand-alone and therefore usable in undergraduate and graduate system engineering classes where safety is just one part of the class contents and the practical design aspects of safety are the most relevant.

Acknowledgments

The research that resulted in this book was partially supported by numerous research grants over many years from NSF and NASA. David Eckhardt at the NASA Langley Research Center provided the early funding that got this work started.

I also am indebted to all my students and colleagues who have helped develop these ideas over the years. There are too many to list, but I have tried to give them credit throughout the book for the ideas they came up with or we worked on together. I apologize in advance if I have inadvertently not given credit where it is due. My students, colleagues, and I engage in frequent discussions and sharing of ideas, and it is sometimes difficult to determine where the ideas originated. Usually the creation involves a process where we each build on what the other has done. Determining who is responsible for what becomes impossible. Needless to say, they provided invaluable input and contributed greatly to my thinking.

I am particularly indebted to the students who were at MIT while I was writing this book and played an important role in developing the ideas: Nicolas Dulac, Margaret Stringfellow, Brandon Owens, Matthieu Couturier, and John Thomas. Several of them assisted with the examples used in this book.

Other former students who provided important input to the ideas in this book are Matt Jaffe, Elwin Ong, Natasha Neogi, Karen Marais, Kathryn Weiss, David Zipkin, Stephen Friedenthal, Michael Moore, Mirna Daouk, John Stealey, Stephanie Chiesi, Brian Wong, Mal Atherton, Shuichiro Daniel Ota, and Polly Allen.

Colleagues who provided assistance and input include Sidney Dekker, John Carroll, Joel Cutcher-Gershenfeld, Joseph Sussman, Betty Barrett, Ed Bachelder, Margaret-Anne Storey, Meghan Dierks, and Stan Finkelstein.

FOUNDATIONS

1 Why Do We Need Something Different?

This book presents a new approach to building safer systems that departs in important ways from traditional safety engineering. While the traditional approaches worked well for the simpler systems of the past for which they were devised, significant changes have occurred in the types of systems we are attempting to build today and the context in which they are being built. These changes are stretching the limits of safety engineering:

- **Fast pace of technological change:** Although learning from past accidents is still an important part of safety engineering, lessons learned over centuries about designing to prevent accidents may be lost or become ineffective when older technologies are replaced with new ones. Technology is changing much faster than our engineering techniques are responding to these changes. New technology introduces unknowns into our systems and creates new paths to losses.

- **Reduced ability to learn from experience:** At the same time that the development of new technology has sprinted forward, the time to market for new products has greatly decreased, and strong pressures exist to decrease this time even further. The average time to translate a basic technical discovery into a commercial product in the early part of this century was thirty years. Today our technologies get to market in two to three years and may be obsolete in five. We no longer have the luxury of carefully testing systems and designs to understand all the potential behaviors and risks before commercial or scientific use.

- **Changing nature of accidents:** As our technology and society change, so do the causes of accidents. System engineering and system safety engineering techniques have not kept up with the rapid pace of technological innovation. Digital technology, in particular, has created a quiet revolution in most fields of engineering. Many of the approaches to prevent accidents that worked on electromechanical components—such as replication of components to protect against individual component failure—are ineffective in controlling accidents that arise from the use of digital systems and software.

• **New types of hazards:**   Advances in science and societal changes have created new hazards. For example, the public is increasingly being exposed to new man-made chemicals or toxins in our food and our environment. Large numbers of people may be harmed by unknown side effects of pharmaceutical products. Misuse or overuse of antibiotics has given rise to resistant microbes. The most common safety engineering strategies have limited impact on many of these new hazards.

• **Increasing complexity and coupling:**   Complexity comes in many forms, most of which are increasing in the systems we are building. Examples include *interactive complexity* (related to interaction among system components), *dynamic complexity* (related to changes over time), *decompositional complexity* (where the structural decomposition is not consistent with the functional decomposition), and *nonlinear complexity* (where cause and effect are not related in a direct or obvious way). The operation of some systems is so complex that it defies the understanding of all but a few experts, and sometimes even they have incomplete information about the system's potential behavior. The problem is that we are attempting to build systems that are beyond our ability to intellectually manage; increased complexity of all types makes it difficult for the designers to consider all the potential system states or for operators to handle all normal and abnormal situations and disturbances safely and effectively. In fact, complexity can be defined as intellectual unmanageability.

  This situation is not new. Throughout history, inventions and new technology have often gotten ahead of their scientific underpinnings and engineering knowledge, but the result has always been increased risk and accidents until science and engineering caught up.[1] We are now in the position of having to catch up with our technological advances by greatly increasing the power of current approaches to controlling risk and creating new improved risk management strategies.

---

1. As an example, consider the introduction of high-pressure steam engines in the first half of the nineteenth century, which transformed industry and transportation but resulted in frequent and disastrous explosions. While engineers quickly amassed scientific information about thermodynamics, the action of steam in the cylinder, the strength of materials in the engine, and many other aspects of steam engine operation, there was little scientific understanding about the buildup of steam pressure in the boiler, the effect of corrosion and decay, and the causes of boiler explosions. High-pressure steam had made the current boiler design obsolete by producing excessive strain on the boilers and exposing weaknesses in the materials and construction. Attempts to add technological safety devices were unsuccessful because engineers did not fully understand what went on in steam boilers: It was not until well after the middle of the century that the dynamics of steam generation was understood [29].

• **Decreasing tolerance for single accidents:** The losses stemming from accidents are increasing with the cost and potential destructiveness of the systems we build. New scientific and technological discoveries have not only created new or increased hazards (such as radiation exposure and chemical pollution) but have also provided the means to harm increasing numbers of people as the scale of our systems increases and to impact future generations through environmental pollution and genetic damage. Financial losses and lost potential for scientific advances are also increasing in an age where, for example, a spacecraft may take ten years and up to a billion dollars to build, but only a few minutes to lose. Financial system meltdowns can affect the world's economy in our increasingly connected and interdependent global economy. Learning from accidents or major losses (the *fly-fix-fly* approach to safety) needs to be supplemented with increasing emphasis on preventing the first one.

• **Difficulty in selecting priorities and making tradeoffs:** At the same time that potential losses from single accidents are increasing, companies are coping with aggressive and competitive environments in which cost and productivity play a major role in short-term decision making. Government agencies must cope with budget limitations in an age of increasingly expensive technology. Pressures are great to take shortcuts and to place higher priority on cost and schedule risks than on safety. Decision makers need the information required to make these tough decisions.

• **More complex relationships between humans and automation:** Humans are increasingly sharing control of systems with automation and moving into positions of higher-level decision making with automation implementing the decisions. These changes are leading to new types of human error—such as various types of mode confusion—and a new distribution of human errors, for example, increasing errors of omission versus commission [182, 183]. Inadequate communication between humans and machines is becoming an increasingly important factor in accidents. Current approaches to safety engineering are unable to deal with these new types of errors.

All human behavior is influenced by the context in which it occurs, and operators in high-tech systems are often at the mercy of the design of the automation they use or the social and organizational environment in which they work. Many recent accidents that have been blamed on operator error could more accurately be labeled as resulting from flaws in the environment in which they operate. New approaches to reducing accidents through improved design of the workplace and of automation are long overdue.

• **Changing regulatory and public views of safety:** In today's complex and interrelated societal structure, responsibility for safety is shifting from the individual to government. Individuals no longer have the ability to control the risks around them and are demanding that government assume greater responsibility for ensuring public safety through laws and various forms of oversight and regulation as companies struggle to balance safety risks with pressure to satisfy time-to-market and budgetary pressures. Ways to design more effective regulatory strategies without impeding economic goals are needed. The alternative is for individuals and groups to turn to the courts for protection, which has many potential downsides, such as stifling innovation through fear of lawsuits as well as unnecessarily increasing costs and decreasing access to products and services.

Incremental improvements in traditional safety engineering approaches over time have not resulted in significant improvement in our ability to engineer safer systems. A paradigm change is needed in the way we engineer and operate the types of systems and hazards we are dealing with today. This book shows how systems theory and systems thinking can be used to extend our understanding of accident causation and provide more powerful (and surprisingly less costly) new accident analysis and prevention techniques. It also allows a broader definition of safety and accidents that go beyond human death and injury and includes all types of major losses including equipment, mission, financial, and information.

Part I of this book presents the foundation for the new approach. The first step is to question the current assumptions and oversimplifications about the cause of accidents that no longer fit today's systems (if they ever did) and create new assumptions to guide future progress. The new, more realistic assumptions are used to create goals to reach for and criteria against which new approaches can be judged. Finally, the scientific and engineering foundations for a new approach are outlined.

Part II presents a new, more inclusive model of causality, followed by part III, which describes how to take advantage of the expanded accident causality model to better manage safety in the twenty-first century.

2 Questioning the Foundations of Traditional Safety Engineering

It's never what we don't know that stops us. It's what we do know that just ain't so.[1]

Paradigm changes necessarily start with questioning the basic assumptions underlying what we do today. Many beliefs about safety and why accidents occur have been widely accepted without question. This chapter examines and questions some of the most important assumptions about the cause of accidents and how to prevent them that "just ain't so." There is, of course, some truth in each of these assumptions, and many were true for the systems of the past. The real question is whether they still fit today's complex sociotechnical systems and what new assumptions need to be substituted or added.

2.1 Confusing Safety with Reliability

Assumption 1: *Safety is increased by increasing system or component reliability. If components or systems do not fail, then accidents will not occur.*

This assumption is one of the most pervasive in engineering and other fields. The problem is that it's not true. Safety and reliability are *different* properties. One does not imply nor require the other: A system can be reliable but unsafe. It can also be safe but unreliable. In some cases, these two properties even conflict, that is, making the system safer may decrease reliability and enhancing reliability may decrease safety. The confusion on this point is exemplified by the primary focus on failure events in most accident and incident analysis. Some researchers in organizational aspects of safety also make this mistake by suggesting that high *reliability* organizations will be safe [107, 175, 177, 205, 206].

---

1. Attributed to Will Rogers (e.g., *New York Times*, 10/7/84, p. B4), Mark Twain, and Josh Billings (*Oxford Dictionary of Quotations*, 1979, p. 49), among others.

Because this assumption about the equivalence between safety and reliability is so widely held, the distinction between these two properties needs to be carefully considered. First, let's consider accidents where none of the system components fail.

Reliable but Unsafe

In complex systems, accidents often result from interactions among components that are all satisfying their individual requirements, that is, they have *not* failed. The loss of the Mars Polar Lander was attributed to noise (spurious signals) generated when the landing legs were deployed during the spacecraft's descent to the planet surface [95]. This noise was normal and expected and did not represent a failure in the landing leg system. The onboard software interpreted these signals as an indication that landing had occurred (which the software engineers were told such signals would indicate) and shut down the descent engines prematurely, causing the spacecraft to crash into the Mars surface. The landing legs and the software performed correctly (as specified in their requirements) and reliably, but the accident occurred because the system designers did not account for all the potential interactions between landing leg deployment and the descent engine control software.

The Mars Polar Lander loss is a *component interaction accident*. Such accidents arise in the interactions among system components (electromechanical, digital, human, and social) rather than in the failure of individual components. In contrast, the other main type of accident, *a component failure accident*, results from component failures, including the possibility of multiple and cascading failures. In component failure accidents, the failures are usually treated as random phenomena. In component interaction accidents, there may be no failures and the system design errors giving rise to unsafe behavior are not random events.

A *failure* in engineering can be defined as the non-performance or inability of a component (or system) to perform its intended function. Intended function (and thus failure) is defined with respect to the component's behavioral requirements. If the behavior of a component satisfies its specified requirements (such as turning off the descent engines when a signal from the landing legs is received), even though the requirements may include behavior that is undesirable from a larger system context, that component has *not* failed.

Component failure accidents have received the most attention in engineering, but component interaction accidents are becoming more common as the complexity of our system designs increases. In the past, our designs were more intellectually manageable, and the potential interactions among components could be thoroughly planned, understood, anticipated, and guarded against [155]. In addition, thorough testing was possible and could be used to eliminate design errors before use. Modern, high-tech systems no longer have these properties, and system design errors are

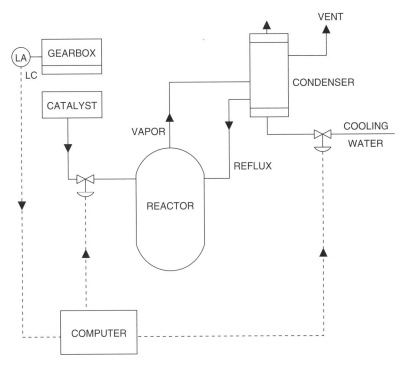

Figure 2.1
A chemical reactor design (adapted from Kletz [103, p. 6]).

increasingly the cause of major accidents, even when all the components have operated reliably—that is, the components have not failed.

Consider another example of a component interaction accident that occurred in a batch chemical reactor in England [103]. The design of this system is shown in figure 2.1. The computer was responsible for controlling the flow of catalyst into the reactor and also the flow of water into the reflux condenser to cool off the reaction. Additionally, sensor inputs to the computer were supposed to warn of any problems in various parts of the plant. The programmers were told that if a fault occurred in the plant, they were to leave all controlled variables as they were and to sound an alarm.

On one occasion, the computer received a signal indicating a low oil level in a gearbox. The computer reacted as the requirements specified: It sounded an alarm and left everything as it was. By coincidence, a catalyst had just been added to the reactor, but the computer had only started to increase the cooling-water flow to the reflux condenser; the flow was therefore kept at a low rate. The reactor over-heated, the relief valve lifted, and the content of the reactor was discharged into the atmosphere.

Note that there were no component failures involved in this accident: the individual components, including the software, worked as specified, but together they created a hazardous system state. The problem was in the overall system design. Merely increasing the reliability of the individual components or protecting against their failure would not have prevented this accident because none of the components failed. Prevention required identifying and eliminating or mitigating unsafe interactions among the system components. High component reliability does not prevent component interaction accidents.

Safe but Unreliable

Accidents like the Mars Polar Lander or the British batch chemical reactor losses, where the cause lies in dysfunctional interactions of non-failing, reliable components—i.e., the problem is in the overall system design—illustrate reliable components in an unsafe system. There can also be safe systems with unreliable components if the system is designed and operated so that component failures do not create hazardous system states. Design techniques to prevent accidents are described in chapter 16 of *Safeware*. One obvious example is systems that are fail-safe, that is, they are designed to fail into a safe state.

For an example of behavior that is unreliable but safe, consider human operators. If operators do not follow the specified procedures, then they are not operating reliably. In some cases, that can lead to an accident. In other cases, it may prevent an accident when the specified procedures turn out to be unsafe under the particular circumstances existing at that time. Examples abound of operators ignoring prescribed procedures in order to prevent an accident [115, 155]. At the same time, accidents have resulted precisely because the operators *did* follow the predetermined instructions provided to them in their training, such as at Three Mile Island [115]. When the results of deviating from procedures are positive, operators are lauded, but when the results are negative, they are punished for being "unreliable." In the successful case (deviating from specified procedures averts an accident), their behavior is unreliable but safe. It satisfies the behavioral safety constraints for the system, but not individual reliability requirements with respect to following specified procedures.

It may be helpful at this point to provide some additional definitions. *Reliability* in engineering is defined as the probability that something satisfies its specified behavioral requirements over time and under given conditions—that is, it does not fail [115]. Reliability is often quantified as *mean time between failure*. Every hardware component (and most humans) can be made to "break" or fail given some set of conditions or a long enough time. The limitations in time and operating conditions in the definition are required to differentiate between (1) unreliability under the assumed operating conditions and (2) situations where no component or component design could have continued to operate.

If a driver engages the brakes of a car too late to avoid hitting the car in front, we would not say that the brakes "failed" because they did not stop the car under circumstances for which they were not designed. The brakes, in this case, were *not* unreliable. They operated reliably but the requirements for safety went beyond the capabilities of the brake design. Failure and reliability are always related to requirements and assumed operating (environmental) conditions. If there are no requirements either specified or assumed, then there can be no failure as any behavior is acceptable and no unreliability.

Safety, in contrast, is defined as the absence of accidents, where an accident is an event involving an unplanned and unacceptable loss [115]. To increase safety, the focus should be on eliminating or preventing hazards, not eliminating failures. Making all the components highly reliable will not necessarily make the system safe.

Conflicts between Safety and Reliability

At this point you may be convinced that reliable *components* are not enough for system safety. But surely, if the *system* as a whole is reliable it will be safe and vice versa, if the system is unreliable it will be unsafe. That is, reliability and safety are the same thing at the system level, aren't they? This common assumption is also untrue. A chemical plant may very reliably manufacture chemicals while occasionally (or even continually) releasing toxic materials into the surrounding environment. The plant is reliable but unsafe.

Not only are safety and reliability not the same thing, but they sometimes conflict: Increasing reliability may decrease safety and increasing safety may decrease reliability. Consider the following simple example in physical design. Increasing the working pressure to burst ratio (essentially the strength) of a tank will make the tank more reliable, that is, it will increase the mean time between failure. When a failure does occur, however, more serious damage may result because of the higher pressure at the time of the rupture.

Reliability and safety may also conflict in engineering design when a choice has to be made between retreating to a fail-safe state (and protecting people and property) versus attempting to continue to achieve the system objectives but with increased risk of an accident.

Understanding the conflicts between reliability and safety requires distinguishing between requirements and constraints. Requirements are derived from the mission or reason for the existence of the organization. The mission of the chemical plant is to produce chemicals. Constraints represent acceptable ways the system or organization can achieve the mission goals. Not exposing bystanders to toxins and not polluting the environment are constraints on the way the mission (producing chemicals) can be achieved.

While in some systems safety is part of the mission or reason for existence, such as air traffic control or healthcare, in others safety is not the mission but instead is a constraint on how the mission can be achieved. The best way to ensure the constraints are enforced in such a system may be not to build or operate the system at all. Not building a nuclear bomb is the surest protection against accidental detonation. We may be unwilling to make that compromise, but some compromise is almost always necessary: The most effective design protections (besides not building the bomb at all) against accidental detonation also decrease the likelihood of detonation when it is required.

Not only do safety constraints sometimes conflict with mission goals, but the safety requirements may even conflict among themselves. One safety constraint on an automated train door system, for example, is that the doors must not open unless the train is stopped and properly aligned with a station platform. Another safety constraint is that the doors must open anywhere for emergency evacuation. Resolving these conflicts is one of the important steps in safety and system engineering.

Even systems with mission goals that include assuring safety, such as air traffic control (ATC), usually have other conflicting goals. ATC systems commonly have the mission to both increase system throughput and ensure safety. One way to increase throughput is to decrease safety margins by operating aircraft closer together. Keeping the aircraft separated adequately to assure acceptable risk may decrease system throughput.

There are always multiple goals and constraints for any system—the challenge in engineering design and risk management is to identify and analyze the conflicts, to make appropriate tradeoffs among the conflicting requirements and constraints, and to find ways to increase system safety without decreasing system reliability.

Safety versus Reliability at the Organizational Level

So far the discussion has focused on safety versus reliability at the physical level. But what about the social and organizational levels above the physical system? Are safety and reliability the same here as implied by High Reliability Organization (HRO) advocates who suggest that High Reliability Organizations (HROs) will be safe? The answer, again, is no [124].

Figure 2.2 shows Rasmussen's analysis of the Zeebrugge ferry mishap [167]. Some background is necessary to understand the figure. On the day the ferry capsized, the *Herald of Free Enterprise* was working the route between Dover and the Belgium port of Bruges–Zeebrugge. This route was not her normal one, and the linkspan[2] at Zeebrugge had not been designed specifically for the Spirit type of ships. The linkspan used spanned a single deck and so could not be used to load decks E and G simultaneously. The ramp could also not be raised high enough to meet the level of

---

2. A *linkspan* is a type of drawbridge used in moving vehicles on and off ferries or other vessels.

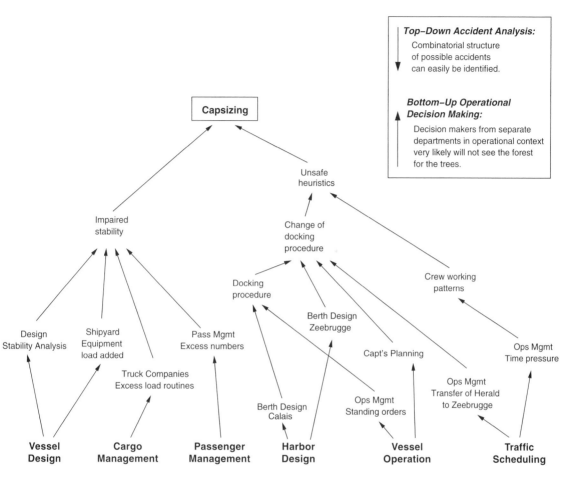

Figure 2.2
The complex interactions in the Zeebrugge accident (adapted from Rasmussen [167, p. 188]).

deck E due to the high spring tides at that time. This limitation was commonly known and was overcome by filling the forward ballast tanks to lower the ferry's bow in the water. The *Herald* was due to be modified during its refit later that year to overcome this limitation in the ship's design.

Before dropping moorings, it was normal practice for a member of the crew, the assistant boatswain, to close the ferry doors. The first officer also remained on deck to ensure they were closed before returning to the wheelhouse. On the day of the accident, in order to keep on schedule, the first officer returned to the wheelhouse before the ship dropped its moorings (which was common practice), leaving the closing of the doors to the assistant boatswain, who had taken a short break after cleaning the car deck upon arrival at Zeebrugge. He had returned to his cabin and was still asleep when the ship left the dock. The captain could only assume that the doors had been closed because he could not see them from the wheelhouse due to their construction, and there was no indicator light in the wheelhouse to show door position. Why nobody else closed the door is unexplained in the accident report.

Other factors also contributed to the loss. One was the depth of the water: if the ship's speed had been below 18 knots (33 km/h) and the ship had not been in shallow water, it was speculated in the accident report that the people on the car deck would probably have had time to notice the bow doors were open and close them [187]. But open bow doors were not alone enough to cause the final capsizing. A few years earlier, one of the *Herald*'s sister ships sailed from Dover to Zeebrugge with the bow doors open and made it to her destination without incident.

Almost all ships are divided into watertight compartments below the waterline so that in the event of flooding, the water will be confined to one compartment, keeping the ship afloat. The *Herald*'s design had an open car deck with no dividers, allowing vehicles to drive in and out easily, but this design allowed water to flood the car deck. As the ferry turned, the water on the car deck moved to one side and the vessel capsized. One hundred and ninety three passengers and crew were killed.

In this accident, those making decisions about vessel design, harbor design, cargo management, passenger management, traffic scheduling, and vessel operation were unaware of the impact (side effects) of their decisions on the others and the overall impact on the process leading to the ferry accident. Each operated "reliably" in terms of making decisions based on the information they had.

Bottom-up decentralized decision making can lead—and has led—to major accidents in complex sociotechnical systems. Each local decision may be "correct" in the limited context in which it was made but lead to an accident when the independent decisions and organizational behaviors interact in dysfunctional ways.

Safety is a system property, not a component property, and must be controlled at the system level, not the component level. We return to this topic in chapter 3.

Assumption 1 is clearly untrue. A new assumption needs to be substituted:

New Assumption 1: *High reliability is neither necessary nor sufficient for safety.*

Building safer systems requires going beyond the usual focus on component failure and reliability to focus on system hazards and eliminating or reducing their occurrence. This fact has important implications for analyzing and designing for safety. Bottom-up reliability engineering analysis techniques, such as failure modes and effects analysis (FMEA), are not appropriate for safety analysis. Even top-down techniques, such as fault trees, if they focus on component failure, are not adequate. Something else is needed.

2.2 Modeling Accident Causation as Event Chains

Assumption 2: *Accidents are caused by chains of directly related events. We can understand accidents and assess risk by looking at the chain of events leading to the loss.*

Some of the most important assumptions in safety lie in our models of how the world works. Models are important because they provide a means for understanding phenomena like accidents or potentially hazardous system behavior and for recording that understanding in a way that can be communicated to others.

A particular type of model, an *accident causality model* (or *accident model* for short) underlies all efforts to engineer for safety. Our accident models provide the foundation for (1) investigating and analyzing the cause of accidents, (2) designing to prevent future losses, and (3) assessing the risk associated with using the systems and products we create. Accident models explain why accidents occur, and they determine the approaches we take to prevent them. While you might not be consciously aware you are using a model when engaged in these activities, some (perhaps subconscious) model of the phenomenon is always part of the process.

All models are abstractions; they simplify the thing being modeled by abstracting away what are assumed to be irrelevant details and focusing on the features of the phenomenon that are judged to be the most relevant. Selecting some factors as relevant and others as irrelevant is, in most cases, arbitrary and entirely the choice of the modeler. That choice, however, is critical in determining the usefulness and accuracy of the model in predicting future events.

An underlying assumption of all accident models is that there are common patterns in accidents and that they are not simply random events. Accident models impose patterns on accidents and influence the factors considered in any safety analysis. Because the accident model influences what cause(s) is ascribed to an accident, the countermeasures taken to prevent future accidents, and the evaluation of the risk in operating a system, the power and features of the accident model used will greatly affect our ability to identify and control hazards and thus prevent accidents.

The earliest formal accident models came from *industrial safety* (sometimes called *occupational safety*) and reflect the factors inherent in protecting workers from injury or illness. Later, these same models or variants of them were applied to the engineering and operation of complex technical and social systems. At the beginning, the focus in industrial accident prevention was on unsafe conditions, such as open blades and unprotected belts. While this emphasis on preventing unsafe conditions was very successful in reducing workplace injuries, the decrease naturally started to slow down as the most obvious hazards were eliminated. The emphasis

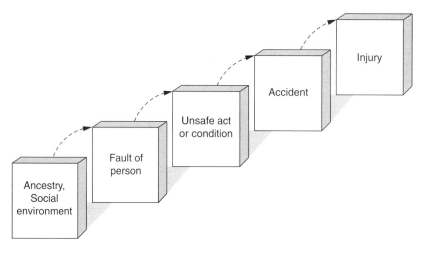

Figure 2.3
Heinrich's Domino Model of Accidents.

then shifted to unsafe acts: Accidents began to be regarded as someone's fault rather than as an event that could have been prevented by some change in the plant or product.

Heinrich's Domino Model, published in 1931, was one of the first published general accident models and was very influential in shifting the emphasis in safety to human error. Heinrich compared the general sequence of accidents to five dominoes standing on end in a line (figure 2.3). When the first domino falls, it automatically knocks down its neighbor and so on until the injury occurs. In any accident sequence, according to this model, ancestry or social environment leads to a fault of a person, which is the proximate reason for an unsafe act or condition (mechanical or physical), which results in an accident, which leads to an injury. In 1976, Bird and Loftus extended the basic Domino Model to include management decisions as a factor in accidents:

1. Lack of control by management, permitting
2. Basic causes (personal and job factors) that lead to
3. Immediate causes (substandard practices/conditions/errors), which are the proximate cause of
4. An accident or incident, which results in
5. A loss

In the same year, Adams suggested a different management-augmented model that included:

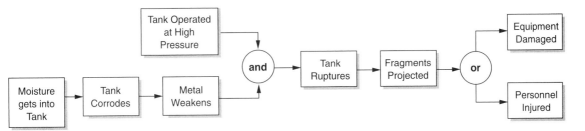

Figure 2.4
A model of the chain of events leading to the rupture of a pressurized tank (adapted from Hammer [79]). Moisture leads to corrosion, which causes weakened metal, which together with high operating pressures causes the tank to rupture, resulting in fragments being projected, and finally leading to personnel injury and/or equipment failure.

1. Management structure (objectives, organization, and operations)

2. Operational errors (management or supervisory behavior)

3. Tactical errors (caused by employee behavior and work conditions)

4. Accident or incident

5. Injury or damage to persons or property

Reason reinvented the Domino Model twenty years later in what he called the Swiss Cheese model, with layers of Swiss cheese substituted for dominos and the layers or dominos labeled as layers of defense[3] that have failed [172, 173].

The basic Domino Model is inadequate for complex systems and other models were developed (see *Safeware* [115], chapter 10), but the assumption that there is a single or *root cause* of an accident unfortunately persists as does the idea of dominos (or layers of Swiss cheese) and chains of failures, each directly causing or leading to the next one in the chain. It also lives on in the emphasis on human error in identifying accident causes.

The most common accident models today explain accidents in terms of multiple events sequenced as a forward chain over time. The events included almost always involve some type of failure" event or human error, or they are energy related (for example, an explosion). The chains may be branching (as in fault trees) or there may be multiple chains synchronized by time or common events. Lots of notations have been developed to represent the events in a graphical form, but the underlying model is the same. Figure 2.4 shows an example for the rupture of a pressurized tank.

---

3. Designing layers of defense is a common safety design approach used primarily in the process industry, particularly for nuclear power. Different design approaches are commonly used in other industries.

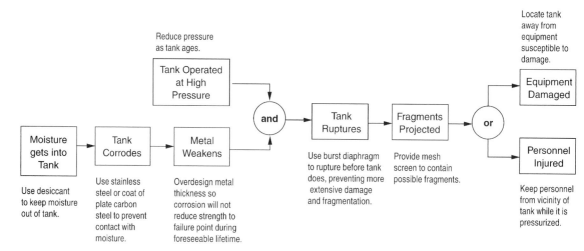

Figure 2.5
The pressurized tank rupture event chain along with measures that could be taken to "break" the chain by preventing individual events in it.

The use of event-chain models of causation has important implications for the way engineers design for safety. If an accident is caused by a chain of events, then the most obvious preventive measure is to break the chain before the loss occurs. Because the most common events considered in these models are component failures, preventive measures tend to be focused on preventing failure events—increasing component integrity or introducing redundancy to reduce the likelihood of the event occurring. If corrosion can be prevented in the tank rupture accident, for example, then the tank rupture is averted.

Figure 2.5 is annotated with mitigation measures designed to break the chain. These mitigation measures are examples of the most common design techniques based on event-chain models of accidents, such as barriers (for example, preventing the contact of moisture with the metal used in the tank by coating it with plate carbon steel or providing mesh screens to contain fragments), interlocks (using a burst diaphragm), overdesign (increasing the metal thickness), and operational procedures (reducing the amount of pressure as the tank ages).

For this simple example involving only physical failures, designing to prevent such failures works well. But even this simple example omits any consideration of factors indirectly related to the events in the chain. An example of a possible indirect or systemic example is competitive or financial pressures to increase efficiency that could lead to not following the plan to reduce the operating pressure as the tank ages. A second factor might be changes over time to the plant design that require workers to spend time near the tank while it is pressurized.

Figure 2.6
Conditions cause events, which lead to new conditions, which cause further events . . .

Formal and informal notations for representing the event chain may contain only the events or they may also contain the conditions that led to the events. Events create conditions that, along with existing conditions, lead to events that create new conditions, and so on (figure 2.6). The *tank corrodes* event leads to a *corrosion exists in tank* condition, which leads to a *metal weakens* event, which leads to a *weakened metal* condition, and so forth.

The difference between events and conditions is that events are limited in time, while conditions persist until some event occurs that results in new or changed conditions. For example, the three conditions that must exist before a flammable mixture will explode (the event) are the flammable gases or vapors themselves, air, and a source of ignition. Any one or two of these may exist for a period of time before the other(s) occurs and leads to the explosion. An event (the explosion) creates new conditions, such as uncontrolled energy or toxic chemicals in the air.

Causality models based on event chains (or dominos or layers of Swiss cheese) are simple and therefore appealing. But they are too simple and do not include what is needed to understand why accidents occur and how to prevent them. Some important limitations include requiring direct causality relationships, subjectivity in selecting the events to include, subjectivity in identifying chaining conditions, and exclusion of systemic factors.

2.2.1   Direct Causality

The causal relationships between the events in event chain models (or between dominoes or Swiss cheese slices) are required to be direct and linear, representing the notion that the preceding event must have occurred and the linking conditions must have been present for the subsequent event to occur: if event A had not occurred then the following event B would not have occurred. As such, event chain models encourage limited notions of linear causality, and it is difficult or impossible to incorporate nonlinear relationships. Consider the statement "Smoking causes lung cancer." Such a statement would not be allowed in the event-chain model of causality because there is no direct relationship between the two. Many smokers do not get lung cancer, and some people who get lung cancer are not smokers. It is widely accepted, however, that there is some relationship between the two, although it may be quite complex and nonlinear.

In addition to limitations in the types of causality considered, the causal factors identified using event-chain models depend on the events that are considered and on the selection of the conditions that link the events. Other than the physical events immediately preceding or directly involved in the loss, however, the choice of events to include is subjective and the conditions selected to explain the events is even more so. Each of these two limitations is considered in turn.

2.2.2   Subjectivity in Selecting Events

The selection of events to include in an event chain is dependent on the stopping rule used to determine how far back the sequence of explanatory events goes. Although the first event in the chain is often labeled the *initiating event* or *root cause*, the selection of an initiating event is arbitrary and previous events and conditions could always be added.

Sometimes the initiating event is selected (the backward chaining stops) because it represents a type of event that is familiar and thus acceptable as an explanation for the accident or it is a deviation from a standard [166]. In other cases, the initiating event or root cause is chosen because it is the first event in the backward chain for which it is felt that something can be done for correction.[4]

The backward chaining may also stop because the causal path disappears due to lack of information. Rasmussen suggests that a practical explanation for why actions by operators actively involved in the dynamic flow of events are so often identified as the cause of an accident is the difficulty in continuing the backtracking "through" a human [166].

A final reason why a "root cause" may be selected is that it is politically acceptable as the identified cause. Other events or explanations may be excluded or not examined in depth because they raise issues that are embarrassing to the organization or its contractors or are politically unacceptable.

The accident report on a friendly fire shootdown of a U.S. Army helicopter over the Iraqi no-fly zone in 1994, for example, describes the chain of events leading to the shootdown. Included in these events is the fact that the helicopter pilots did not change to the radio frequency required in the no-fly zone when they entered it (they stayed on the enroute frequency). Stopping at this event in the chain (which the official report does), it appears that the helicopter pilots were partially at fault for the loss by not following radio procedures. An independent account of the accident [159], however, notes that the U.S. commander of the operation had made

---

4. As an example, a NASA Procedures and Guidelines document (NPG 8621 Draft 1) defines a root cause as: "Along a chain of events leading to an mishap, the first causal action or failure to act that could have been controlled systematically either by policy/practice/procedure or individual adherence to policy/practice/procedure."

an exception about the radio frequency to be used by the helicopters in order to mitigate a different safety concern (see chapter 5), and therefore the pilots were simply following orders when they did not switch to the "required" frequency. The command to the helicopter pilots not to follow official radio procedures is not included in the chain of events provided in the official government accident report, but it suggests a very different understanding of the role of the helicopter pilots in the loss.

In addition to a *root* cause or causes, some events or conditions may be identified as *proximate* or *direct* causes while others are labeled as *contributory*. There is no more basis for this distinction than the selection of a root cause.

Making such distinctions between causes or limiting the factors considered can be a hindrance in learning from and preventing future accidents. Consider the following aircraft examples.

In the crash of an American Airlines DC-10 at Chicago's O'Hare Airport in 1979, the U.S. National Transportation Safety Board (NTSB) blamed only a "maintenance-induced crack," and not also a design error that allowed the slats to retract if the wing was punctured. Because of this omission, McDonnell Douglas was not required to change the design, leading to future accidents related to the same design flaw [155].

Similar omissions of causal factors in aircraft accidents have occurred more recently. One example is the crash of a China Airlines A300 on April 26, 1994, while approaching the Nagoya, Japan, airport. One of the factors involved in the accident was the design of the flight control computer software. Previous incidents with the same type of aircraft had led to a Service Bulletin being issued for a modification of the two flight control computers to fix the problem. But because the computer problem had not been labeled a "cause" of the previous incidents (for perhaps at least partially political reasons), the modification was labeled *recommended* rather than *mandatory*. China Airlines concluded, as a result, that the implementation of the changes to the computers was not urgent and decided to delay modification until the next time the flight computers on the plane needed repair [4]. Because of that delay, 264 passengers and crew died.

In another DC-10 saga, explosive decompression played a critical role in a near miss over Windsor, Ontario. An American Airlines DC-10 lost part of its passenger floor, and thus all of the control cables that ran through it, when a cargo door opened in flight in June 1972. Thanks to the extraordinary skill and poise of the pilot, Bryce McCormick, the plane landed safely. In a remarkable coincidence, McCormick had trained himself to fly the plane using only the engines because he had been concerned about a decompression-caused collapse of the floor. After this close call, McCormick recommended that every DC-10 pilot be informed of the consequences of explosive decompression and trained in the flying techniques that he and his crew had used to save their passengers and aircraft. FAA investigators, the National Transportation Safety Board, and engineers at a subcontractor to McDonnell Douglas that designed the fuselage of the plane, all recommended changes in the design of the aircraft. Instead, McDonnell Douglas attributed the Windsor incident totally to human error on the part of the baggage handler responsible for closing the cargo compartment door (a convenient event in the event chain) and not to any error on the part of their designers or engineers and decided all they had to do was to come up with a fix that would prevent baggage handlers from forcing the door.

One of the discoveries after the Windsor incident was that the door could be improperly closed but the external signs, such as the position of the external handle, made it appear to be closed properly. In addition, this incident proved that the cockpit warning system could fail, and the crew would then not know that the plane was taking off without a properly closed door:

> The aviation industry does not normally receive such manifest warnings of basic design flaws in an aircraft without cost to human life. Windsor deserved to be celebrated as an exceptional case when every life was saved through a combination of crew skill and the sheer luck that the plane was so lightly loaded. If there had been more passengers and thus more weight, damage to the control cables would undoubtedly have been more severe, and it is highly questionable if any amount of skill could have saved the plane [61].

Almost two years later, in March 1974, a fully loaded Turkish Airlines DC-10 crashed near Paris, resulting in 346 deaths—one of the worst accidents in aviation history. Once again, the cargo door had opened in flight, causing the cabin floor to collapse, severing the flight control cables. Immediately after the accident, Sanford McDonnell stated the official McDonnell-Douglas position that once again placed the blame on the baggage handler and the ground crew. This time, however, the FAA finally ordered modifications to all DC-10s that eliminated the hazard. In addition, an FAA regulation issued in July 1975 required all wide-bodied jets to be able to tolerate a hole in the fuselage of twenty square feet. By labeling the root cause in the event chain as baggage handler error and attempting only to eliminate that event or link in the chain rather than the basic engineering design flaws, fixes that could have prevented the Paris crash were not made.

Until we do a better job of identifying causal factors in accidents, we will continue to have unnecessary repetition of incidents and accidents.

2.2.3 Subjectivity in Selecting the Chaining Conditions

In addition to subjectivity in selecting the events and the root cause event, the links between the events that are chosen to explain them are subjective and subject to bias. Leplat notes that the links are justified by knowledge or rules of different types, including physical and organizational knowledge. The same event can give rise to different types of links according to the mental representations the analyst has of the production of this event. When several types of rules are possible, the analyst will apply those that agree with his or her mental model of the situation [111].

Consider, for example, the loss of an American Airlines B757 near Cali, Colombia, in 1995 [2]. Two significant events in this loss were

(1) Pilot asks for clearance to take the rozo approach

followed later by

(2) Pilot types R into the FMS.[5]

In fact, the pilot should have typed the four letters ROZO instead of R—the latter was the symbol for a different radio beacon (called ROMEO) near Bogota. As a result, the aircraft incorrectly turned toward mountainous terrain. While these events are noncontroversial, the link between the two events could be explained by any of the following:

- *Pilot Error:*   In the rush to start the descent, the pilot executed a change of course without verifying its effect on the flight path.

- *Crew Procedure Error:*   In the rush to start the descent, the captain entered the name of the waypoint without normal verification from the other pilot.

- *Approach Chart and FMS Inconsistencies:*   The identifier used to identify rozo on the approach chart (R) did not match the identifier used to call up rozo in the FMS.

- *FMS Design Deficiency:*   The FMS did not provide the pilot with feedback that choosing the first identifier listed on the display was not the closest beacon having that identifier.

- *American Airlines Training Deficiency:*   The pilots flying into South America were not warned about duplicate beacon identifiers nor adequately trained on the logic and priorities used in the FMS on the aircraft.

- *Manufacturer Deficiency:*   Jeppesen-Sanderson did not inform airlines operating FMS-equipped aircraft of the differences between navigation information provided by Jeppesen-Sanderson Flight Management System navigation databases and Jeppesen-Sanderson approach charts or the logic and priorities employed in the display of electronic FMS navigation information.

- *International Standards Deficiency:*   No single worldwide standard provides unified criteria for the providers of electronic navigation databases used in Flight Management Systems.

---

5. An FMS is an automated flight management system that assists the pilots in various ways. In this case, it was being used to provide navigation information.

The selection of the linking condition (or events) will greatly influence the cause ascribed to the accident yet in the example all are plausible and each could serve as an explanation of the event sequence. The choice may reflect more on the person or group making the selection than on the accident itself. In fact, understanding this accident and learning enough from it to prevent future accidents requires identifying *all* of these factors to explain the incorrect input: The accident model used should encourage and guide a comprehensive analysis at multiple technical and social system levels.

2.2.4 Discounting Systemic Factors

The problem with event chain models is not simply that the selection of the events to include and the labeling of some of them as causes are arbitrary or that the selection of which conditions to include is also arbitrary and usually incomplete. Even more important is that viewing accidents as chains of events and conditions may limit understanding and learning from the loss and omit causal factors that cannot be included in an event chain.

Event chains developed to explain an accident usually concentrate on the proximate events immediately preceding the loss. But the foundation for an accident is often laid years before. One event simply triggers the loss, but if that event had not happened, another one would have led to a loss. The Bhopal disaster provides a good example.

The release of methyl isocyanate (MIC) from the Union Carbide chemical plant in Bhopal, India, in December 1984 has been called the worst industrial accident in history: Conservative estimates point to 2,000 fatalities, 10,000 permanent disabilities (including blindness), and 200,000 injuries [38]. The Indian government blamed the accident on human error—the improper cleaning of a pipe at the plant. A relatively new worker was assigned to wash out some pipes and filters, which were clogged. MIC produces large amounts of heat when in contact with water, and the worker properly closed the valves to isolate the MIC tanks from the pipes and filters being washed. Nobody, however, inserted a required safety disk (called a *slip blind*) to back up the valves in case they leaked [12].

A chain of events describing the accident mechanism for Bhopal might include:

E1 Worker washes pipes without inserting a slip blind.

E2 Water leaks into MIC tank.

E3 Explosion occurs.

E4 Relief valve opens.

E5 MIC vented into air.

E6 Wind carries MIC into populated area around plant.

Both Union Carbide and the Indian government blamed the worker washing the pipes for the accident.[6] A different operator error might be identified as the root cause (initiating event) if the chain is followed back farther. The worker who had been assigned the task of washing the pipes reportedly knew that the valves leaked, but he did not check to see whether the pipe was properly isolated because, he said, it was not his job to do so. Inserting the safety disks was the job of the maintenance department, but the maintenance sheet contained no instruction to insert this disk. The pipe-washing operation should have been supervised by the second shift supervisor, but that position had been eliminated in a cost-cutting effort. So the root cause might instead have been assigned to the person responsible for inserting the slip blind or to the lack of a second shift supervisor.

But the selection of a stopping point and the specific operator action to label as the root cause—and operator actions are almost always selected as root causes—is not the real problem here. The problem is the oversimplification implicit in using a chain of events to understand why this accident occurred. Given the design and operating conditions of the plant, an accident was waiting to happen:

> However [water] got in, it would not have caused the severe explosion had the refrigeration unit not been disconnected and drained of freon, or had the gauges been properly working and monitored, or had various steps been taken at the first smell of MIC instead of being put off until after the tea break, or had the scrubber been in service, or had the water sprays been designed to go high enough to douse the emissions, or had the flare tower been working and been of sufficient capacity to handle a large excursion. [156, p. 349]

It is not uncommon for a company to turn off passive safety devices, such as refrigeration units, to save money. The operating manual specified that the refrigeration unit *must* be operating whenever MIC was in the system: The chemical has to be maintained at a temperature no higher than 5° Celsius to avoid uncontrolled reactions. A high temperature alarm was to sound if the MIC reached 11°. The refrigeration unit was turned off, however, to save money, and the MIC was usually stored at nearly 20°. The plant management adjusted the threshold of the alarm, accordingly, from 11° to 20° and logging of tank temperatures was halted, thus eliminating the possibility of an early warning of rising temperatures.

Gauges at plants are frequently out of service [23]. At the Bhopal facility, there were few alarms or interlock devices in critical locations that might have warned operators of abnormal conditions—a system design deficiency.

---

6. Union Carbide lawyers argued that the introduction of water into the MIC tank was an act of sabotage rather than a maintenance worker's mistake. While this differing interpretation of the initiating event has important implications with respect to legal liability, it makes no difference in the argument presented here regarding the limitations of event-chain models of accidents or even, as will be seen, understanding why this accident occurred.

Other protection devices at the plant had inadequate design thresholds. The vent scrubber, had it worked, was designed to neutralize only small quantities of gas at fairly low pressures and temperatures: The pressure of the escaping gas during the accident exceeded the scrubber's design by nearly two and a half times, and the temperature of the escaping gas was at least 80° Celsius more than the scrubber could handle. Similarly, the flare tower (which was supposed to burn off released vapor) was totally inadequate to deal with the estimated 40 tons of MIC that escaped during the accident. In addition, the MIC was vented from the vent stack 108 feet above the ground, well above the height of the water curtain intended to knock down the gas: The water curtain reached only 40 to 50 feet above the ground. The water jets could reach as high as 115 feet, but only if operated individually.

Leaks were routine occurrences and the reasons for them were seldom investigated: Problems were either fixed without further examination or were ignored. A safety audit two years earlier by a team from Union Carbide had noted many safety problems at the plant, including several involved in the accident, such as filter-cleaning operations without using slip blinds, leaking valves, the possibility of contaminating the tank with material from the vent gas scrubber, and bad pressure gauges. The safety auditors had recommended increasing the capability of the water curtain and had pointed out that the alarm at the flare tower from which the MIC leaked was nonoperational, and thus any leak could go unnoticed for a long time. None of the recommended changes were made [23]. There is debate about whether the audit information was fully shared with the Union Carbide India subsidiary and about who was responsible for making sure changes were made. In any event, there was no follow-up to make sure that the problems identified in the audit had been corrected.

A year before the accident, the chemical engineer managing the MIC plant resigned because he disapproved of falling safety standards, and still no changes were made. He was replaced by an electrical engineer. Measures for dealing with a chemical release once it occurred were no better. Alarms at the plant sounded so often (the siren went off twenty to thirty times a week for various purposes) that an actual alert could not be distinguished from routine events or practice alerts. Ironically, the warning siren was not turned on until two hours after the MIC leak was detected (and after almost all the injuries had occurred) and then was turned off after only five minutes—which was company policy [12]. Moreover, the numerous practice alerts did not seem to be effective in preparing for an emergency: When the danger during the release became known, many employees ran from the contaminated areas of the plant, totally ignoring the buses that were sitting idle ready to evacuate workers and nearby residents. Plant workers had only a bare minimum of emergency equipment—a shortage of oxygen masks, for example, was discovered after the accident started—and they had almost no knowledge or training about how to handle nonroutine events.

The police were not notified when the chemical release began. In fact, when called by police and reporters, plant spokesmen first denied the accident and then claimed that MIC was not dangerous. Nor was the surrounding community warned of the dangers, before or during the release, or informed of the simple precautions that could have saved them from lethal exposure, such as putting a wet cloth over their face and closing their eyes. If the community had been alerted and provided with this simple information, many (if not most) lives would have been saved and injuries prevented [106].

Some of the reasons why the poor conditions in the plant were allowed to persist are financial. Demand for MIC had dropped sharply after 1981, leading to reductions in production and pressure on the company to cut costs. The plant was operating at less than half capacity when the accident occurred. Union Carbide put pressure on the Indian management to reduce losses, but gave no specific details on how to achieve the reductions. In response, the maintenance and operating personnel were cut in half. Maintenance procedures were severely cut back and the shift relieving system was suspended—if no replacement showed up at the end of the shift, the following shift went unmanned. The person responsible for inserting the slip blind in the pipe had not showed up for his shift. Top management justified the cuts as merely reducing avoidable and wasteful expenditures without affecting overall safety.

As the plant lost money, many of the skilled workers left for more secure jobs. They either were not replaced or were replaced by unskilled workers. When the plant was first built, operators and technicians had the equivalent of two years of college education in chemistry or chemical engineering. In addition, Union Carbide provided them with six months training. When the plant began to lose money, educational standards and staffing levels were reportedly reduced. In the past, UC flew plant personnel to West Virginia for intensive training and had teams of U.S. engineers make regular on-site safety inspections. But by 1982, financial pressures led UC to give up direct supervision of safety at the plant, even though it retained general financial and technical control. No American advisors were resident at Bhopal after 1982.

Management and labor problems followed the financial losses. Morale at the plant was low. "There was widespread belief among employees that the management had taken drastic and imprudent measures to cut costs and that attention to details that ensure safe operation were absent" [127].

These are only a few of the factors involved in this catastrophe, which also include other technical and human errors within the plant, design errors, management negligence, regulatory deficiencies on the part of the U.S. and Indian governments, and general agricultural and technology transfer policies related to the reason they were making such a dangerous chemical in India in the first place. Any one of these perspectives or "causes" is inadequate by itself to understand the accident and to prevent future ones. In particular, identifying only operator error or sabotage as the root cause of the accident ignores most of the opportunities for the prevention of similar accidents in the future. Many of the systemic causal factors are only indirectly related to the proximate events and conditions preceding the loss.

When all the factors, including indirect and systemic ones, are considered, it becomes clear that the maintenance worker was, in fact, only a minor and somewhat irrelevant player in the loss. Instead, degradation in the safety margin occurred over time and without any particular single decision to do so but simply as a series of decisions that moved the plant slowly toward a situation where any slight error would lead to a major accident. Given the overall state of the Bhopal Union Carbide plant and its operation, if the action of inserting the slip disk had not been left out of the pipe washing operation that December day in 1984, something else would have triggered an accident. In fact, a similar leak had occurred the year before, but did not have the same catastrophic consequences and the true root causes of that incident were neither identified nor fixed.

To label one event (such as a maintenance worker leaving out the slip disk) or even several events as the root cause or the start of an event chain leading to the Bhopal accident is misleading at best. Rasmussen writes:

> The stage for an accidental course of events very likely is prepared through time by the normal efforts of many actors in their respective daily work context, responding to the standing request to be more productive and less costly. Ultimately, a quite normal variation in somebody's behavior can then release an accident. Had this "root cause" been avoided by some additional safety measure, the accident would very likely be released by another cause at another point in time. In other words, an explanation of the accident in terms of events, acts, and errors is not very useful for design of improved systems [167].

In general, event-based models are poor at representing systemic accident factors such as structural deficiencies in the organization, management decision making, and flaws in the safety culture of the company or industry. An accident model should encourage a broad view of accident mechanisms that expands the investigation beyond the proximate events: A narrow focus on technological components and pure engineering activities or a similar narrow focus on operator errors may lead to ignoring some of the most important factors in terms of preventing future accidents. The accident model used to explain why the accident occurred should not only encourage the inclusion of all the causal factors but should provide guidance in identifying these factors.

2.2.5 Including Systems Factors in Accident Models

Large-scale engineered systems are more than just a collection of technological artifacts: They are a reflection of the structure, management, procedures, and culture of the engineering organization that created them. They are usually also a reflection of the society in which they were created. Ralph Miles Jr., in describing the basic concepts of systems theory, notes,

> Underlying every technology is at least one basic science, although the technology may be well developed long before the science emerges. Overlying every technical or civil system is a social system that provides purpose, goals, and decision criteria. [137, p. 1]

Effectively preventing accidents in complex systems requires using accident models that include that social system as well as the technology and its underlying science. Without understanding the purpose, goals, and decision criteria used to construct and operate systems, it is not possible to completely understand and most effectively prevent accidents.

Awareness of the importance of social and organizational aspects of safety goes back to the early days of System Safety.[7] In 1968, Jerome Lederer, then the director of the NASA Manned Flight Safety Program for Apollo, wrote:

> System safety covers the total spectrum of risk management. It goes *beyond the hardware* and associated procedures of system safety engineering. It involves: attitudes and motivation of designers and production people, employee/management rapport, the relation of industrial associations among themselves and with government, human factors in supervision and quality control, documentation on the interfaces of industrial and public safety with design and operations, the interest and attitudes of top management, the effects of the legal system on accident investigations and exchange of information, the certification of critical workers, political considerations, resources, public sentiment and many other non-technical but vital influences on the attainment of an acceptable level of risk control. These non-technical aspects of system safety cannot be ignored. [109]

Too often, however, these non-technical aspects *are* ignored.

At least three types of factors need to be considered in accident causation. The first is the proximate event chain, which for the *Herald of Free Enterprise* includes the assistant boatswain's not closing the doors and the return of the first officer to the wheelhouse prematurely. Note that there was a redundant design here, with the first officer checking the work of the assistant boatswain, but it did not prevent the accident, as is often the case with redundancy [115, 155].

The second type of information includes the conditions that allowed the events to occur: the high spring tides, the inadequate design of the ferry loading ramp for this harbor, and the desire of the first officer to stay on schedule (thus leaving the car deck before the doors were closed). All of these conditions can be directly mapped to the events.

---

7. When this term is capitalized in this book, it denotes the specific form of safety engineering developed originally by the Defense Department and its contractors for the early ICBM systems and defined by MIL-STD-882. System safety (uncapitalized) or safety engineering denotes all the approaches to engineering for safety.

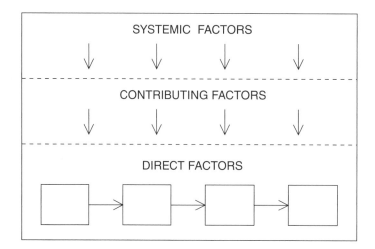

Figure 2.7
Johnson's three-level model of accidents.

Systemic factors

   The third set of causal factors is only indirectly related to the events and conditions, but these indirect factors are critical in fully understanding why the accident occurred and thus how to prevent future accidents. In this case, the systemic factors include the owner of the ferry (Townsend Thoresen) needing ships that were designed to permit fast loading and unloading and quick acceleration in order to remain competitive in the ferry business, and pressure by company management on the captain and first officer to strictly adhere to schedules, also related to competitive factors.

   Several attempts have been made to graft systemic factors onto event models, but all have important limitations. The most common approach has been to add hierarchical levels above the event chain. In the seventies, Johnson proposed a model and sequencing method that described accidents as chains of direct events and causal factors arising from contributory factors, which in turn arise from systemic factors (figure 2.7) [93].

   Johnson also tried to put management factors into fault trees (a technique called MORT, or Management Oversight Risk Tree), but ended up simply providing a general checklist for auditing management practices. While such a checklist can be very useful, it presupposes that every error can be predefined and put into a checklist form. The checklist is comprised of a set of questions that should be asked during an accident investigation. Examples of the questions from a DOE MORT User's Manual are: Was there sufficient training to update and improve needed supervisory skills? Did the supervisors have their own technical staff or access to such individuals? Was there technical support of the right discipline(s) sufficient for the needs of supervisory programs and review functions? Were there established methods for measuring performance that permitted the effectiveness of supervisory programs to be evaluated? Was a maintenance plan provided before startup? Was all relevant information provided to planners and managers? Was it used? Was concern for safety displayed by vigorous, visible personal action by top executives? And so forth.

Johnson originally provided hundreds of such questions, and additions have been made to his checklist since Johnson created it in the 1970s so it is now even larger. The use of the MORT checklist is feasible because the items are so general, but that same generality also limits its usefulness. Something more effective than checklists is needed.

The most sophisticated of the hierarchical add-ons to event chains is Rasmussen and Svedung's model of the sociotechnical system involved in risk management [167]. As shown in figure 2.8, at the social and organizational levels they use a hierarchical control structure, with levels for government, regulators and associations, company, management, and staff. At all levels they map information flow. The model concentrates on operations; information from the system design and analysis process is treated as input to the operations process. At each level, they model the factors involved using event chains, with links to the event chains at the level below. Notice that they still assume there is a root cause and causal chain of events. A generalization of the Rasmussen and Svedung model, which overcomes these limitations, is presented in chapter 4.

Once again, a new assumption is needed to make progress in learning how to design and operate safer systems:

New Assumption 2: *Accidents are complex processes involving the entire socio-technical system. Traditional event-chain models cannot describe this process adequately.*

Most of the accident models underlying safety engineering today stem from the days when the types of systems we were building and the context in which they were built were much simpler. As noted in chapter 1, new technology and social factors are making fundamental changes in the etiology of accidents, requiring changes in the explanatory mechanisms used to understand them and in the engineering techniques applied to prevent them. *[handwritten: Study of Causation]*

Event-based models are limited in their ability to represent accidents as complex processes, particularly at representing systemic accident factors such as structural deficiencies in the organization, management deficiencies, and flaws in the safety culture of the company or industry. We need to understand how the whole system, including the organizational and social components, operating together, led to the loss. While some extensions to event-chain models have been proposed, all are unsatisfactory in important ways.

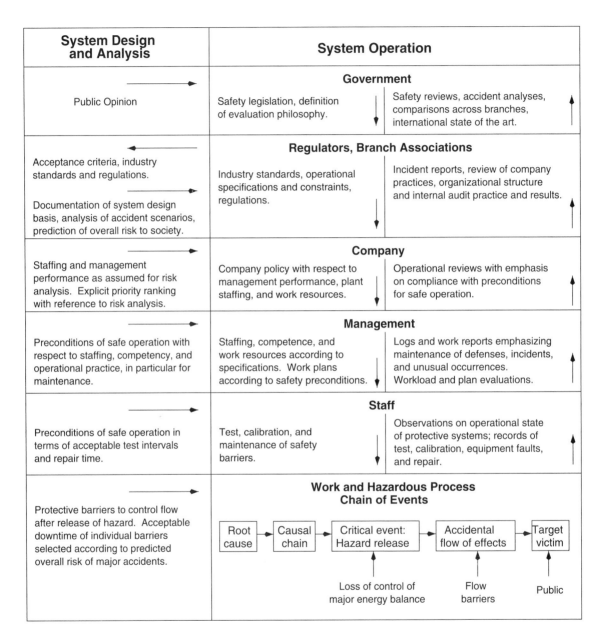

Figure 2.8
The Rasmussen/Svedung model of risk management.

An accident model should encourage a broad view of accident mechanisms that expands the investigation beyond the proximate events: A narrow focus on operator actions, physical component failures, and technology may lead to ignoring some of the most important factors in terms of preventing future accidents. The whole concept of "root cause" needs to be reconsidered.

2.3   Limitations of Probabilistic Risk Assessment

Assumption 3: *Probabilistic risk analysis based on event chains is the best way to assess and communicate safety and risk information.*

The limitations of event-chain models are reflected in the current approaches to quantitative risk assessment, most of which use trees or other forms of event chains. Probabilities (or probability density functions) are assigned to the events in the chain and an overall likelihood of a loss is calculated.

In performing a probabilistic risk assessment (PRA), initiating events in the chain are usually assumed to be mutually exclusive. While this assumption simplifies the mathematics, it may not match reality. As an example, consider the following description of an accident chain for an offshore oil platform:

> An initiating event is an event that triggers an accident sequence—e.g., a wave that exceeds the jacket's capacity that, in turn, triggers a blowout that causes failures of the foundation. As initiating events, they are mutually exclusive; only one of them starts the accident sequence. A catastrophic platform failure can start by failure of the foundation, failure of the jacket, or failure of the deck. These initiating failures are also (by definition) mutually exclusive and constitute the basic events of the [probabilistic risk assessment] model in its simplest form. [152, p. 121]

The selection of the failure of the foundation, jacket, or deck as the initiating event is arbitrary, as we have seen, and eliminates from consideration prior events leading to them such as manufacturing or construction problems. The failure of the foundation, for example, might be related to the use of inferior construction materials, which in turn might be related to budget deficiencies or lack of government oversight.

In addition, there does not seem to be any reason for assuming that initiating failures are mutually exclusive and that only one starts the accident, except perhaps again to simplify the mathematics. In accidents, seemingly independent failures may have a common systemic cause (often not a failure) that results in coincident failures. For example, the same pressures to use inferior materials in the foundation may result in their use in the jacket and the deck, leading to a wave causing coincident, dependent failures in all three. Alternatively, the design of the foundation—a systemic factor rather than a failure event—may lead to pressures on the jacket and deck when stresses cause deformities in the foundation. Treating such events as independent may lead to unrealistic risk assessments.

In the Bhopal accident, the vent scrubber, flare tower, water spouts, refrigeration unit, and various monitoring instruments were all out of operation simultaneously. Assigning probabilities to all these seemingly unrelated events and assuming independence would lead one to believe that this accident was merely a matter of a once-in-a-lifetime coincidence. A probabilistic risk assessment based on an event chain model most likely would have treated these conditions as independent failures and then calculated their coincidence as being so remote as to be beyond consideration. Reason, in his popular Swiss Cheese Model of accident causation based on defense in depth, does the same, arguing that in general "the chances of such a trajectory of opportunity finding loopholes in all the defences at any one time is very small indeed" [172, p. 208]. As suggested earlier, a closer look at Bhopal and, indeed, most accidents paints a quite different picture and shows these were not random failure events but were related to engineering and management decisions stemming from common systemic factors.

Most accidents in well-designed systems involve two or more low-probability events occurring in the worst possible combination. When people attempt to predict system risk, they explicitly or implicitly multiply events with low probability—assuming independence—and come out with impossibly small numbers, when, in fact, the events are dependent. This dependence may be related to common systemic factors that do not appear in an event chain. Machol calls this phenomenon the *Titanic coincidence* [131].[8]

A number of "coincidences" contributed to the *Titanic* accident and the subsequent loss of life. For example, the captain was going far too fast for existing conditions, a proper watch for icebergs was not kept, the ship was not carrying enough lifeboats, lifeboat drills were not held, the lifeboats were lowered properly but arrangements for manning them were insufficient, and the radio operator on a nearby ship was asleep and so did not hear the distress call. Many of these events or conditions may be considered independent but appear less so when we consider that overconfidence due to incorrect engineering analyses about the safety and unsinkability of the ship most likely contributed to the excessive speed, the lack of a proper watch, and the insufficient number of lifeboats and drills. That the collision occurred at night contributed to the iceberg not being easily seen, made abandoning ship more difficult than it would have been during the day, and was a factor in why

---

8. Watt defined a related phenomenon he called the *Titanic effect* to explain the fact that major accidents are often preceded by a belief that they cannot happen. The Titanic effect says that the magnitude of disasters decreases to the extent that people believe that disasters are possible and plan to prevent them or to minimize their effects [204].

the nearby ship's operator was asleep [135]. Assuming independence here leads to a large underestimate of the true risk.

Another problem in probabilistic risk assessment (PRA) is the emphasis on failure events—design errors are usually omitted and only come into the calculation indirectly through the probability of the failure event. Accidents involving dysfunctional interactions among non-failing (operational) components—that is, component interaction accidents—are usually not considered. Systemic factors also are not reflected. In the offshore oil platform example at the beginning of this section, the true probability density function for the failure of the deck might reflect a poor design for the conditions the deck must withstand (a human design error) or, as noted earlier, the use of inadequate construction materials due to lack of government oversight or project budget limitations.

When historical data are used to determine the failure probabilities used in the PRA, non-failure factors, such as design errors or unsafe management decisions, may differ between the historic systems from which the data was derived and the system under consideration. It is possible (and obviously desirable) for each PRA to include a description of the conditions under which the probabilities were derived. If such a description is not included, it may not be possible to determine whether conditions in the platform being evaluated differ from those built previously that might significantly alter the risk. The introduction of a new design feature or of active control by a computer might greatly affect the probability of failure and the usefulness of data from previous experience then becomes highly questionable.

The most dangerous result of using PRA arises from considering only immediate physical failures. Latent design errors may be ignored and go uncorrected due to overconfidence in the risk assessment. An example, which is a common but dangerous practice judging from its implication in a surprising number of accidents, is wiring a valve to detect only that power has been applied to open or close it and not that the valve position has actually changed. In one case, an Air Force system included a relief valve to be opened by the operator to protect against overpressurization [3]. A second, backup relief valve was installed in case the primary valve failed. The operator needed to know that the first valve had not opened, however, in order to determine that the backup valve must be activated. One day, the operator issued a command to open the primary valve. The position indicator and open indicator lights both illuminated but the primary relief valve was *not* open. The operator, thinking the primary valve had opened, did not activate the backup valve and an explosion occurred.

A post-accident investigation discovered that the indicator light circuit was wired to indicate *presence of power* at the valve, but it did not indicate valve *position*. Thus, the indicator showed only that the activation button had been pushed, not that the valve had operated. An extensive probabilistic risk assessment of this design had correctly assumed a low probability of simultaneous failure for the two relief valves, but had ignored the possibility of a design error in the electrical wiring: The probability of that design error was not quantifiable. If it had been identified, of course, the proper solution would have been to eliminate the design error, not to assign a probability to it. The same type of design flaw was a factor in the Three Mile Island accident: An indicator misleadingly showed that a discharge valve had been ordered closed but not that it had actually closed. In fact, the valve was blocked in an open position.

In addition to these limitations of PRA for electromechanical systems, current methods for quantifying risk that are based on combining probabilities of individual component failures and mutually exclusive events are not appropriate for systems controlled by software and by humans making cognitively complex decisions, and there is no effective way to incorporate management and organizational factors, such as flaws in the safety culture, despite many well-intentioned efforts to do so. As a result, these critical factors in accidents are often omitted from risk assessment because analysts do not know how to obtain a "failure" probability, or alternatively, a number is pulled out of the air for convenience. If we knew enough to measure these types of design flaws, it would be better to fix them than to try to measure them.

Another possibility for future progress is usually not considered:

New Assumption 3:   *Risk and safety may be best understood and communicated in ways other than probabilistic risk analysis.*

Understanding risk is important in decision making. Many people assume that risk information is most appropriately communicated in the form of a probability. Much has been written, however, about the difficulty people have in interpreting probabilities [97]. Even if people could use such values appropriately, the tools commonly used to compute these quantities, which are based on computing probabilities of failure events, have serious limitations. An accident model that is not based on failure events, such as the one introduced in this book, could provide an entirely new basis for understanding and evaluating safety and, more generally, risk.

2.4   The Role of Operators in Accidents

Assumption 4:   *Most accidents are caused by operator error. Rewarding safe behavior and punishing unsafe behavior will eliminate or reduce accidents significantly.*

As we have seen, the definition of "caused by" is debatable. But the fact remains that if there are operators in the system, they are most likely to be blamed for an accident. This phenomenon is not new. In the nineteenth century, coupling accidents on railroads were one of the main causes of injury and death to railroad workers [79]. In the seven years between 1888 and 1894, 16,000 railroad workers were killed in coupling accidents and 170,000 were crippled. Managers claimed that such accidents were due only to worker error and negligence, and therefore nothing could be done aside from telling workers to be more careful. The government finally stepped in and required that automatic couplers be installed. As a result, fatalities dropped sharply. According to the June 1896 (three years after Congress acted on the problem) issue of *Scientific American*:

> Few battles in history show so ghastly a fatality. A large percentage of these deaths were caused by the use of imperfect equipment by the railroad companies; twenty years ago it was practically demonstrated that cars could be automatically coupled, and that it was no longer necessary for a railroad employee to imperil his life by stepping between two cars about to be connected. In response to appeals from all over, the U.S. Congress passed the Safety Appliance Act in March 1893. It has or will cost the railroads $50,000,000 to fully comply with the provisions of the law. Such progress has already been made that the death rate has dropped by 35 per cent.

2.4.1 Do Operators Cause Most Accidents?

The tendency to blame the operator is not simply a nineteenth century problem, but persists today. During and after World War II, the Air Force had serious problems with aircraft accidents: From 1952 to 1966, for example, 7,715 aircraft were lost and 8,547 people killed [79]. Most of these accidents were blamed on pilots. Some aerospace engineers in the 1950s did not believe the cause was so simple and argued that safety must be designed and built into aircraft just as are performance, stability, and structural integrity. Although a few seminars were conducted and papers written about this approach, the Air Force did not take it seriously until they began to develop intercontinental ballistic missiles: there were no pilots to blame for the frequent and devastating explosions of these liquid-propellant missiles. In having to confront factors other than pilot error, the Air Force began to treat safety as a system problem, and System Safety programs were developed to deal with them. Similar adjustments in attitude and practice may be forced in the future by the increasing use of unmanned autonomous aircraft and other automated systems.

It is still common to see statements that 70 percent to 80 percent of aircraft accidents are caused by pilot error or that 85 percent of work accidents are due to unsafe acts by workers rather than unsafe conditions. However, closer examination shows that the data may be biased and incomplete: the less that is known about an accident, the most likely it will be attributed to operator error [93]. Thorough investigation of serious accidents almost invariably finds other factors.

Part of the problem stems from the use of the chain-of-events model in accident investigation because it is difficult to find an *event* preceding and causal to the operator behavior, as mentioned earlier. If the problem is in the system design, there is no proximal event to explain the error, only a flawed decision during system design.

Even if a technical failure precedes the human action, the tendency is to put the blame on an inadequate response to the failure by an operator. Perrow claims that even in the best of industries, there is rampant attribution of accidents to operator error, to the neglect of errors by designers or managers [155]. He cites a U.S. Air Force study of aviation accidents demonstrating that the designation of human error (pilot error in this case) is a convenient classification for mishaps whose real cause is uncertain, complex, or embarrassing to the organization.

Beside the fact that operator actions represent a convenient stopping point in an event chain, other reasons for the operator error statistics include: (1) operator actions are generally reported only when they have a negative impact on safety and not when they are responsible for preventing accidents; (2) blame may be based on unrealistic expectations that operators can overcome every emergency; (3) operators may have to intervene at the limits of system behavior when the consequences of not succeeding are likely to be serious and often involve a situation the designer never anticipated and was not covered by the operator's training; and (4) hindsight often allows us to identify a better decision in retrospect, but detecting and correcting potential errors before they have been made obvious by an accident is far more difficult.[9]

2.4.2 Hindsight Bias

The psychological phenomenon called *hindsight bias* plays such an important role in attribution of causes to accidents that it is worth spending time on it. The report on the Clapham Junction railway accident in Britain concluded:

> There is almost no human action or decision that cannot be made to look flawed and less sensible in the misleading light of hindsight. It is essential that the critic should keep himself constantly aware of that fact. [82, pg. 147]

After an accident, it is easy to see where people went wrong, what they should have done or not done, to judge people for missing a piece of information that turned out to be critical, and to see exactly the kind of harm that they should have foreseen or prevented [51]. Before the event, such insight is difficult and, perhaps, impossible.

---

9. The attribution of operator error as the cause of accidents is discussed more thoroughly in *Safeware* (chapter 5).

Dekker [51] points out that hindsight allows us to:

- Oversimplify causality because we can start from the outcome and reason backward to presumed or plausible "causes."

- Overestimate the likelihood of the outcome—and people's ability to foresee it—because we already know what the outcome is.

- Overrate the role of rule or procedure "violations." There is always a gap between written guidance and actual practice, but this gap almost never leads to trouble. It only takes on causal significance once we have a bad outcome to look at and reason about.

- Misjudge the prominence or relevance of data presented to people at the time.

- Match outcome with the actions that went before it. If the outcome was bad, then the actions leading up to it must have also been bad—missed opportunities, bad assessments, wrong decisions, and misperceptions.

Avoiding hindsight bias requires changing our emphasis in analyzing the role of humans in accidents from what they did wrong to why it made sense for them to act the way they did.

2.4.3 The Impact of System Design on Human Error

All human activity takes place within and is influenced by the environment, both physical and social, in which it takes place. It is, therefore, often very difficult to separate system design error from operator error: In highly automated systems, the operator is often at the mercy of the system design and operational procedures. One of the major mistakes made by the operators at Three Mile Island was following the procedures provided to them by the utility. The instrumentation design also did not provide the information they needed to act effectively in recovering from the hazardous state [99].

In the lawsuits following the 1995 B757 Cali accident, American Airlines was held liable for the crash based on the Colombian investigators blaming crew error entirely for the accident. The official accident investigation report cited the following four causes for the loss [2]:

1. The flightcrew's failure to adequately plan and execute the approach to runway 19 and their inadequate use of automation.

2. Failure of the flightcrew to discontinue their approach, despite numerous cues alerting them of the inadvisability of continuing the approach.

3. The lack of situational awareness of the flightcrew regarding vertical navigation, proximity to terrain, and the relative location of critical radio aids.

4. Failure of the flightcrew to revert to basic radio navigation at a time when the FMS-assisted navigation became confusing and demanded an excessive workload in a critical phase of the flight.

Look in particular the fourth identified cause: the blame is placed on the pilots when the automation became confusing and demanded an excessive workload rather than on the design of the automation. To be fair, the report also identifies two "contributory factors"—but *not* causes—as:

• FMS logic that dropped all intermediate fixes from the display(s) in the event of execution of a direct routing.

• FMS-generated navigational information that used a different naming convention from that published in navigational charts.

These two "contributory factors" are highly related to the third cause—the pilots' "lack of situational awareness." Even using an event-chain model of accidents, the FMS-related events preceded and contributed to the pilot errors. There seems to be no reason why, at the least, they should be treated any different than the labeled "causes." There were also many other factors in this accident that were not reflected in either the identified causes or contributory factors.

In this case, the Cali accident report conclusions were challenged in court. A U.S. appeals court rejected the conclusion of the report about the four causes of the accident [13], which led to a lawsuit by American Airlines in a federal court in which American alleged that components of the automated aircraft system made by Honeywell Air Transport Systems and Jeppesen Sanderson helped cause the crash. American blamed the software, saying Jeppesen stored the location of the Cali airport beacon in a different file from most other beacons. Lawyers for the computer companies argued that the beacon code could have been properly accessed and that the pilots were in error. The jury concluded that the two companies produced a defective product and that Jeppesen was 17 percent responsible, Honeywell was 8 percent at fault, and American was held to be 75 percent responsible [7]. While such distribution of responsibility may be important in determining how much each company will have to pay, it is arbitrary and does not provide any important information with respect to accident prevention in the future. The verdict is interesting, however, because the jury rejected the oversimplified notion of causality being argued. It was also one of the first cases not settled out of court where the role of software in the loss was acknowledged.

This case, however, does not seem to have had much impact on the attribution of pilot error in later aircraft accidents.

Part of the problem is engineers' tendency to equate people with machines. Human "failure" usually is treated the same as a physical component failure—a deviation from the performance of a specified or prescribed sequence of actions. This definition is equivalent to that of machine failure. Alas, human behavior is much more complex than machines.

As many human factors experts have found, instructions and written procedures are almost never followed exactly as operators try to become more efficient and productive and to deal with time pressures [167]. In studies of operators, even in such highly constrained and high-risk environments as nuclear power plants, modification of instructions is repeatedly found [71, 201, 213]. When examined, these violations of rules appear to be quite rational, given the workload and timing constraints under which the operators must do their job. The explanation lies in the basic conflict between error viewed as a deviation from *normative procedure* and error viewed as a deviation from the rational and normally used *effective procedure* [169].

One implication is that following an accident, it will be easy to find someone involved in the dynamic flow of events that has violated a formal rule by following *established practice* rather than *specified practice*. Given the frequent deviation of established practice from normative work instructions and rules, it is not surprising that operator "error" is found to be the cause of 70 percent to 80 percent of accidents. As noted in the discussion of assumption 2, a root cause is often selected because that event involves a deviation from a standard.

2.4.4   The Role of Mental Models
The updating of human mental models plays a significant role here (figure 2.9). Both the designer and the operator will have their own mental models of the plant. It is quite natural for the designer's and operator's models to differ and even for both to have significant differences from the actual plant as it exists. During development, the designer evolves a model of the plant to the point where it can be built. The *designer's model* is an idealization formed *before* the plant is constructed. Significant differences may exist between this ideal model and the actual constructed system. Besides construction variances, the designer always deals with ideals or averages, not with the actual components themselves. Thus, a designer may have a model of a valve with an average closure time, while real valves have closure times that fall somewhere along a continuum of timing behavior that reflects manufacturing and material differences. The designer's idealized model is used to develop operator work instructions and training. But the actual system may differ from the designer's model because of manufacturing and construction variances and evolution and changes over time.

The *operator's model* of the system will be based partly on formal training created from the designer's model and partly on experience with the system. The operator must cope with the system as it is constructed and not as it may have been

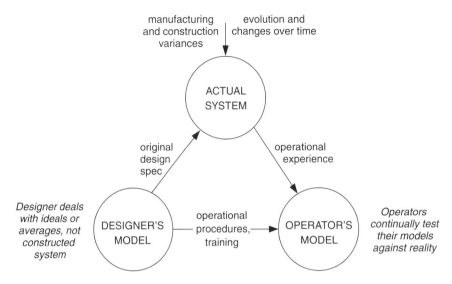

Figure 2.9
The relationship between mental models.

envisioned. As the physical system changes and evolves over time, the operator's model and operational procedures must change accordingly. While the formal procedures, work instructions, and training will be updated periodically to reflect the current operating environment, there is necessarily always a time lag. In addition, the operator may be working under time and productivity pressures that are not reflected in the idealized procedures and training.

Operators use feedback to update their mental models of the system as the system evolves. The only way for the operator to determine that the system has changed and that his or her mental model must be updated is through experimentation: To learn where the boundaries of safe behavior currently are, occasionally they must be crossed.

Experimentation is important at all levels of control [166]. For manual tasks where the optimization criteria are speed and smoothness, the limits of acceptable adaptation and optimization can only be known from the error experienced when occasionally crossing a limit. Errors are an integral part of maintaining a skill at an optimal level and a necessary part of the feedback loop to achieve this goal. The role of such experimentation in accidents cannot be understood by treating human errors as events in a causal chain separate from the feedback loops in which they operate.

At higher levels of cognitive control and supervisory decision making, experimentation is needed for operators to update procedures to handle changing conditions or to evaluate hypotheses while engaged in reasoning about the best response to unexpected situations. Actions that are quite rational and important during the search for information and test of hypotheses may appear to be unacceptable mistakes in hindsight, without access to the many details of a "turbulent" situation [169].

The ability to adapt mental models through experience in interacting with the operating system is what makes the human operator so valuable. For the reasons discussed, the operators' actual behavior may differ from the prescribed procedures because it is based on current inputs and feedback. When the deviation is correct (the designers' models are less accurate than the operators' models at that particular instant in time), then the operators are considered to be doing their job. When the operators' models are incorrect, they are often blamed for any unfortunate results, even though their incorrect mental models may have been reasonable given the information they had at the time.

Providing feedback and allowing for experimentation in system design, then, is critical in allowing operators to optimize their control ability. In the less automated system designs of the past, operators naturally had this ability to experiment and update their mental models of the current system state. Designers of highly automated systems sometimes do not understand this requirement and design automation that takes operators "out of the loop." Everyone is then surprised when the operator makes a mistake based on an incorrect mental model. Unfortunately, the reaction to such a mistake is to add even more automation and to marginalize the operators even more, thus exacerbating the problem [50].

Flawed decisions may also result from limitations in the boundaries of the operator's or designer's model. Decision makers may simply have too narrow a view of the system their decisions will impact. Recall figure 2.2 and the discussion of the *Herald of Free Enterprise* accident. The boundaries of the system model relevant to a particular decision maker may depend on the activities of several other decision makers found within the total system [167]. Accidents may result from the interaction and side effects of their decisions based on their limited model. Before an accident, it will be difficult for the individual decision makers to see the full picture during their daily operational decision making and to judge the current state of the multiple defenses and safety margins that are partly dependent on decisions made by other people in other departments and organizations [167].

Rasmussen stresses that most decisions are sound using local judgment criteria and given the time and budget pressures and short-term incentives that shape behavior. Experts do their best to meet local conditions and in the busy daily flow of activities may be unaware of the potentially dangerous side effects of their behavior. Each individual decision may appear safe and rational within the context of the individual work environments and local pressures, but may be unsafe when considered as a whole: It is difficult—if not impossible—for any individual to judge the safety of their decisions when it is dependent on the decisions made by other people in other departments and organizations.

Decentralized decision making is, of course, required in some time-critical situations. But like all safety-critical decision making, the decentralized decisions must be made in the context of system-level information and from a total systems perspective in order to be effective in reducing accidents. One way to make distributed decision making safe is to decouple the system components in the overall system design, if possible, so that decisions do not have systemwide repercussions. Another common way to deal with the problem is to specify and train standard emergency responses. Operators may be told to sound the evacuation alarm any time an indicator reaches a certain level. In this way, safe procedures are determined at the system level and operators are socialized and trained to provide uniform and appropriate responses to crisis situations.

There are situations, of course, when unexpected conditions occur and avoiding losses requires the operators to violate the specified (and in such cases unsafe) procedures. If the operators are expected to make decisions in real time and not just follow a predetermined procedure, then they usually must have the relevant *system-level* information about the situation in order to make safe decisions. This is not required, of course, if the system design decouples the components and thus allows operators to make independent safe decisions. Such decoupling must be designed into the system, however.

Some high reliability organization (HRO) theorists have argued just the opposite. They have asserted that HROs are safe because they allow professionals at the front lines to use their knowledge and judgment to maintain safety. During crises, they argue, decision making in HROs migrates to the frontline workers who have the necessary judgment to make decisions [206]. The problem is that the assumption that frontline workers will have the necessary knowledge and judgment to make decisions is not necessarily true. One example is the friendly fire accident analyzed in chapter 5 where the pilots ignored the rules of engagement they were told to follow and decided to make real-time decisions on their own based on the inadequate information they had.

Many of the HRO theories were derived from studying safety-critical systems, such as aircraft carrier flight operations. La Porte and Consolini [107], for example, argue that while the operation of aircraft carriers is subject to the Navy's chain of command, even the lowest-level seaman can abort landings. Clearly, this local authority is necessary in the case of aborted landings because decisions must be made too quickly to go up a chain of command. But note that such low-level personnel can only make decisions in one direction, that is, they may only abort landings. In essence, they are allowed to change to an inherently safe state (a go-around) with respect to the hazard involved. System-level information is not needed because a safe state exists that has no conflicts with other hazards, and the actions governed by these decisions and the conditions for making them are relatively simple. Aircraft carriers are usually operating in areas containing little traffic—they are decoupled from the larger system—and therefore localized decisions to abort are almost always safe and can be allowed from a larger system safety viewpoint.

Consider a slightly different situation, however, where a pilot makes a decision to go-around (abort a landing) at a busy urban airport. While executing a go-around when a clear danger exists if the pilot lands is obviously the right decision, there have been near misses when a pilot executed a go-around and came too close to another aircraft that was taking off on a perpendicular runway. The solution to this problem is not at the decentralized level—the individual pilot lacks the system-level information to avoid hazardous system states in this case. Instead, the solution must be at the system level, where the danger must be reduced by instituting different landing and takeoff procedures, building new runways, redistributing air traffic, or by making other system-level changes. We want pilots to be able to execute a go-around if they feel it is necessary, but unless the encompassing system is designed to prevent collisions, the action decreases one hazard while increasing a different one. Safety is a system property.

2.4.5 An Alternative View of Human Error

Traditional decision-making research views decisions as discrete processes that can be separated from the context in which the decisions are made and studied as an isolated phenomenon. This view is starting to be challenged. Instead of thinking of operations as predefined sequences of actions, human interaction with a system is increasingly being considered to be a continuous control task in which separate "decisions" or errors are difficult to identify.

Edwards, back in 1962, was one of the first to argue that decisions can only be understood as part of an ongoing process [63]. The state of the system is perceived in terms of possible actions, one of these actions is chosen, and the resulting response from the controlled system acts as a background for the next actions. Errors then are difficult to localize in the stream of behavior; the effects of less successful actions are a natural part of the search by the operator for optimal performance. As an example, consider steering a boat. The helmsman of ship A may see an obstacle ahead (perhaps another ship) and decide to steer the boat to the left to avoid it. The wind, current, and wave action may require the helmsman to make continual adjustments in order to hold the desired course. At some point, the other ship may also change course, making the helmsman's first decision about what would be a safe course no longer correct and needing to be revised. Steering then can be perceived as a continuous control activity or process with what is the correct and safe behavior changing over time and with respect to the results of prior behavior. The helmsman's mental model of the effects of the actions of the sea and the assumed behavior of the other ship has to be continually adjusted.

Not only are individual unsafe actions difficult to identify in this nontraditional control model of human decision making, but the study of decision making cannot be separated from a simultaneous study of the social context, the value system in which it takes place, and the dynamic work process it is intended to control [166]. This view is the foundation of some modern trends in decision-making research, such as *dynamic decision making* [25], the new field of *naturalistic decision making* [217, 102], and the approach to safety described in this book.

As argued by Rasmussen and others, devising more effective accident models that go beyond the simple event chain and human failure models requires shifting the emphasis in explaining the role of humans in accidents from error (that is, deviations from normative procedures) to focus instead on the mechanisms and factors that shape human behavior, that is, the performance-shaping context in which human actions take place and decisions are made. Modeling human behavior by decomposing it into decisions and actions and studying it as a phenomenon isolated from the context in which the behavior takes place is not an effective way to understand behavior [167].

The alternative view requires a new approach to representing and understanding human behavior, focused not on human error and violation of rules but on the mechanisms generating behavior in the actual, dynamic context. Such as approach must take into account the work system constraints, the boundaries of acceptable performance, the need for experimentation, and the subjective criteria guiding adaptation to change. In this approach, traditional task analysis is replaced or augmented with *cognitive work analysis* [169, 202] or *cognitive task analysis* [75]. Behavior is modeled in terms of the objectives of the decision maker, the boundaries of acceptable performance, the behavior-shaping constraints of the environment (including the value system and safety constraints), and the adaptive mechanisms of the human actors.

Such an approach leads to new ways of dealing with the human contribution to accidents and human "error." Instead of trying to control human behavior by fighting deviations from specified procedures, focus should be on controlling behavior by identifying the boundaries of safe performance (the behavioral safety constraints), by making the boundaries explicit and known, by giving opportunities to develop coping skills at the boundaries, by designing systems to support safe optimization and adaptation of performance in response to contextual influences and pressures, by providing means for identifying potentially dangerous side effects of individual decisions in the network of decisions over the entire system, by designing for error tolerance (making errors observable and reversible before safety constraints are violated) [167], and by counteracting the pressures that drive operators and decision makers to violate safety constraints.

Once again, future progress in accident reduction requires tossing out the old assumption and substituting a new one:

New Assumption 4: *Operator behavior is a product of the environment in which it occurs. To reduce operator "error" we must change the environment in which the operator works.*

Human behavior is always influenced by the environment in which it takes place. Changing that environment will be much more effective in changing operator error than the usual behaviorist approach of using reward and punishment. Without changing the environment, human error cannot be reduced for long. We design systems in which operator error is inevitable, and then blame the operator and not the system design.

As argued by Rasmussen and others, devising more effective accident causality models requires shifting the emphasis in explaining the role that humans play in accidents from error (deviations from normative procedures) to focus on the mechanisms and factors that shape human behavior, that is the performance-shaping features and context in which human actions take place and decisions are made. Modeling behavior by decomposing it into decisions and actions or events, which most all current accident models do, and studying it as a phenomenon isolated from the context in which the behavior takes place is not an effective way to understand behavior [167].

2.5 The Role of Software in Accidents

Assumption 5: *Highly reliable software is safe.*

The most common approach to ensuring safety when the system includes software is to try to make the software highly reliable. To help readers who are not software professionals see the flaws in this assumption, a few words about software in general may be helpful.

The uniqueness and power of the digital computer over other machines stems from the fact that, for the first time, we have a general-purpose machine:

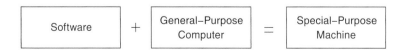

We no longer need to build a mechanical or analog autopilot from scratch, for example, but simply to write down the "design" of an autopilot in the form of instructions or steps to accomplish the desired goals. These steps are then loaded into the computer, which, while executing the instructions, in effect *becomes* the special-purpose machine (the autopilot). If changes are needed, the instructions can be changed and the same physical machine (the computer hardware) is used instead of having to build a different physical machine from scratch. Software in essence is the *design of a machine abstracted from its physical realization.* In other words, the logical design of a machine (the software) is separated from the physical design of that machine (the computer hardware).

Machines that previously were physically impossible or impractical to build become feasible, and the design of a machine can be changed quickly without going through an entire retooling and manufacturing process. In essence, the manufacturing phase is eliminated from the lifecycle of these machines: the physical parts of the machine (the computer hardware) can be reused, leaving only the design and verification phases. The design phase also has changed: The designer can concentrate on identifying the steps to be achieved without having to worry about how those steps will be realized physically.

These advantages of using computers (along with others specific to particular applications, such as reduced size and weight) have led to an explosive increase in their use, including their introduction into potentially dangerous systems. There are, however, some potential disadvantages of using computers and some important changes that their use introduces into the traditional engineering process that are leading to new types of accidents as well as creating difficulties in investigating accidents and preventing them.

One of the most important changes is that with computers, the design of the special purpose machine is usually created by someone who is not an expert on designing such machines. The autopilot design expert, for example, decides how the autopilot should work, and then provides that information to a software engineer, who is an expert in software design but not autopilots. It is the software engineer who then creates the detailed design of the autopilot. The extra communication step between the engineer and the software developer is the source of the most serious problems with software today.

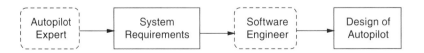

It should not be surprising, then, that most errors found in operational software can be traced to requirements flaws, particularly incompleteness. Completeness is a quality often associated with requirements but rarely defined. The most appropriate definition in the context of this book has been proposed by Jaffe: Software requirements specifications are complete if they are sufficient to distinguish the desired behavior of the software from that of any other undesired program that might be designed [91].

Nearly all the serious accidents in which software has been involved in the past twenty years can be traced to requirements flaws, not coding errors. The requirements may reflect incomplete or wrong assumptions

- About the operation of the system components being controlled by the software (for example, how quickly the component can react to a software-generated control command) or
- About the required operation of the computer itself

In the Mars Polar Lander loss, the software requirements did not include information about the potential for the landing leg sensors to generate noise or, alternatively, to ignore any inputs from the sensors while the spacecraft was more than forty meters above the planet surface. In the batch chemical reactor accident, the software engineers were never told to open the water valve before the catalyst valve and apparently thought the ordering was therefore irrelevant.

The problems may also stem from unhandled controlled-system states and environmental conditions. An F-18 was lost when a mechanical failure in the aircraft led to the inputs arriving faster than expected, which overwhelmed the software [70]. Another F-18 loss resulted from the aircraft getting into an attitude that the engineers had assumed was impossible and that the software was not programmed to handle.

In these cases, simply trying to get the software "correct" in terms of accurately implementing the requirements will not make it safer. Software may be highly reliable and correct and still be unsafe when:

- The software correctly implements the requirements, but the specified behavior is unsafe from a system perspective.
- The software requirements do not specify some particular behavior required for system safety (that is, they are incomplete).
- The software has unintended (and unsafe) behavior beyond what is specified in the requirements.

If the problems stem from the software doing what the software engineer thought it should do when that is not what the original design engineer wanted, the use of integrated product teams and other project management schemes to help with communication are useful. The most serious problems arise, however, when *nobody* understands what the software should do or even what it should not do. We need better techniques to assist in determining these requirements.

There is not only anecdotal but some hard data to support the hypothesis that safety problems in software stem from requirements flaws and not coding errors. Lutz examined 387 software errors uncovered during integration and system testing of the Voyager and Galileo spacecraft [130]. She concluded that the software errors identified as potentially hazardous to the system tended to be produced by different error mechanisms than non-safety-related software errors. She showed that for these two spacecraft, the safety-related software errors arose most commonly from (1) discrepancies between the documented requirements specifications and the requirements needed for correct functioning of the system and (2) misunderstandings about the software's interface with the rest of the system. They did not involve coding errors in implementing the documented requirements.

Many software requirements problems arise from what could be called the *curse of flexibility.* The computer is so powerful and so useful because it has eliminated many of the physical constraints of previous machines. This is both its blessing and its curse: We no longer have to worry about the physical realization of our designs, but we also no longer have physical laws that limit the complexity of our designs. Physical constraints enforce discipline on the design, construction, and modification of our design artifacts. Physical constraints also control the complexity of what we build. With software, the limits of what is *possible* to accomplish are different than the limits of what can be accomplished *successfully* and *safely* — the limiting factors change from the structural integrity and physical constraints of our materials to limits on our intellectual capabilities.

It is possible and even quite easy to build software that we cannot understand in terms of being able to determine how it will behave under all conditions. We can construct software (and often do) that goes beyond human intellectual limits. The result has been an increase in component interaction accidents stemming from intellectual unmanageability that allows potentially unsafe interactions to go undetected during development. The software often controls the interactions among the system components so its close relationship with component interaction accidents should not be surprising. But this fact has important implications for how software must be engineered when it controls potentially unsafe systems or products: Software or system engineering techniques that simply ensure software reliability or correctness (consistency of the code with the requirements) will have little or no impact on safety.

Techniques that *are* effective will rest on a new assumption:

New Assumption 5: *Highly reliable software is not necessarily safe. Increasing software reliability or reducing implementation errors will have little impact on safety.*

2.6   Static versus Dynamic Views of Systems

Assumption 6: *Major accidents occur from the chance simultaneous occurrence of random events.*

Most current safety engineering techniques suffer from the limitation of considering only the events underlying an accident and not the entire accident *process*. Accidents are often viewed as some unfortunate coincidence of factors that come together at one particular point in time and lead to the loss. This belief arises from too narrow a view of the causal time line. Looking only at the immediate time of the Bhopal MIC release, it does seem to be a coincidence that the refrigeration system, flare tower, vent scrubber, alarms, water curtain, and so on had all been inoperable at the same time. But viewing the accident through a larger lens makes it clear that the causal factors were all related to systemic causes that had existed for a long time.

Systems are not static. Rather than accidents being a chance occurrence of multiple independent events, they tend to involve a migration to a state of increasing risk over time [167]. A point is reached where an accident is inevitable unless the high risk is detected and reduced. The particular events involved at the time of the loss are somewhat irrelevant: if those events had not occurred, something else would have led to the loss. This concept is reflected in the common observation that a loss was "an accident waiting to happen." The proximate cause of the *Columbia* Space Shuttle loss was the foam coming loose from the external tank and damaging the reentry heat control structure. But many potential problems that could have caused the loss of the Shuttle had preceded this event and an accident was avoided by luck or unusual circumstances. The economic and political pressures led the Shuttle program to migrate to a state where any slight deviation could have led to a loss [117].

Any approach to enhancing safety that includes the social system and humans must account for adaptation. To paraphrase a familiar saying, the only constant is that nothing ever remains constant. Systems and organizations continually experience change as adaptations are made in response to local pressures and short-term productivity and cost goals. People adapt to their environment or they change their environment to better suit their purposes. A corollary to this propensity for systems and people to adapt over time is that safety defenses are likely to degenerate systematically through time, particularly when pressure toward cost-effectiveness and increased productivity is the dominant element in decision making. Rasmussen noted that the critical factor here is that such adaptation is not a random process—it is an optimization process depending on search strategies—and thus should be predictable and potentially controllable [167].

Woods has stressed the importance of adaptation in accidents. He describes organizational and human failures as breakdowns in adaptations directed at coping with complexity, and accidents as involving a "drift toward failure as planned defenses erode in the face of production pressures and change" [214].

Similarly, Rasmussen has argued that major accidents are often caused not by a coincidence of independent failures but instead reflect a systematic migration of organizational behavior to the boundaries of safe behavior under pressure toward cost-effectiveness in an aggressive, competitive environment [167]. One implication of this viewpoint is that the struggle for a good safety culture will never end because it must continually fight against the functional pressures of the work environment. Improvement of the safety culture will therefore require an analytical approach directed toward the behavior-shaping factors in the environment. A way of achieving this goal is described in part III.

Humans and organizations can adapt and still maintain safety as long as they stay within the area bounded by safety constraints. But in the search for optimal operations, humans and organizations will close in on and explore the boundaries of established practice. Such exploration implies the risk of occasionally crossing the limits of safe practice unless the constraints on safe behavior are enforced.

The natural migration toward the boundaries of safe behavior, according to Rasmussen, is complicated by the fact that it results from the decisions of multiple people, in different work environments and contexts within the overall sociotechnical system, all subject to competitive or budgetary stresses and each trying to optimize their decisions within their own immediate context. Several decision makers at different times, in different parts of the company or organization, all striving locally to optimize cost effectiveness may be preparing the stage for an accident, as illustrated by the Zeebrugge ferry accident (see figure 2.2) and the friendly fire accident described in chapter 5. The dynamic flow of events can then be released by a single act.

Our new assumption is therefore:

New Assumption 6: *Systems will tend to migrate toward states of higher risk. Such migration is predictable and can be prevented by appropriate system design or detected during operations using leading indicators of increasing risk.*

To handle system adaptation over time, our causal models and safety techniques must consider the *processes* involved in accidents and not simply events and conditions: Processes control a sequence of events and describe system and human behavior as it changes and adapts over time rather than considering individual events and human actions. To talk about the cause or causes of an accident makes no sense in this systems or process view of accidents. As Rasmussen argues, deterministic causal models are inadequate to explain the organizational and social factors in highly adaptive sociotechnical systems. Instead, accident causation must be viewed as a complex process involving the entire sociotechnical system including legislators, government agencies, industry associations and insurance companies, company management, technical and engineering personnel, operations, and so on [167].

2.7   The Focus on Determining Blame

Assumption 7:   *Assigning blame is necessary to learn from and prevent accidents or incidents.*

Beyond the tendency to blame operators described under assumption 3, other types of subjectivity in ascribing cause exist. Rarely are all the causes of an accident perceived identically by everyone involved, including engineers, managers, operators, union officials, insurers, lawyers, politicians, the press, the state, and the victims and their families. Such conflicts are typical in situations that involve normative, ethical, and political considerations about which people may legitimately disagree. Some conditions may be considered unnecessarily hazardous by one group yet adequately safe and necessary by another. In addition, judgments about the cause of an accident may be affected by the threat of litigation or by conflicting interests.

Research data validates this hypothesis. Various studies have found the selection of a cause(s) depends on characteristics of the victim and of the analyst (e.g., hierarchical status, degree of involvement, and job satisfaction) as well as on the relationships between the victim and the analyst and on the severity of the accident [112].

For example, one study found that workers who were satisfied with their jobs and who were integrated into and participating in the enterprise attributed accidents mainly to personal causes. In contrast, workers who were not satisfied and who had a low degree of integration and participation more often cited nonpersonal causes that implied that the enterprise was responsible [112]. Another study found differences in the attribution of accident causes among victims, safety managers, and general managers. Other researchers have suggested that accidents are attributed to factors in which the individuals are less directly involved. A further consideration may be position in the organization: The lower the position in the hierarchy, the greater the tendency to blame accidents on factors linked to the organization; individuals who have a high position in the hierarchy tend to blame workers for accidents [112].

There even seem to be differences in causal attribution between accidents and incidents: Accident investigation data on near-miss (incident) reporting suggest that causes for these events are mainly attributed to technical deviations while similar events that result in losses are more often blamed on operator error [62, 100].

Causal identification may also be influenced by the data collection methods. Data are usually collected in the form of textual descriptions of the sequence of events of the accident, which, as we have seen, tend to concentrate on obvious conditions or events closely preceding the accident in time and tend to leave out less obvious or indirect events and factors. There is no simple solution to this inherent bias: On one hand, report forms that do not specifically ask for nonproximal factors often do not elicit them while, on the other hand, more directive report forms that do request particular information may limit the categories or conditions considered [101].

Other factors affecting causal filtering in accident and incident reports may be related to the design of the reporting system itself. For example, the NASA Aviation Safety Reporting System (ASRS) has a category that includes nonadherence to FARs (Federal Aviation Regulations). In a NASA study of reported helicopter incidents and accidents over a nine-year period, this category was by far the largest category cited [81]. The NASA study concluded that the predominance of FAR violations in the incident data may reflect the motivation of the ASRS reporters to obtain immunity from perceived or real violations of FARs and not necessarily the true percentages.

A final complication is that human actions always involve some interpretation of the person's goals and motives. The individuals involved may be unaware of their actual goals and motivation or may be subject to various types of pressures to reinterpret their actions. Explanations by accident analysts after the fact may be influenced by their own mental models or additional goals and pressures.

Note the difference between an explanation based on goals and one based on motives: a goal represents an end state while a motive explains *why* that end state was chosen. Consider the hypothetical case where a car is driven too fast during a snowstorm and slides into a telephone pole. An explanation based on goals for this chain of events might include the fact that the driver wanted to get home quickly. An explanation based on motives might include the fact that guests were coming for dinner and the driver had to prepare the food before they arrived.

Explanations based on goals and motives depend on assumptions that cannot be directly measured or observed by the accident investigator. Leplat illustrates this dilemma by describing three different motives for the event *"operator sweeps the floor"*: (1) the floor is dirty, (2) the supervisor is present, or (3) the machine is broken and the operator needs to find other work [113]. Even if the people involved survive the accident, true goals and motives may not be revealed for a variety of reasons.

Where does all this leave us? There are two possible reasons for conducting an accident investigation: (1) to assign blame for the accident and (2) to understand why it happened so that future accidents can be prevented. When the goal is to assign blame, the backward chain of events considered often stops when someone or something appropriate to blame is found, such as the baggage handler in the

DC-10 case or the maintenance worker at Bhopal. As a result, the selected initiating event may provide too superficial an explanation of why the accident occurred to prevent similar losses in the future.

As another example, stopping at the O-ring failure in the *Challenger* accident and fixing that particular design flaw would not have eliminated the systemic flaws that could lead to accidents in the future. For *Challenger*, examples of those systemic problems include flawed decision making and the political and economic pressures that led to it, poor problem reporting, lack of trend analysis, a "silent" or ineffective safety program, communication problems, etc. None of these are "events" (although they may be manifested in particular events) and thus do not appear in the chain of events leading to the accident. Wisely, the authors of the *Challenger* accident report used an event chain only to identify the proximate physical cause and not the reasons those events occurred, and the report's recommendations led to many important changes at NASA or at least attempts to make such changes.

Twenty years later, another Space Shuttle was lost. While the proximate cause for the *Columbia* accident (foam hitting the wing of the orbiter) was very different than that for *Challenger*, many of the systemic causal factors were similar and reflected either inadequate fixes of these factors after the *Challenger* accident or their reemergence in the years between these losses [117].

Blame is not an engineering concept; it is a legal or moral one. Usually there is no objective criterion for distinguishing one factor or several factors from other factors that contribute to an accident. While lawyers and insurers recognize that many factors contribute to a loss event, for practical reasons and particularly for establishing liability, they often oversimplify the causes of accidents and identify what they call the *proximate* (immediate or direct) cause. The goal is to determine the parties in a dispute that have the legal liability to pay damages, which may be affected by the ability to pay or by public policy considerations, such as discouraging company management or even an entire industry from acting in a particular way in the future.

When learning how to engineer safer systems is the goal rather than identifying who to punish and establishing liability, then the emphasis in accident analysis needs to shift from *cause* (in terms of events or errors), which has a limiting, blame orientation, to understanding accidents in terms of *reasons*, that is, why the events and errors occurred. In an analysis by the author of recent aerospace accidents involving software, most of the reports stopped after assigning blame—usually to the operators who interacted with the software—and never got to the root of why the accident occurred, e.g., why the operators made the errors they did and how to prevent such errors in the future (perhaps by changing the software) or why the software requirements specified unsafe behavior, why that requirements error was introduced, and why it was not detected and fixed before the software was used [116].

When trying to understand operator contributions to accidents, just as with overcoming hindsight bias, it is more helpful in learning how to prevent future accidents to focus *not* on what the operators did "wrong" but on why it made sense for them to behave that way under those conditions [51]. Most people are not malicious but are simply trying to do the best they can under the circumstances and with the information they have. Understanding why those efforts were not enough will help in changing features of the system and environment so that sincere efforts are more successful in the future. Focusing on assigning blame contributes nothing toward achieving this goal and may impede it by reducing openness during accident investigations, thereby making it more difficult to find out what really happened.

A focus on blame can also lead to a lot of finger pointing and arguments that someone or something else was more to blame. Much effort is usually spent in accident investigations on determining which factors were the most important and assigning them to categories such as root cause, primary cause, contributory cause. In general, determining the relative importance of various factors to an accident may not be useful in preventing future accidents. Haddon [77] argues, reasonably, that countermeasures to accidents should *not* be determined by the relative importance of the causal factors; instead, priority should be given to the measures that will be most effective in reducing future losses. Explanations involving events in an event chain often do not provide the information necessary to prevent future losses, and spending a lot of time determining the relative contributions of events or conditions to accidents (such as arguing about whether an event is the root cause or a contributory cause) is not productive outside the legal system. Rather, Haddon suggests that engineering effort should be devoted to identifying the factors (1) that are easiest or most feasible to change, (2) that will prevent large classes of accidents, and (3) over which we have the greatest control.

Because the goal of this book is to describe a new approach to understanding and preventing accidents rather than assigning blame, the emphasis is on identifying *all* the factors involved in an accident and understanding the relationship among these causal factors in order to provide an explanation of why the accident occurred. That explanation can then be used to generate recommendations for preventing losses in the future. Building safer systems will be more effective when we consider all causal factors, both direct and indirect. In the new approach presented in this book, there is no attempt to determine which factors are more "important" than others but rather how they all relate to each other and to the final loss event or near miss.

One final new assumption is needed to complete the foundation for future progress:

New Assumption 7:   *Blame is the enemy of safety. Focus should be on understanding how the system behavior as a whole contributed to the loss and not on who or what to blame for it.*

Table 2.1
The basis for a new foundation for safety engineering

Old Assumption	New Assumption
Safety is increased by increasing system or component reliability; if components do not fail, then accidents will not occur.	High reliability is neither necessary nor sufficient for safety.
Accidents are caused by chains of directly related events. We can understand accidents and assess risk by looking at the chains of events leading to the loss.	Accidents are complex processes involving the entire sociotechnical system. Traditional event-chain models cannot describe this process adequately.
Probabilistic risk analysis based on event chains is the best way to assess and communicate safety and risk information.	Risk and safety may be best understood and communicated in ways other than probabilistic risk analysis.
Most accidents are caused by operator error. Rewarding safe behavior and punishing unsafe behavior will eliminate or reduce accidents significantly.	Operator error is a product of the environment in which it occurs. To reduce operator "error" we must change the environment in which the operator works.
Highly reliable software is safe.	Highly reliable software is not necessarily safe. Increasing software reliability will have only minimal impact on safety.
Major accidents occur from the chance simultaneous occurrence of random events.	Systems will tend to migrate toward states of higher risk. Such migration is predictable and can be prevented by appropriate system design or detected during operations using leading indicators of increasing risk.
Assigning blame is necessary to learn from and prevent accidents or incidents.	Blame is the enemy of safety. Focus should be on understanding how the system behavior as a whole contributed to the loss and not on who or what to blame for it.

We will be more successful in enhancing safety by focusing on why accidents occur rather than on blame.

Updating our assumptions about accident causation will allow us to make greater progress toward building safer systems in the twenty-first century. The old and new assumptions are summarized in table 2.1. The new assumptions provide the foundation for a new view of accident causation.

2.8   Goals for a New Accident Model

Event-based models work best for accidents where one or several components fail, leading to a system failure or hazard. Accident models and explanations involving only simple chains of failure events, however, can easily miss subtle and complex

couplings and interactions among failure events and omit entirely accidents involving no component failure at all. The event-based models developed to explain
physical phenomena (which they do well) are inadequate to explain accidents
involving organizational and social factors and human decisions and software design
errors in highly adaptive, tightly-coupled, interactively complex sociotechnical systems—namely, those accidents related to the new factors (described in chapter 1)
in the changing environment in which engineering is taking place.

The search for a new model, resulting in the accident model presented in part II,
was driven by the following goals:

- *Expand accident analysis by forcing consideration of factors other than component failures and human errors.* The model should encourage a broad
  view of accident mechanisms, expanding the investigation from simply considering proximal events to considering the entire sociotechnical system.
  Such a model should include societal, regulatory, and cultural factors. While
  some accident reports do this well, for example the space shuttle *Challenger*
  report, such results appear to be ad hoc and dependent on the personalities
  involved in the investigation rather than being guided by the accident
  model itself.

- *Provide a more scientific way to model accidents that produces a better and less
  subjective understanding of why the accident occurred and how to prevent
  future ones.* Event-chain models provide little guidance in the selection of
  events to include in the accident explanation or the conditions to investigate.
  The model should provide more assistance in identifying and understanding a
  comprehensive set of factors involved, including the adaptations that led to
  the loss.

- *Include system design errors and dysfunctional system interactions.* The models
  used widely were created before computers and digital components and do not
  handle them well. In fact, many of the event-based models were developed to
  explain industrial accidents, such as workers falling into holes or injuring themselves during the manufacturing process, and do not fit system safety at all. A
  new model must be able to account for accidents arising from dysfunctional
  interactions among the system components.

- *Allow for and encourage new types of hazard analyses and risk assessments that
  go beyond component failures and can deal with the complex role software and
  humans are assuming in high-tech systems.* Traditional hazard analysis techniques, such as fault tree analysis and the various other types of failure analysis
  techniques, do not work well for human errors and for software and other
  system design errors. An appropriate model should suggest hazard analysis
  techniques to augment these failure-based methods and encourage a wider

variety of risk reduction measures than redundancy and monitoring. In addition, risk assessment is currently firmly rooted in the probabilistic analysis of failure events. Attempts to extend current probabilistic risk assessment techniques to software and other new technology, to management, and to cognitively complex human control activities have been disappointing. This way forward may lead to a dead end, but starting from a different theoretical foundation may allow significant progress in finding new, more comprehensive approaches to risk assessment for complex systems.

- *Shift the emphasis in the role of humans in accidents from errors (deviations from normative behavior) to focus on the mechanisms and factors that shape human behavior (i.e., the performance-shaping mechanisms and context in which human actions take place and decisions are made).* A new model should account for the complex role that human decisions and behavior are playing in the accidents occurring in high-tech systems and handle not simply individual decisions but also sequences of decisions and the interactions among decisions by multiple, interacting decision makers [167]. The model must include examining the possible goals and motives behind human behavior as well as the contextual factors that influenced that behavior.

- *Encourage a shift in the emphasis in accident analysis from "cause"—which has a limiting, blame orientation—to understanding accidents in terms of reasons, that is, why the events and errors occurred* [197]. Learning how to engineer safer systems is the goal here, not identifying whom to punish.

- *Examine the processes involved in accidents and not simply events and conditions* Processes control a sequence of events and describe changes and adaptations over time rather than considering events and human actions individually.

- *Allow for and encourage multiple viewpoints and multiple interpretations when appropriate* Operators, managers, and regulatory agencies may all have different views of the flawed processes underlying an accident, depending on the hierarchical level of the sociotechnical control structure from which the process is viewed. At the same time, the factual data should be separated from the interpretation of that data.

- *Assist in defining operational metrics and analyzing performance data.* Computers allow the collection of massive amounts of operational data, but analyzing that data to determine whether the system is moving toward the boundaries of safe behavior is difficult. A new accident model should provide directions for identifying appropriate safety metrics and operational auditing procedures to evaluate decisions made during design and development, to determine whether controls over hazards are adequate, to detect erroneous operational and environmental assumptions underlying the hazard analysis and design process, to identify leading indicators and dangerous trends and changes in operations before they lead to accidents, and to identify any maladaptive system or environment changes over time that could increase accident risk to unacceptable levels.

These goals are achievable if models based on systems theory, rather than reliability theory, underlie our safety engineering activities.

3   Systems Theory and Its Relationship to Safety

To achieve the goals set at the end of the last chapter, a new theoretical under-pinning is needed for system safety. Systems theory provides that foundation. This chapter introduces some basic concepts in systems theory, how this theory is reflected in system engineering, and how all of this relates to system safety.

3.1   An Introduction to Systems Theory

Systems theory dates from the 1930s and 1940s and was a response to limitations of the classic analysis techniques in coping with the increasingly complex systems start-ing to be built at that time [36]. Norbert Wiener applied the approach to control and communications engineering [210], while Ludwig von Bertalanffy developed similar ideas for biology [21]. Bertalanffy suggested that the emerging ideas in various fields could be combined into a general theory of systems.

In the traditional scientific method, sometimes referred to as *divide and conquer*, systems are broken into distinct parts so that the parts can be examined separately: Physical aspects of systems are decomposed into separate physical components, while behavior is decomposed into discrete events over time.

Physical aspects → Separate physical components

Behavior → Discrete events over time

This decomposition (formally called *analytic reduction*) assumes that the separation is feasible: that is, each component or subsystem operates independently, and analy-sis results are not distorted when these components are considered separately. This assumption in turn implies that the components or events are not subject to feed-back loops and other nonlinear interactions and that the behavior of the compo-nents is the same when examined singly as when they are playing their part in the whole. A third fundamental assumption is that the principles governing the assem-bling of the components into the whole are straightforward, that is, the interactions

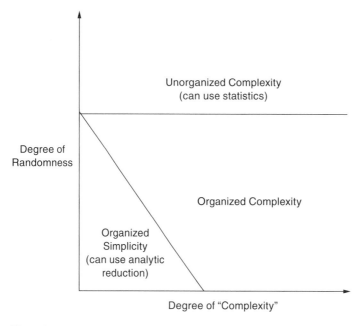

Figure 3.1
Three categories of systems (adapted from Gerald Weinberg, *An Introduction to General Systems Thinking* [John Wiley, 1975]).

among the subsystems are simple enough that they can be considered separate from the behavior of the subsystems themselves.

These are reasonable assumptions, it turns out, for many of the physical regularities of the universe. System theorists have described these systems as displaying *organized simplicity* (figure 3.1) [207]. Such systems can be separated into non-interacting subsystems for analysis purposes: the precise nature of the component interactions is known and interactions can be examined pairwise. Analytic reduction has been highly effective in physics and is embodied in structural mechanics.

Other types of systems display what systems theorists have labeled *unorganized complexity* — that is, they lack the underlying structure that allows reductionism to be effective. They can, however, often be treated as aggregates: They are complex, but regular and random enough in their behavior that they can be studied statistically. This study is simplified by treating them as a structureless mass with interchangeable parts and then describing them in terms of averages. The basis of this approach is the *law of large numbers*: The larger the population, the more likely that observed values are close to the predicted average values. In physics, this approach is embodied in statistical mechanics.

A third type of system exhibits what system theorists call *organized complexity*. These systems are too complex for complete analysis and too organized for statistics; the averages are deranged by the underlying structure [207]. Many of the complex engineered systems of the post–World War II era, as well as biological systems and social systems, fit into this category. Organized complexity also represents particularly well the problems that are faced by those attempting to build complex software, and it explains the difficulty computer scientists have had in attempting to apply analysis and statistics to software.

Systems theory was developed for this third type of system. The systems approach focuses on systems taken as a whole, not on the parts taken separately. It assumes that some properties of systems can be treated adequately only in their entirety, taking into account all facets relating the social to the technical aspects [161]. These system properties derive from the relationships between the parts of systems: how the parts interact and fit together [1]. Concentrating on the analysis and design of the whole as distinct from the components or parts provides a means for studying systems exhibiting organized complexity.

The foundation of systems theory rests on two pairs of ideas: (1) *emergence* and *hierarchy* and (2) *communication* and *control* [36].

3.2   Emergence and Hierarchy

A general model of complex systems can be expressed in terms of a *hierarchy* of levels of organization, each more complex than the one below, where a level is characterized by having *emergent* properties. Emergent properties do not exist at lower levels; they are meaningless in the language appropriate to those levels. The shape of an apple, although eventually explainable in terms of the cells of the apple, has no meaning at that lower level of description. The operation of the processes at the lower levels of the hierarchy result in a higher level of complexity—that of the whole apple itself—that has emergent properties, one of them being the apple's shape [36]. The concept of emergence is the idea that at a given level of complexity, some properties characteristic of that level (emergent at that level) are irreducible.

Hierarchy theory deals with the fundamental differences between one level of complexity and another. Its ultimate aim is to explain the relationships between different levels: what generates the levels, what separates them, and what links them. Emergent properties associated with a set of components at one level in a hierarchy are related to *constraints upon the degree of freedom* of those components. Describing the emergent properties resulting from the imposition of constraints requires a language at a higher level (a metalevel) different than that describing the components themselves. Thus, different languages of description are appropriate at different levels.

Reliability is a component property.[1] Conclusions can be reached about the reliability of a valve in isolation, where reliability is defined as the probability that the behavior of the valve will satisfy its specification over time and under given conditions.

Safety, on the other hand, is clearly an emergent property of systems: Safety can be determined only in the context of the whole. Determining whether a plant is acceptably safe is not possible, for example, by examining a single valve in the plant. In fact, statements about the "safety of the valve" without information about the context in which that valve is used are meaningless. Safety is determined by the relationship between the valve and the other plant components. As another example, pilot procedures to execute a landing might be safe in one aircraft or in one set of circumstances but unsafe in another.

Although they are often confused, reliability and safety are different properties. The pilots may reliably execute the landing procedures on a plane or at an airport in which those procedures are unsafe. A gun when discharged out on a desert with no other humans or animals for hundreds of miles may be both safe and reliable. When discharged in a crowded mall, the reliability will not have changed, but the safety most assuredly has.

Because safety is an emergent property, it is not possible to take a single system component, like a software module or a single human action, in isolation and assess its safety. A component that is perfectly safe in one system or in one environment may not be when used in another.

The new model of accidents introduced in part II of this book incorporates the basic systems theory idea of hierarchical levels, where constraints or lack of constraints at the higher levels control or allow lower-level behavior. Safety is treated as an emergent property at each of these levels. Safety depends on the enforcement of constraints on the behavior of the components in the system, including constraints on their potential interactions. Safety in the batch chemical reactor in the previous chapter, for example, depends on the enforcement of a constraint on the relationship between the state of the catalyst valve and the water valve.

3.3   Communication and Control

The second major pair of ideas in systems theory is *communication* and *control*. An example of regulatory or *control* action is the imposition of *constraints* upon the

---

1. This statement is somewhat of an oversimplification, because the reliability of a system component can, under some conditions (e.g., magnetic interference or excessive heat) be impacted by its environment. The basic reliability of the component, however, can be defined and measured in isolation, whereas the safety of an individual component is undefined except in a specific environment.

activity at one level of a hierarchy, which define the "laws of behavior" at that level. Those laws of behavior yield activity meaningful at a higher level. Hierarchies are characterized by control processes operating at the interfaces between levels [36].

The link between control mechanisms studied in natural systems and those engineered in man-made systems was provided by a part of systems theory known as cybernetics. Checkland writes:

> Control is always associated with the imposition of constraints, and an account of a control process necessarily requires our taking into account at least two hierarchical levels. At a given level, it is often possible to describe the level by writing dynamical equations, on the assumption that one particle is representative of the collection and that the forces at other levels do not interfere. But any description of a control process entails an upper level imposing constraints upon the lower. The upper level is a source of an alternative (simpler) description of the lower level in terms of specific functions that are emergent as a result of the imposition of constraints [36, p. 87].

Note Checkland's statement about control always being associated with the imposition of constraints. Imposing *safety constraints* plays a fundamental role in the approach to safety presented in this book. The limited focus on avoiding failures, which is common in safety engineering today, is replaced by the larger concept of imposing constraints on system behavior to avoid unsafe events or conditions, that is, hazards.

Control in open systems (those that have inputs and outputs from their environment) implies the need for *communication*. Bertalanffy distinguished between *closed systems*, in which unchanging components settle into a state of equilibrium, and *open systems*, which can be thrown out of equilibrium by exchanges with their environment.

In control theory, open systems are viewed as interrelated components that are kept in a state of dynamic equilibrium by feedback loops of information and control. The plant's overall performance has to be controlled in order to produce the desired product while satisfying cost, safety, and general quality constraints.

In order to control a process, four conditions are required [10]:

- *Goal Condition:* The controller must have a goal or goals (for example, to maintain the setpoint).
- *Action Condition:* The controller must be able to affect the state of the system. In engineering, control actions are implemented by *actuators*.
- *Model Condition:* The controller must be (or contain) a model of the system (see section 4.3).
- *Observability Condition:* The controller must be able to ascertain the state of the system. In engineering terminology, observation of the state of the system is provided by *sensors*.

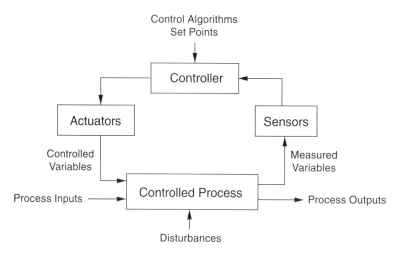

Figure 3.2
A standard control loop.

Figure 3.2 shows a typical control loop. The plant controller obtains information about (observes) the process state from measured variables (*feedback*) and uses this information to initiate action by manipulating *controlled variables* to keep the process operating within predefined limits or *set points* (the goal) despite disturbances to the process. In general, the maintenance of any open-system hierarchy (either biological or man-made) will require a set of processes in which there is communication of information for regulation or control [36].

Control actions will generally lag in their effects on the process because of delays in signal propagation around the control loop: an actuator may not respond immediately to an external command signal (called *dead time*); the process may have delays in responding to manipulated variables (*time constants*); and the sensors may obtain values only at certain sampling intervals (*feedback delays*). Time lags restrict the speed and extent with which the effects of disturbances, both within the process itself and externally derived, can be reduced. They also impose extra requirements on the controller, for example, the need to infer delays that are not directly observable.

The model condition plays an important role in accidents and safety. In order to create effective control actions, the controller must know the current state of the controlled process and be able to estimate the effect of various control actions on that state. As discussed further in section 4.3, many accidents have been caused by the controller incorrectly assuming the controlled system was in a particular state and imposing a control action (or not providing one) that led to a loss: the Mars Polar Lander descent engine controller, for example, assumed that the spacecraft was on the surface of the planet and shut down the descent engines. The captain of the *Herald of Free Enterprise* thought the car deck doors were shut and left the mooring.

3.4 Using Systems Theory to Understand Accidents

Safety approaches based on systems theory consider accidents as arising from the interactions among system components and usually do not specify single causal variables or factors [112]. Whereas industrial (occupational) safety models and event chain models focus on unsafe acts or conditions, classic system safety models instead look at what went wrong with the system's operation or organization to allow the accident to take place.

This systems approach treats safety as an emergent property that arises when the system components interact within an environment. Emergent properties like safety are controlled or enforced by a set of constraints (control laws) related to the behavior of the system components. For example, the spacecraft descent engines must remain on until the spacecraft reaches the surface of the planet and the car deck doors on the ferry must be closed before leaving port. Accidents result from interactions among components that violate these constraints—in other words, from a lack of appropriate constraints on the interactions. Component interaction accidents, as well as component failure accidents, can be explained using these concepts.

Safety then can be viewed as a control problem. Accidents occur when component failures, external disturbances, and/or dysfunctional interactions among system components are not adequately controlled. In the space shuttle *Challenger* loss, the O-rings did not adequately control propellant gas release by sealing a tiny gap in the field joint. In the Mars Polar Lander loss, the software did not adequately control the descent speed of the spacecraft—it misinterpreted noise from a Hall effect sensor (feedback of a measured variable) as an indication the spacecraft had reached the surface of the planet. Accidents such as these, involving engineering design errors, may in turn stem from inadequate control over the development process. A Milstar satellite was lost when a typo in the software load tape was not detected during the development and testing. Control is also imposed by the management functions in an organization—the *Challenger* and *Columbia* losses, for example, involved inadequate controls in the launch-decision process.

While events reflect the *effects* of dysfunctional interactions and inadequate enforcement of safety constraints, the inadequate control itself is only indirectly reflected by the events—the events are the *result* of the inadequate control. The control structure itself must be examined to determine why it was inadequate to maintain the constraints on safe behavior and why the events occurred.

As an example, the unsafe behavior (hazard) in the *Challenger* loss was the release of hot propellant gases from the field joint. The miscreant O-ring was used to control the hazard—that is, its role was to seal a tiny gap in the field joint created by pressure at ignition. The loss occurred because the system design, including the O-ring, did not effectively impose the required constraint on the propellant gas release. Starting from here, there are then several questions that need to be answered to understand why the accident occurred and to obtain the information necessary to prevent future accidents. Why was this particular design unsuccessful in imposing the constraint, why was it chosen (what was the decision process), why was the flaw not found during development, and was there a different design that might have been more successful? These questions and others consider the original *design process.*

Understanding the accident also requires examining the contribution of the *operations process.* Why were management decisions made to launch despite warnings that it might not be safe to do so? One constraint that was violated during operations was the requirement to correctly handle feedback about any potential violation of the safety design constraints, in this case, feedback during operations that the control by the O-rings of the release of hot propellant gases from the field joints was not being adequately enforced by the design. There were several instances of feedback that was not adequately handled, such as data about O-ring blowby and erosion during previous shuttle launches and feedback by engineers who were concerned about the behavior of the O-rings in cold weather. Although the lack of redundancy provided by the second O-ring was known long before the loss of *Challenger*, that information was never incorporated into the NASA Marshall Space Flight Center database and was unknown by those making the launch decision. In addition, there was missing feedback about changes in the design and testing procedures during operations, such as the use of a new type of putty and the introduction of new O-ring leak checks without adequate verification that they satisfied system safety constraints on the field joints. As a final example, the control processes that ensured unresolved safety concerns were fully considered before each flight, that is, the flight readiness reviews and other feedback channels to project management making flight decisions, were flawed.

Systems theory provides a much better foundation for safety engineering than the classic analytic reduction approach underlying event-based models of accidents. It provides a way forward to much more powerful and effective safety and risk analysis and management procedures that handle the inadequacies and needed extensions to current practice described in chapter 2.

Combining a systems-theoretic approach to safety with system engineering processes will allow designing safety into the system as it is being developed or reengineered. System engineering provides an appropriate vehicle for this process because it rests on the same systems theory foundation and involves engineering the system as a whole.

3.5 Systems Engineering and Safety

The emerging theory of systems, along with many of the historical forces noted in chapter 1, gave rise after World War II to a new emphasis in engineering, eventually called systems engineering. During and after the war, technology expanded rapidly and engineers were faced with designing and building more complex systems than had been attempted previously. Much of the impetus for the creation of this new discipline came from military programs in the 1950s and 1960s, particularly intercontinental ballistic missile (ICBM) systems. *Apollo* was the first nonmilitary government program in which systems engineering was recognized from the beginning as an essential function [24].

System Safety, as defined in MIL-STD-882, is a subdiscipline of system engineering. It was created at the same time and for the same reasons. The defense community tried using the standard safety engineering techniques on their complex new systems, but the limitations became clear when interface and component interaction problems went unnoticed until it was too late, resulting in many losses and near misses. When these early aerospace accidents were investigated, the causes of a large percentage of them were traced to deficiencies in design, operations, and management. Clearly, big changes were needed. System engineering along with its subdiscipline, System Safety, were developed to tackle these problems.

Systems theory provides the theoretical foundation for systems engineering, which views each system as an integrated whole even though it is composed of diverse, specialized components. The objective is to integrate the subsystems into the most effective system possible to achieve the overall objectives, given a prioritized set of design criteria. Optimizing the system design often requires making tradeoffs between these design criteria (goals).

The development of systems engineering as a discipline enabled the solution of enormously more complex and difficult technological problems than previously [137]. Many of the elements of systems engineering can be viewed merely as good engineering: It represents more a shift in emphasis than a change in content. In addition, while much of engineering is based on technology and science, systems engineering is equally concerned with overall management of the engineering process.

A systems engineering approach to safety starts with the basic assumption that some properties of systems, in this case safety, can only be treated adequately in the context of the social and technical system as a whole. A basic assumption of systems engineering is that optimization of individual components or subsystems will not in general lead to a system optimum; in fact, improvement of a particular subsystem may actually worsen the overall system performance because of complex, nonlinear interactions among the components. When each aircraft tries to optimize its path from its departure point to its destination, for example, the overall air transportation system throughput may not be optimized when they all arrive at a popular hub at the same time. One goal of the air traffic control system is to optimize the overall air transportation system throughput while, at the same time, trying to allow as much flexibility for the individual aircraft and airlines to achieve their goals. In the end, if system engineering is successful, everyone gains. Similarly, each pharmaceutical company acting to optimize its profits, which is a legitimate and reasonable company goal, will not necessarily optimize the larger societal *system* goal of producing safe and effective pharmaceutical and biological products to enhance public health. These system engineering principles are applicable even to systems beyond those traditionally thought of as in the engineering realm. The financial system and its meltdown starting in 2007 is an example of a social system that could benefit from system engineering concepts.

Another assumption of system engineering is that individual component behavior (including events or actions) cannot be understood without considering the components' role and interaction within the system as a whole. This basis for systems engineering has been stated as the principle that a system is more than the sum of its parts. Attempts to improve long-term safety in complex systems by analyzing and changing individual components have often proven to be unsuccessful over the long term. For example, Rasmussen notes that over many years of working in the field of nuclear power plant safety, he found that attempts to improve safety from models of local features were compensated for by people adapting to the change in an unpredicted way [167].

Approaches used to enhance safety in complex systems must take these basic systems engineering principles into account. Otherwise, our safety engineering approaches will be limited in the types of accidents and systems they can handle. At the same time, approaches that include them, such as those described in this book, have the potential to greatly improve our ability to engineer safer and more complex systems.

3.6   Building Safety into the System Design

System Safety, as practiced by the U.S. defense and aerospace communities as well as the new approach outlined in this book, fit naturally within the general systems engineering process and the problem-solving approach that a system view provides. This problem-solving process entails several steps. First, a need or problem is specified in terms of objectives that the system must satisfy along with criteria that can be used to rank alternative designs. For a system that has potential hazards, the objectives will include safety objectives and criteria along with high-level requirements and safety design constraints. The hazards for an automated train system, for example, might include the train doors closing while a passenger is in the doorway. The safety-related design constraint might be that obstructions in the path of a closing door must be detected and the door closing motion reversed.

After the high-level requirements and constraints on the system design are identified, a process of system synthesis takes place that results in a set of alternative designs. Each of these alternatives is analyzed and evaluated in terms of the stated objectives and design criteria, and one alternative is selected to be implemented. In practice, the process is highly iterative: The results from later stages are fed back to early stages to modify objectives, criteria, design alternatives, and so on. Of course, the process described here is highly simplified and idealized.

The following are some examples of basic systems engineering activities and the role of safety within them:

- *Needs analysis:* The starting point of any system design project is a perceived need. This need must first be established with enough confidence to justify the commitment of resources to satisfy it and understood well enough to allow appropriate solutions to be generated. Criteria must be established to provide a means to evaluate both the evolving and final system. If there are hazards associated with the operation of the system, safety should be included in the needs analysis.

- *Feasibility studies:* The goal of this step in the design process is to generate a set of realistic designs. This goal is accomplished by identifying the principal constraints and design criteria—including safety constraints and safety design criteria—for the specific problem being addressed and then generating plausible solutions to the problem that satisfy the requirements and constraints and are physically and economically feasible.

- *Trade studies:* In trade studies, the alternative feasible designs are evaluated with respect to the identified design criteria. A hazard might be controlled by any one of several safeguards: A trade study would determine the relative desirability of each safeguard with respect to effectiveness, cost, weight, size, safety, and any other relevant criteria. For example, substitution of one material for another may reduce the risk of fire or explosion, but may also reduce reliability or efficiency. Each alternative design may have its own set of safety constraints (derived from the system hazards) as well as other performance goals and constraints that need to be assessed. Although decisions ideally should be based upon mathematical analysis, quantification of many of the key factors is often difficult, if not impossible, and subjective judgment often has to be used.

• *System architecture development and analysis:* In this step, the system engineers break down the system into a set of subsystems, together with the functions and constraints, including safety constraints, imposed upon the individual subsystem designs, the major system interfaces, and the subsystem interface topology. These aspects are analyzed with respect to desired system performance characteristics and constraints (again including safety constraints) and the process is iterated until an acceptable system design results. The preliminary design at the end of this process must be described in sufficient detail that subsystem implementation can proceed independently.

• *Interface analysis:* The interfaces define the functional boundaries of the system components. From a management standpoint, interfaces must (1) optimize visibility and control and (2) isolate components that can be implemented independently and for which authority and responsibility can be delegated [158]. From an engineering standpoint, interfaces must be designed to separate independent functions and to facilitate the integration, testing, and operation of the overall system. One important factor in designing the interfaces is safety, and safety analysis should be a part of the system interface analysis. Because interfaces tend to be particularly susceptible to design error and are implicated in the majority of accidents, a paramount goal of interface design is simplicity. Simplicity aids in ensuring that the interface can be adequately designed, analyzed, and tested prior to integration and that interface responsibilities can be clearly understood.

Any specific realization of this general systems engineering process depends on the engineering models used for the system components and the desired system qualities. For safety, the models commonly used to understand why and how accidents occur have been based on events, particularly failure events, and the use of reliability engineering techniques to prevent them. Part II of this book further details the alternative systems approach to safety introduced in this chapter, while part III provides techniques to perform many of these safety and system engineering activities.

II STAMP: AN ACCIDENT MODEL BASED ON SYSTEMS THEORY

Part II introduces an expanded accident causality model based on the new assumptions in chapter 2 and satisfying the goals stemming from them. The theoretical foundation for the new model is systems theory, as introduced in chapter 3. Using this new causality model, called STAMP (Systems-Theoretic Accident Model and Processes), changes the emphasis in system safety from preventing failures to enforcing behavioral safety constraints. Component failure accidents are still included, but our conception of causality is extended to include component interaction accidents. Safety is reformulated as a control problem rather than a reliability problem. This change leads to much more powerful and effective ways to engineer safer systems, including the complex sociotechnical systems of most concern today.

The three main concepts in this model—safety constraints, hierarchical control structures, and process models—are introduced first in chapter 4. Then the STAMP causality model is described, along with a classification of accident causes implied by the new model.

To provide additional understanding of STAMP, it is used to describe the causes of several very different types of losses—a friendly fire shootdown of a U.S. Army helicopter by a U.S. Air Force fighter jet over northern Iraq, the contamination of a public water system with *E. coli* bacteria in a small town in Canada, and the loss of a Milstar satellite. Chapter 5 presents the friendly fire accident analysis. The other accident analyses are contained in appendixes B and C.

4   A Systems-Theoretic View of Causality

In the traditional causality models, accidents are considered to be caused by chains of failure events, each failure directly causing the next one in the chain. Part I explained why these simple models are no longer adequate for the more complex sociotechnical systems we are attempting to build today. The definition of accident causation needs to be expanded beyond failure events so that it includes component interaction accidents and indirect or systemic causal mechanisms.

The first step is to generalize the definition of an accident.[1] An *accident* is an unplanned and undesired loss event. That loss may involve human death and injury, but it may also involve other major losses, including mission, equipment, financial, and information losses.

Losses result from component failures, disturbances external to the system, interactions among system components, and behavior of individual system components that lead to hazardous system states. Examples of hazards include the release of toxic chemicals from an oil refinery, a patient receiving a lethal dose of medicine, two aircraft violating minimum separation requirements, and commuter train doors opening between stations.[2]

In systems theory, emergent properties, such as safety, arise from the interactions among the system components. The emergent properties are controlled by imposing constraints on the behavior of and interactions among the components. Safety then becomes a *control* problem where the goal of the control is to enforce the safety constraints. Accidents result from inadequate control or enforcement of safety-related constraints on the development, design, and operation of the system.

At Bhopal, the safety constraint that was violated was that the MIC must not come in contact with water. In the Mars Polar Lander, the safety constraint was that the spacecraft must not impact the planet surface with more than a maximum force.

---

1. A set of definitions used in this book can be found in appendix A.

2. Hazards are more carefully defined in chapter 7.

In the batch chemical reactor accident described in chapter 2, one safety constraint is a limitation on the temperature of the contents of the reactor.

The problem then becomes one of control where the goal is to control the behavior of the system by enforcing the safety constraints in its design and operation. Controls must be established to accomplish this goal. These controls need not necessarily involve a human or automated controller. Component behavior (including failures) and unsafe interactions may be controlled through physical design, through process (such as manufacturing processes and procedures, maintenance processes, and operations), or through social controls. Social controls include organizational (management), governmental, and regulatory structures, but they may also be cultural, policy, or individual (such as self-interest). As an example of the latter, one explanation that has been given for the 2009 financial crisis is that when investment banks went public, individual controls to reduce personal risk and long-term profits were eliminated and risk shifted to shareholders and others who had few and weak controls over those taking the risks.

In this framework, understanding why an accident occurred requires determining why the control was ineffective. Preventing future accidents requires shifting from a focus on preventing failures to the broader goal of designing and implementing controls that will enforce the necessary constraints.

The STAMP (System-Theoretic Accident Model and Processes) accident model is based on these principles. Three basic constructs underlie STAMP: safety constraints, hierarchical safety control structures, and process models.

4.1 Safety Constraints

The most basic concept in STAMP is not an event, but a constraint. Events leading to losses occur only because safety constraints were not successfully enforced.

The difficulty in identifying and enforcing safety constraints in design and operations has increased from the past. In many of our older and less automated systems, physical and operational constraints were often imposed by the limitations of technology and of the operational environments. Physical laws and the limits of our materials imposed natural constraints on the complexity of physical designs and allowed the use of passive controls.

In engineering, *passive controls* are those that maintain safety by their presence — basically, the system fails into a safe state or simple interlocks are used to limit the interactions among system components to safe ones. Some examples of passive controls that maintain safety by their presence are shields or barriers such as containment vessels, safety harnesses, hardhats, passive restraint systems in vehicles, and fences. Passive controls may also rely on physical principles, such as gravity, to fail into a safe state. An example is an old railway semaphore that used weights to ensure that if the cable (controlling the semaphore) broke, the arm would auto-matically drop into the STOP position. Other examples include mechanical relays designed to fail with their contacts open, and retractable landing gear for aircraft in which the wheels drop and lock in the landing position if the pressure system that raises and lowers them fails. For the batch chemical reactor example in chapter 2, where the order valves are opened is crucial, designers might have used a physical interlock that did not allow the catalyst valve to be opened while the water valve was closed.

In contrast, *active controls* require some action(s) to provide protection: (1) *detection* of a hazardous event or condition (monitoring), (2) *measurement* of some variable(s), (3) interpretation of the measurement (*diagnosis*), and (4) *response* (recovery or fail-safe procedures), all of which must be completed before a loss occurs. These actions are usually implemented by a control system, which now commonly includes a computer.

Consider the simple passive safety control where the circuit for a high-power outlet is run through a door that shields the power outlet. When the door is opened, the circuit is broken and the power disabled. When the door is closed and the power enabled, humans cannot touch the high power outlet. Such a design is simple and foolproof. An active safety control design for the same high power source, requires some type of sensor to detect when the access door to the power outlet is opened and an active controller to issue a control command to cut the power. The failure modes for the active control system are greatly increased over the passive design, as is the complexity of the system component interactions. In the railway semaphore example, there must be a way to detect that the cable has broken (probably now a digital system is used instead of a cable so the failure of the digital signaling system must be detected) and some type of active controls used to warn operators to stop the train. The design of the batch chemical reactor described in chapter 2 used a computer to control the valve opening and closing order instead of a simple mechanical interlock.

While simple examples are used here for practical reasons, the complexity of our designs is reaching and exceeding the limits of our intellectual manageability with a resulting increase in component interaction accidents and lack of enforcement of the system safety constraints. Even the relatively simple computer-based batch chemical reactor valve control design resulted in a component interaction accident. There are often very good reasons to use active controls instead of passive ones, including increased functionality, more flexibility in design, ability to operate over large distances, weight reduction, and so on. But the difficulty of the engineering problem is increased and more potential for design error is introduced.

A similar argument can be made for the interactions between operators and the processes they control. Cook [40] suggests that when controls were primarily

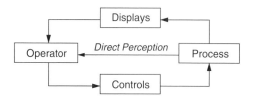

Figure 4.1
Operator has direct perception of process and mechanical controls.

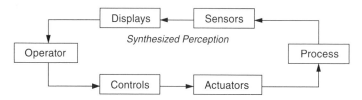

Figure 4.2
Operator has indirect information about process state and indirect controls.

mechanical and were operated by people located close to the operating process, proximity allowed sensory perception of the status of the process via direct physical feedback such as vibration, sound, and temperature (figure 4.1). Displays were directly linked to the process and were essentially a physical extension of it. For example, the flicker of a gauge needle in the cab of a train indicated that (1) the engine valves were opening and closing in response to slight pressure fluctuations, (2) the gauge was connected to the engine, (3) the pointing indicator was free, and so on. In this way, the displays provided a rich source of information about the controlled process and the state of the displays themselves.

The introduction of electromechanical controls allowed operators to control processes from a greater distance (both physical and conceptual) than possible with pure mechanically linked controls (figure 4.2). That distance, however, meant that operators lost a lot of direct information about the process—they could no longer sense the process state directly and the control and display surfaces no longer provided as rich a source of information about the process or the state of the controls themselves. The system designers had to synthesize and provide an image of the process state to the operators. An important new source of design errors was introduced by the need for the designers to determine beforehand what information the operator would need under all conditions to safely control the process. If the designers had not anticipated a particular situation could occur and provided for it in the original system design, they might also not anticipate the need of the operators for information about it during operations.

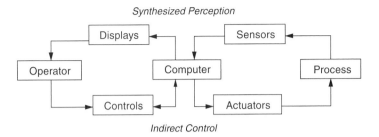

Figure 4.3
Operator has computer-generated displays and controls the process through a computer.

Designers also had to provide feedback on the actions of the operators and on any failures that might have occurred. The controls could now be operated without the desired effect on the process, and the operators might not know about it. Accidents started to occur due to incorrect feedback. For example, major accidents (including Three Mile Island) have involved the operators commanding a valve to open and receiving feedback that the valve had opened, when in reality it had not. In this case and others, the valves were wired to provide feedback indicating that power had been applied to the valve, but not that the valve had actually opened. Not only could the design of the feedback about success and failures of control actions be misleading in these systems, but the return links were also subject to failure.

Electromechanical controls relaxed constraints on the system design allowing greater functionality (figure 4.3). At the same time, they created new possibilities for designer and operator error that had not existed or were much less likely in mechanically controlled systems. The later introduction of computer and digital controls afforded additional advantages and removed even more constraints on the control system design—and introduced more possibility for error. Proximity in our old mechanical systems provided rich sources of feedback that involved almost all of the senses, enabling early detection of potential problems. We are finding it hard to capture and provide these same qualities in new systems that use automated controls and displays.

It is the freedom from constraints that makes the design of such systems so difficult. Physical constraints enforced discipline and limited complexity in system design, construction, and modification. The physical constraints also shaped system design in ways that efficiently transmitted valuable physical component and process information to operators and supported their cognitive processes.

The same argument applies to the increasing complexity in organizational and social controls and in the interactions among the components of sociotechnical systems. Some engineering projects today employ thousands of engineers. The Joint

Strike Fighter, for example, has eight thousand engineers spread over most of the United States. Corporate operations have become global, with greatly increased interdependencies and producing a large variety of products. A new holistic approach to safety, based on control and enforcing safety constraints in the entire sociotechnical system, is needed to ensure safety.

To accomplish this goal, system-level constraints must be identified, and responsibility for enforcing them must be divided up and allocated to appropriate groups. For example, the members of one group might be responsible for performing hazard analyses. The manager of this group might be assigned responsibility for ensuring that the group has the resources, skills, and authority to perform such analyses and for ensuring that high-quality analyses result. Higher levels of management might have responsibility for budgets, for establishing corporate safety policies, and for providing oversight to ensure that safety policies and activities are being carried out successfully and that the information provided by the hazard analyses is used in design and operations.

During system and product design and development, the safety constraints will be broken down and sub-requirements or constraints allocated to the components of the design as it evolves. In the batch chemical reactor, for example, the system safety requirement is that the temperature in the reactor must always remain below a particular level. A design decision may be made to control this temperature using a reflux condenser. This decision leads to a new constraint: "Water must be flowing into the reflux condenser whenever catalyst is added to the reactor." After a decision is made about what component(s) will be responsible for operating the catalyst and water valves, additional requirements will be generated. If, for example, a decision is made to use software rather than (or in addition to) a physical interlock, the software must be assigned the responsibility for enforcing the constraint: "The water valve must always be open when the catalyst valve is open."

In order to provide the level of safety demanded by society today, we first need to identify the safety constraints to enforce and then to design effective controls to enforce them. This process is much more difficult for today's complex and often high-tech systems than in the past and new techniques, such as those described in part III, are going to be required to solve it, for example, methods to assist in generating the component safety constraints from the system safety constraints. The alternative—building only the simple electromechanical systems of the past or living with higher levels of risk—is for the most part not going to be considered an acceptable solution.

4.2   The Hierarchical Safety Control Structure

In systems theory (see section 3.3), systems are viewed as hierarchical structures, where each level imposes constraints on the activity of the level beneath it—that is, constraints or lack of constraints at a higher level allow or control lower-level behavior.

Control processes operate between levels to control the processes at lower levels in the hierarchy. These control processes enforce the safety constraints for which the control process is responsible. Accidents occur when these processes provide inadequate control and the safety constraints are violated in the behavior of the lower-level components.

By describing accidents in terms of a hierarchy of control based on adaptive feedback mechanisms, adaptation plays a central role in the understanding and prevention of accidents.

At each level of the hierarchical structure, inadequate control may result from missing constraints (unassigned responsibility for safety), inadequate safety control commands, commands that were not executed correctly at a lower level, or inadequately communicated or processed feedback about constraint enforcement. For example, an operations manager may provide unsafe work instructions or procedures to the operators, or the manager may provide instructions that enforce the safety constraints, but the operators may ignore them. The operations manager may not have the feedback channels established to determine that unsafe instructions were provided or that his or her safety-related instructions are not being followed.

Figure 4.4 shows a typical sociotechnical hierarchical safety control structure common in a regulated, safety-critical industry in the United States, such as air transportation. Each system, of course, must be modeled to include its specific features. Figure 4.4 has two basic hierarchical control structures—one for system development (on the left) and one for system operation (on the right)—with interactions between them. An aircraft manufacturer, for example, might have only system development under its immediate control, but safety involves both development and operational use of the aircraft, and neither can be accomplished successfully in isolation: Safety during operation depends partly on the original design and development and partly on effective control over operations. Communication channels may be needed between the two structures.[3] For example, aircraft manufacturers must communicate to their customers the assumptions about the operational environment upon which the safety analysis was based, as well as information about safe operating procedures. The operational environment (e.g., the commercial airline industry), in turn, provides feedback to the manufacturer about the performance of the system over its lifetime.

Between the hierarchical levels of each safety control structure, effective communication channels are needed, both a downward *reference channel* providing the

---

3. Not all interactions between the two control structures are shown in the figure to simplify it and make it more readable.

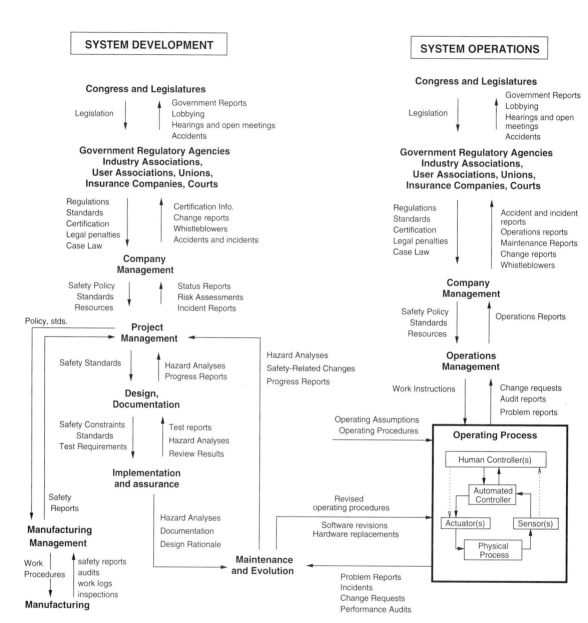

Figure 4.4
General form of a model of sociotechnical control.

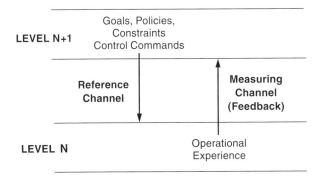

Figure 4.5
Communication channels between control levels.

information necessary to impose safety constraints on the level below and an upward *measuring channel* to provide feedback about how effectively the constraints are being satisfied (figure 4.5). Feedback is critical in any open system in order to provide adaptive control. The controller uses the feedback to adapt future control commands to more readily achieve its goals.

Government, general industry groups, and the court system are the top two levels of each of the generic control structures shown in figure 4.4. The government control structure in place to control development may differ from that controlling operations—responsibility for certifying the aircraft developed by aircraft manufacturers is assigned to one group at the FAA, while responsibility for supervising airline operations is assigned to a different group. The appropriate constraints in each control structure and at each level will vary but in general may include technical design and process constraints, management constraints, manufacturing constraints, and operational constraints.

At the highest level in both the system development and system operation hierarchies are Congress and state legislatures.[4] Congress controls safety by passing laws and by establishing and funding government regulatory structures. Feedback as to the success of these controls or the need for additional ones comes in the form of government reports, congressional hearings and testimony, lobbying by various interest groups, and, of course, accidents.

The next level contains government regulatory agencies, industry associations, user associations, insurance companies, and the court system. Unions have always played an important role in ensuring safe operations, such as the air traffic controllers union in the air transportation system, or in ensuring worker safety in

---

4. Obvious changes are required in the model for countries other than the United States. The United States is used in the example because of the author's familiarity with it.

manufacturing. The legal system tends to be used when there is no regulatory authority and the public has no other means to encourage a desired level of concern for safety in company management. The constraints generated at this level and imposed on companies are usually in the form of policy, regulations, certification, standards (by trade or user associations), or threat of litigation. Where there is a union, safety-related constraints on operations or manufacturing may result from union demands and collective bargaining.

Company management takes the standards, regulations, and other general controls on its behavior and translates them into specific policy and standards for the company. Many companies have a general safety policy (it is required by law in Great Britain) as well as more detailed standards documents. Feedback may come in the form of status reports, risk assessments, and incident reports.

In the development control structure (shown on the left of figure 4.4), company policies and standards are usually tailored and perhaps augmented by each engineering project to fit the needs of the particular project. The higher-level control process may provide only general goals and constraints and the lower levels may then add many details to operationalize the general goals and constraints given the immediate conditions and local goals. For example, while government or company standards may require a hazard analysis be performed, the system designers and documenters (including those designing the operational procedures and writing user manuals) may have control over the actual hazard analysis process used to identify specific safety constraints on the design and operation of the system. These detailed procedures may need to be approved by the level above.

The design constraints identified as necessary to control system hazards are passed to the implementers and assurers of the individual system components along with standards and other requirements. Success is determined through feedback provided by test reports, reviews, and various additional hazard analyses. At the end of the development process, the results of the hazard analyses as well as documentation of the safety-related design features and design rationale should be passed on to the maintenance group to be used in the system evolution and sustainment process.

A similar process involving layers of control is found in the system operation control structure. In addition, there will be (or at least should be) interactions between the two structures. For example, the safety design constraints used during development should form the basis for operating procedures and for performance and process auditing.

As in any control loop, time lags may affect the flow of control actions and feedback and may impact the effectiveness of the control loop in enforcing the safety constraints. For example, standards can take years to develop or change—a time scale that may keep them behind current technology and practice. At the physical level, new technology may be introduced in different parts of the system at different rates, which may result in *asynchronous evolution* of the control structure. In the accidental shootdown of two U.S. Army Black Hawk helicopters by two U.S. Air Force F-15s in the no-fly zone over northern Iraq in 1994, for example, the fighter jet aircraft and the helicopters were inhibited in communicating by radio because the F-15 pilots used newer jam-resistant radios that could not communicate with the older-technology Army helicopter radios. Hazard analysis needs to include the influence of these time lags and potential changes over time.

A common way to deal with time lags leading to delays is to delegate responsibility to lower levels that are not subject to as great a delay in obtaining information or feedback from the measuring channels. In periods of quickly changing technology, time lags may make it necessary for the lower levels to augment the control processes passed down from above or to modify them to fit the current situation. Time lags at the lowest levels, as in the Black Hawk shootdown example, may require the use of feedforward control to overcome lack of feedback or may require temporary controls on behavior: Communication between the F-15s and the Black Hawks would have been possible if the F-15 pilots had been told to use an older radio technology available to them, as they were commanded to do for other types of friendly aircraft.

More generally, control structures always change over time, particularly those that include humans and organizational components. Physical devices also change with time, but usually much slower and in more predictable ways. If we are to handle social and human aspects of safety, then our accident causality models must include the concept of change. In addition, controls and assurance that the safety control structure remains effective in enforcing the constraints over time are required.

Control does not necessarily imply rigidity and authoritarian management styles. Rasmussen notes that control at each level may be enforced in a very prescriptive command and control structure or it may be loosely implemented as performance objectives with many degrees of freedom in how the objectives are met [165]. Recent trends from management by *oversight* to management by *insight* reflect differing levels of feedback control that are exerted over the lower levels and a change from prescriptive management control to management by objectives, where the objectives are interpreted and satisfied according to the local context.

Management insight, however, does not mean abdication of safety-related responsibility. In a Milstar satellite loss [151] and both the Mars Climate Orbiter [191] and Mars Polar Lander [95, 213] losses, the accident reports all note that a poor transition from oversight to insight was a factor in the losses. Attempts to delegate decisions and to manage by objectives require an explicit formulation of the value criteria to be used and an effective means for communicating the values down through society and organizations. In addition, the impact of specific decisions at each level on the objectives and values passed down need to be adequately and formally evaluated. Feedback is required to measure how successfully the functions are being performed.

Although regulatory agencies are included in the figure 4.4 example, there is no implication that government regulation is required for safety. The only requirement is that responsibility for safety is distributed in an appropriate way throughout the sociotechnical system. In aircraft safety, for example, manufacturers play the major role while the FAA type certification authority simply provides oversight that safety is being successfully engineered into aircraft at the lower levels of the hierarchy. If companies or industries are unwilling or incapable of performing their public safety responsibilities, then government has to step in to achieve the overall public safety goals. But a much better solution is for company management to take responsibility, as it has direct control over the system design and manufacturing and over operations.

The safety-control structure will differ among industries and examples are spread among the following chapters. Figure C.1 in appendix C shows the control structure and safety constraints for the hierarchical water safety control system in Ontario, Canada. The structure is drawn on its side (as is more common for control diagrams) so that the top of the hierarchy is on the left side of the figure. The system hazard is exposure of the public to *E. coli* or other health-related contaminants through the public drinking water system; therefore, the goal of the safety control structure is to prevent such exposure. This goal leads to two system safety constraints:

1. Water quality must not be compromised.
2. Public health measures must reduce the risk of exposure if water quality is somehow compromised (such as notification and procedures to follow).

The physical processes being controlled by this control structure (shown at the right of the figure) are the water system, the wells used by the local public utilities, and public health. Details of the control structure are discussed in appendix C, but appropriate responsibility, authority, and accountability must be assigned to each component with respect to the role it plays in the overall control structure. For example, the responsibility of the Canadian federal government is to establish a nationwide public health system and ensure that it is operating effectively. The provincial government must establish regulatory bodies and codes, provide resources to the regulatory bodies, provide oversight and feedback loops to ensure that the regulators are doing their job adequately, and ensure that adequate risk assessment is conducted and effective risk management plans are in place. Local public utility operations must apply adequate doses of chlorine to kill bacteria, measure the chlorine residuals, and take further steps if evidence of bacterial contamination is found. While chlorine residuals are a quick way to get feedback about possible contamination, more accurate feedback is provided by analyzing water samples but takes longer (it has a greater time lag). Both have their uses in the overall safety control structure of the public water supply.

Safety control structures may be very complex: Abstracting and concentrating on parts of the overall structure may be useful in understanding and communicating about the controls. In examining different hazards, only subsets of the overall structure may be relevant and need to be considered in detail and the rest can be treated as the inputs to or the environment of the substructure. The only critical part is that the hazards must first be identified at the system level and the process must then proceed top-down and not bottom-up to identify the safety constraints for the parts of the overall control structure.

The operation of sociotechnical safety control structures at all levels is facing the stresses noted in chapter 1, such as rapidly changing technology, competitive and time-to-market pressures, and changing public and regulatory views of responsibility for safety. These pressures can lead to a need for new procedures or new controls to ensure that required safety constraints are not ignored.

4.3   Process Models

The third concept used in STAMP, along with safety constraints and hierarchical safety control structures, is process models. Process models are an important part of control theory. The four conditions required to control a process are described in chapter 3. The first is a *goal*, which in STAMP is the safety constraints that must be enforced by each controller in the hierarchical safety control structure. The *action condition* is implemented in the (downward) control channels and the *observability condition* is embodied in the (upward) feedback or measuring channels. The final condition is the *model condition*: *Any* controller—human or automated—needs a model of the process being controlled to control it effectively (figure 4.6).

At one extreme, this process model may contain only one or two variables, such as the model required for a simple thermostat, which contains the current temperature and the setpoint and perhaps a few control laws about how temperature is changed. At the other extreme, effective control may require a very complex model with a large number of state variables and transitions, such as the model needed to control air traffic.

Whether the model is embedded in the control logic of an automated controller or in the mental model maintained by a human controller, it must contain the same type of information: the required relationship among the system variables (the control laws), the current state (the current values of the system variables), and the ways the process can change state. This model is used to determine what control

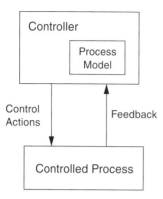

Figure 4.6
Every controller must contain a model of the process being controlled. Accidents can occur when the controller's process model does not match the system being controlled and the controller issues unsafe commands.

actions are needed, and it is updated through various forms of feedback. If the model of the room temperature shows that the ambient temperature is less than the setpoint, then the thermostat issues a control command to start a heating element. Temperature sensors provide feedback about the (hopefully rising) temperature. This feedback is used to update the thermostat's model of the current room temperature. When the setpoint is reached, the thermostat turns off the heating element. In the same way, human operators also require accurate process or mental models to provide safe control actions.

Component interaction accidents can usually be explained in terms of incorrect process models. For example, the Mars Polar Lander software thought the spacecraft had landed and issued a control instruction to shut down the descent engines. The captain of the *Herald of Free Enterprise* thought the ferry doors were closed and ordered the ship to leave the mooring. The pilots in the Cali Colombia B757 crash thought *R* was the symbol denoting the radio beacon near Cali.

In general, accidents often occur, particularly component interaction accidents and accidents involving complex digital technology or human error, when the process model used by the controller (automated or human) does not match the process and, as a result:

1. Incorrect or unsafe control commands are given

2. Required control actions (for safety) are not provided

3. Potentially correct control commands are provided at the wrong time (too early or too late), or

4. Control is stopped too soon or applied too long.

These four types of inadequate control actions are used in the new hazard analysis technique described in chapter 8.

A model of the process being controlled is required not just at the lower physical levels of the hierarchical control structure, but at all levels. In order to make proper decisions, the manager of an oil refinery may need to have a model of the current maintenance level of the safety equipment of the refinery, the state of safety training of the workforce, and the degree to which safety requirements are being followed or are effective, among other things. The CEO of the global oil conglomerate has a much less detailed model of the state of the refineries he controls but at the same time requires a broader view of the state of safety of all the corporate assets in order to make appropriate corporate-level decisions impacting safety.

Process models are not only used during operations but also during system development activities. Designers use both models of the system being designed and models of the development process itself. The developers may have an incorrect model of the system or software behavior necessary for safety or the physical laws controlling the system. Safety may also be impacted by developers' incorrect models of the development process itself.

As an example of the latter, a Titan/Centaur satellite launch system, along with the Milstar satellite it was transporting into orbit, was lost due to a typo in a load tape used by the computer to determine the attitude change instructions to issue to the engines. The information on the load tape was essentially part of the process model used by the attitude control software. The typo was not caught during the development process partly because of flaws in the developers' models of the testing process—each thought someone else was testing the software using the actual load tape when, in fact, nobody was (see appendix B).

In summary, process models play an important role (1) in understanding why accidents occur and why humans provide inadequate control over safety-critical systems and (2) in designing safer systems.

4.4   STAMP

The STAMP (Systems-Theoretic Accident Model and Process) model of accident causation is built on these three basic concepts—safety constraints, a hierarchical safety control structure, and process models—along with basic systems theory concepts. All the pieces for a new causation model have been presented. It is now simply a matter of putting them together.

In STAMP, systems are viewed as interrelated components kept in a state of dynamic equilibrium by feedback control loops. Systems are not treated as static but as dynamic processes that are continually adapting to achieve their ends and to react to changes in themselves and their environment.

Safety is an emergent property of the system that is achieved when appropriate constraints on the behavior of the system and its components are satisfied. The original design of the system must not only enforce appropriate constraints on behavior to ensure safe operation, but the system must continue to enforce the safety constraints as changes and adaptations to the system design occur over time.

Accidents are the result of flawed processes involving interactions among people, societal and organizational structures, engineering activities, and physical system components that lead to violating the system safety constraints. The process leading up to an accident is described in STAMP in terms of an adaptive feedback function that fails to maintain safety as system performance changes over time to meet a complex set of goals and values.

Instead of defining safety management in terms of preventing component failures, it is defined as creating a safety control structure that will enforce the behavioral safety constraints and ensure its continued effectiveness as changes and adaptations occur over time. Effective safety (and risk) management may require limiting the types of changes that occur but the goal is to allow as much flexibility and performance enhancement as possible while enforcing the safety constraints.

Accidents can be understood, using STAMP, by identifying the safety constraints that were violated and determining why the controls were inadequate in enforcing them. For example, understanding the Bhopal accident requires determining not simply why the maintenance personnel did not insert the slip blind, but also why the controls that had been designed into the system to prevent the release of hazardous chemicals and to mitigate the consequences of such occurrences—including maintenance procedures and oversight of maintenance processes, refrigeration units, gauges and other monitoring units, a vent scrubber, water spouts, a flare tower, safety audits, alarms and practice alerts, emergency procedures and equipment, and others—were not successful.

STAMP not only allows consideration of more accident causes than simple component failures, but it also allows more sophisticated analysis of failures and component failure accidents. Component failures may result from inadequate constraints on the manufacturing process; inadequate engineering design such as missing or incorrectly implemented fault tolerance; lack of correspondence between individual component capacity (including human capacity) and task requirements; unhandled environmental disturbances (e.g., electromagnetic interference or EMI); inadequate maintenance; physical degradation (wearout); and so on.

Component failures may be prevented by increasing the integrity or resistance of the component to internal or external influences or by building in safety margins or safety factors. They may also be avoided by operational controls, such as

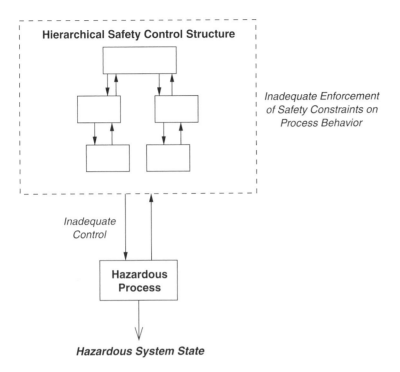

Figure 4.7
Accidents result from inadequate enforcement of the behavioral safety constraints on the process.

operating the component within its design envelope and by periodic inspections and preventive maintenance. Manufacturing controls can reduce deficiencies or flaws introduced during the manufacturing process. The effects of physical component failure on system behavior may be eliminated or reduced by using redundancy. The important difference from other causality models is that STAMP goes beyond simply blaming component failure for accidents by requiring that the reasons be identified for why those failures occurred (including systemic factors) and led to an accident, that is, why the controls instituted for preventing such failures or for minimizing their impact on safety were missing or inadequate. And it includes other types of accident causes, such as component interaction accidents, which are becoming more frequent with the introduction of new technology and new roles for humans in system control.

STAMP does not lend itself to a simple graphic representation of accident causality (see figure 4.7). While dominoes, event chains, and holes in Swiss cheese are very compelling because they are easy to grasp, they oversimplify causality and thus the approaches used to prevent accidents.

4.5   A General Classification of Accident Causes

Starting from the basic definitions in STAMP, the general causes of accidents can be identified using basic systems and control theory. The resulting classification is useful in accident analysis and accident prevention activities.

Accidents in STAMP are the result of a complex process that results in the system behavior violating the safety constraints. The safety constraints are enforced by the control loops between the various levels of the hierarchical control structure that are in place during design, development, manufacturing, and operations.

Using the STAMP causality model, if there is an accident, one or more of the following must have occurred:

1. The safety constraints were not enforced by the controller.

   a. The control actions necessary to enforce the associated safety constraint at each level of the sociotechnical control structure for the system were not provided.

   b. The necessary control actions were provided but at the wrong time (too early or too late) or stopped too soon.

   c. Unsafe control actions were provided that caused a violation of the safety constraints.

2. Appropriate control actions were provided but not followed.

These same general factors apply at each level of the sociotechnical control structure, but the interpretation (application) of the factor at each level may differ.

Classification of accident causal factors starts by examining each of the basic components of a control loop (see figure 3.2) and determining how their improper operation may contribute to the general types of inadequate control.

Figure 4.8 shows the classification. The causal factors in accidents can be divided into three general categories: (1) the controller operation, (2) the behavior of actuators and controlled processes, and (3) communication and coordination among controllers and decision makers. When humans are involved in the control structure, context and behavior-shaping mechanisms also play an important role in causality.

4.5.1   Controller Operation

Controller operation has three primary parts: control inputs and other relevant external information sources, the control algorithms, and the process model. Inadequate, ineffective, or missing control actions necessary to enforce the safety constraints and ensure safety can stem from flaws in each of these parts. For human controllers and actuators, context is also an important factor.

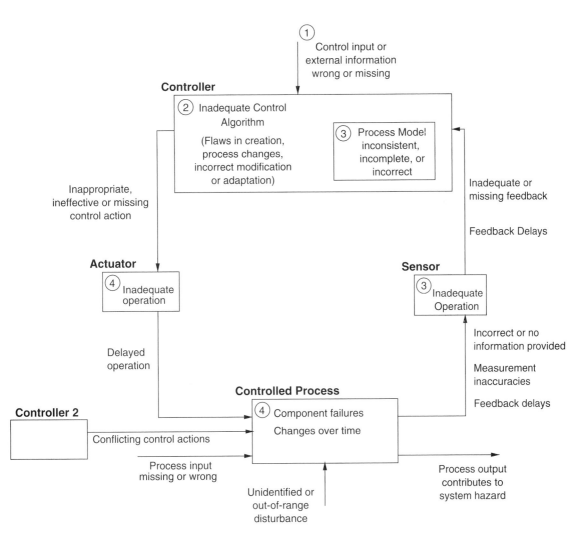

Figure 4.8
A classification of control flaws leading to hazards.

Unsafe Inputs (① in figure 4.8)

Each controller in the hierarchical control structure is itself controlled by higher-level controllers. The control actions and other information provided by the higher level and required for safe behavior may be missing or wrong. Using the Black Hawk friendly fire example again, the F-15 pilots patrolling the no-fly zone were given instructions to switch to a non-jammed radio mode for a list of aircraft types that did not have the ability to interpret jammed broadcasts. Black Hawk helicopters had not been upgraded with new anti-jamming technology but were omitted from the list and so could not hear the F-15 radio broadcasts. Other types of missing or wrong noncontrol inputs may also affect the operation of the controller.

Unsafe Control Algorithms (② in figure 4.8)

Algorithms in this sense are both the procedures designed by engineers for hardware controllers and the procedures that human controllers use. Control algorithms may not enforce safety constraints because the algorithms are inadequately designed originally, the process may change and the algorithms become unsafe, or the control algorithms may be inadequately modified by maintainers if the algorithms are automated or through various types of natural adaptation if they are implemented by humans. Human control algorithms are affected by initial training, by the procedures provided to the operators to follow, and by feedback and experimentation over time (see figure 2.9).

Time delays are an important consideration in designing control algorithms. Any control loop includes time lags, such as the time between the measurement of process parameters and receiving those measurements or between issuing a command and the time the process state actually changes. For example, pilot response delays are important time lags that must be considered in designing the control function for TCAS[5] or other aircraft systems, as are time lags in the controlled process—the aircraft trajectory, for example—caused by aircraft performance limitations.

Delays may not be directly observable, but may need to be inferred. Depending on where in the feedback loop the delay occurs, different control algorithms are required to cope with the delays [25]: dead time and time constants require an algorithm that makes it possible to predict when an action is needed before the need. Feedback delays generate requirements to predict when a prior control action has taken effect and when resources will be available again. Such requirements may impose the need for some type of open loop or feedforward strategy to cope with

---

5. TCAS (Traffic alert and Collision Avoidance System) is an airborne system used to avoid collisions between aircraft. More details about TCAS can be found in chapter 10.

delays. When time delays are not adequately considered in the control algorithm, accidents can result.

Leplat has noted that many accidents relate to *asynchronous evolution* [112], where one part of a system (in this case the hierarchical safety control structure) changes without the related necessary changes in other parts. Changes to subsystems may be carefully designed, but consideration of their effects on other parts of the system, including the safety control aspects, may be neglected or inadequate. Asynchronous evolution may also occur when one part of a properly designed system deteriorates.

In both these cases, the erroneous expectations of users or system components about the behavior of the changed or degraded subsystem may lead to accidents. The Ariane 5 trajectory changed from that of the Ariane 4, but the inertial reference system software was not changed. As a result, an assumption of the inertial reference software was violated and the spacecraft was lost shortly after launch. One factor in the loss of contact with SOHO (SOlar Heliospheric Observatory), a scientific spacecraft, in 1998 was the failure to communicate to operators that a functional change had been made in a procedure to perform gyro spin down. The Black Hawk friendly fire accident (analyzed in chapter 5) had several examples of asynchronous evolution, for example the mission changed and an individual key to communication between the Air Force and Army left, leaving the safety control structure without an important component.

Communication is a critical factor here as well as monitoring for changes that may occur and feeding back this information to the higher-level control. For example, the safety analysis process that generates constraints always involves some basic assumptions about the operating environment of the process. When the environment changes such that those assumptions are no longer true, as in the Ariane 5 and SOHO examples, the controls in place may become inadequate. Embedded pacemakers provide another example. These devices were originally assumed to be used only in adults, who would lie quietly in the doctor's office while the pacemaker was being "programmed." Later these devices began to be used in children, and the assumptions under which the hazard analysis was conducted and the controls were designed no longer held and needed to be revisited. A requirement for effective updating of the control algorithms is that the assumptions of the original (and subsequent) analysis are recorded and retrievable.

Inconsistent, Incomplete, or Incorrect Process Models (③ in figure 4.8)
Section 4.3 stated that effective control is based on a model of the process state. Accidents, particularly component interaction accidents, most often result from inconsistencies between the models of the process used by the controllers (both

human and automated) and the actual process state. When the controller's model of the process (either the human mental model or the software or hardware model) diverges from the process state, erroneous control commands (based on the incorrect model) can lead to an accident: for example, (1) the software does not know that the plane is on the ground and raises the landing gear, or (2) the controller (automated or human) does not identify an object as friendly and shoots a missile at it, or (3) the pilot thinks the aircraft controls are in *speed* mode but the computer has changed the mode to *open descent* and the pilot behaves inappropriately for that mode, or (4) the computer does not think the aircraft has landed and overrides the pilots' attempts to operate the braking system. All of these examples have actually occurred.

The mental models of the system developers are also important. During software development, for example, the programmers' models of required behavior may not match the engineers' models (commonly referred to as a software requirements error), or the software may be executed on computer hardware or may control physical systems during operations that differ from what was assumed by the programmer and used during testing. The situation becomes more even complicated when there are multiple controllers (both human and automated) because each of their process models must also be kept consistent.

The most common form of inconsistency occurs when one or more process models is incomplete in terms of not defining appropriate behavior for all possible process states or all possible disturbances, including unhandled or incorrectly handled component failures. Of course, no models are complete in the absolute sense: The goal is to make them complete enough that no safety constraints are violated when they are used. Criteria for completeness in this sense are presented in *Safeware*, and completeness analysis is integrated into the new hazard analysis method as described in chapter 9.

How does the process model become inconsistent with the actual process state? The process model designed into the system (or provided by training if the controller is human) may be wrong from the beginning, there may be missing or incorrect feedback for updating the process model as the controlled process changes state, the process model may be updated incorrectly (an error in the algorithm of the controller), or time lags may not be accounted for. The result can be uncontrolled disturbances, unhandled process states, inadvertent commanding of the system into a hazardous state, unhandled or incorrectly handled controlled process component failures, and so forth.

Feedback is critically important to the safe operation of the controller. A basic principle of system theory is that no control system will perform better than its measuring channel. Feedback may be missing or inadequate because such feedback is not included in the system design, flaws exist in the monitoring or feedback communication channel, the feedback is not timely, or the measuring instrument operates inadequately.

A contributing factor cited in the Cali B757 accident report, for example, was the omission of the waypoints[6] behind the aircraft from cockpit displays, which contributed to the crew not realizing that the waypoint for which they were searching was behind them (missing feedback). The model of the Ariane 501 attitude used by the attitude control software became inconsistent with the launcher attitude when an error message sent by the inertial reference system was interpreted by the attitude control system as data (incorrect processing of feedback), causing the spacecraft onboard computer to issue an incorrect and unsafe command to the booster and main engine nozzles.

Other reasons for the process models to diverge from the true system state may be more subtle. Information about the process state has to be inferred from measurements. For example, in the TCAS II aircraft collision avoidance system, relative range positions of other aircraft are computed based on round-trip message propagation time. The theoretical control function (control law) uses the true values of the controlled variables or component states (e.g., true aircraft positions). However, at any time, the controller has only measured values, which may be subject to time lags or inaccuracies. The controller must use these measured values to infer the true conditions in the process and, if necessary, to derive corrective actions to maintain the required process state. In the TCAS example, sensors include on-board devices such as altimeters that provide measured altitude (not necessarily true altitude) and antennas for communicating with other aircraft. The primary TCAS actuator is the pilot, who may or may not respond to system advisories. The mapping between the measured or assumed values and the true values can be flawed.

To summarize, process models can be incorrect from the beginning—where correct is defined in terms of consistency with the current process state and with the models being used by other controllers—or they can become incorrect due to erroneous or missing feedback or measurement inaccuracies. They may also be incorrect only for short periods of time due to time lags in the process loop.

4.5.2 Actuators and Controlled Processes (④ in figure 4.8)
The factors discussed so far have involved inadequate control. The other case occurs when the control commands maintain the safety constraints, but the controlled process may not implement these commands. One reason might be a failure or flaw in the reference channel, that is, in the transmission of control commands. Another reason might be an actuator or controlled component fault or failure. A third is that

---

6. A *waypoint* is a set of coordinates that identify a point in physical space.

the safety of the controlled process may depend on inputs from other system components, such as power, for the execution of the control actions provided. If these process inputs are missing or inadequate in some way, the controller process may be unable to execute the control commands and accidents may result. Finally, there may be external disturbances that are not handled by the controller.

In a hierarchical control structure, the actuators and controlled process may themselves be a controller of a lower-level process. In this case, the flaws in executing the control are the same described earlier for a controller.

Once again, these types of flaws do not simply apply to operations or to the technical system but also to system design and development. For example, a common flaw in system development is that the safety information gathered or created by the system safety engineers (the hazards and the necessary design constraints to control them) is inadequately communicated to the system designers and testers, or that flaws exist in the use of this information in the system development process.

4.5.3   Coordination and Communication among Controllers and Decision Makers
When there are multiple controllers (human and/or automated), control actions may be inadequately coordinated, including unexpected side effects of decisions or actions or conflicting control actions. Communication flaws play an important role here.

Leplat suggests that accidents are most likely in *overlap areas* or in *boundary areas* or where two or more controllers (human or automated) control the same process or processes with common boundaries (figure 4.9) [112]. In both boundary and overlap areas, the potential exists for ambiguity and for conflicts among independent decisions.

Responsibility for the control functions in boundary areas is often poorly defined. For example, Leplat cites an iron and steel plant where frequent accidents occurred at the boundary of the blast furnace department and the transport department. One conflict arose when a signal informing transport workers of the state of the blast

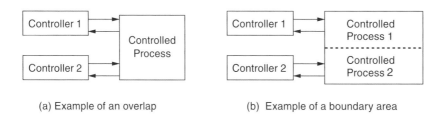

(a) Example of an overlap   (b) Example of a boundary area

Figure 4.9
Problems often occur when there is shared control over the same process or at the boundary areas between separately controlled processes.

furnace did not work and was not repaired because each department was waiting for the other to fix it. Faverge suggests that such dysfunction can be related to the number of management levels separating the workers in the departments from a common manager: The greater the distance, the more difficult the communication, and thus the greater the uncertainty and risk.

Coordination problems in the control of boundary areas are rife. As mentioned earlier, a Milstar satellite was lost due to inadequate attitude control of the Titan/ Centaur launch vehicle, which used an incorrect process model based on erroneous inputs on a software load tape. After the accident, it was discovered that nobody had tested the software using the actual load tape—each group involved in testing and assurance had assumed some other group was doing so. In the system development process, system engineering and mission assurance activities were missing or ineffective, and a common control or management function was quite distant from the individual development and assurance groups (see appendix B). One factor in the loss of the Black Hawk helicopters to friendly fire over northern Iraq was that the helicopters normally flew only in the boundary areas of the no-fly zone and procedures for handling aircraft in those areas were ill defined. Another factor was that an Army base controlled the flights of the Black Hawks, while an Air Force base controlled all the other components of the airspace. A common control point once again was high above where the accident occurred in the control structure. In addition, communication problems existed between the Army and Air Force bases at the intermediate control levels.

Overlap areas exist when a function is achieved by the cooperation of two controllers or when two controllers exert influence on the same object. Such overlap creates the potential for conflicting control actions (dysfunctional interactions among control actions). Leplat cites a study of the steel industry that found 67 percent of technical incidents with material damage occurred in areas of co-activity, although these represented only a small percentage of the total activity areas. In an A320 accident in Bangalore, India, the pilot had disconnected his flight director during approach and assumed that the copilot would do the same. The result would have been a mode configuration in which airspeed is automatically controlled by the autothrottle (the *speed* mode), which is the recommended procedure for the approach phase. However, the copilot had not turned off his flight director, which meant that *open descent* mode became active when a lower altitude was selected instead of *speed* mode, eventually contributing to the crash of the aircraft short of the runway [181]. In the Black Hawks' shootdown by friendly fire, the aircraft surveillance officer (ASO) thought she was responsible only for identifying and tracking aircraft south of the 36th Parallel, while the air traffic controller for the area north of the 36th Parallel thought the ASO was also tracking and identifying aircraft in his area and acted accordingly.

In 2002, two aircraft collided over southern Germany. An important factor in the accident was the lack of coordination between the airborne TCAS (collision avoidance) system and the ground air traffic controller. They each gave different and conflicting advisories on how to avoid a collision. If both pilots had followed one or the other, the loss would have been avoided, but one followed the TCAS advisory and the other followed the ground air traffic control advisory.

4.5.4   Context and Environment

Flawed human decision making can result from incorrect information and inaccurate process models, as described earlier. But human behavior is also greatly impacted by the context and environment in which the human is working. These factors have been called "behavior shaping mechanisms." While value systems and other influences on decision making can be considered to be inputs to the controller, describing them in this way oversimplifies their role and origin. A classification of the contextual and behavior-shaping mechanisms is premature at this point, but relevant principles and heuristics are elucidated throughout the rest of the book.

4.6   Applying the New Model

To summarize, STAMP focuses particular attention on the role of constraints in safety management. Accidents are seen as resulting from inadequate control or enforcement of constraints on safety-related behavior at each level of the system development and system operations control structures. Accidents can be understood in terms of why the controls that were in place did not prevent or detect maladaptive changes.

Accident causal analysis based on STAMP starts with identifying the safety constraints that were violated and then determines why the controls designed to enforce the safety constraints were inadequate or, if they were potentially adequate, why the system was unable to exert appropriate control over their enforcement.

In this conception of safety, there is no "root cause." Instead, the accident "cause" consists of an inadequate safety control structure that under some circumstances leads to the violation of a behavioral safety constraint. Preventing future accidents requires reengineering or designing the safety control structure to be more effective.

Because the safety control structure and the behavior of the individuals in it, like any physical or social system, changes over time, accidents must be viewed as dynamic processes. Looking only at the time of the proximal loss events distorts and omits from view the most important aspects of the larger accident process that are needed to prevent reoccurrences of losses from the same causes in the future. Without that view, we see and fix only the symptoms, that is, the results of the flawed processes and inadequate safety control structure without getting to the sources of those symptoms.

To understand the dynamic aspects of accidents, the process leading to the loss can be viewed as an adaptive feedback function where the safety control system performance degrades over time as the system attempts to meet a complex set of goals and values. Adaptation is critical in understanding accidents, and the adaptive feedback mechanism inherent in the model allows a STAMP analysis to incorporate adaptation as a fundamental system property.

We have found in practice that using this model helps us to separate factual data from the interpretations of that data: While the events and physical data involved in accidents may be clear, their importance and the explanations for why the factors were present are often subjective as is the selection of the events to consider.

STAMP models are also more complete than most accident reports and other models, for example see [9, 89, 140]. Each of the explanations for the incorrect FMS input of R in the Cali American Airlines accident described in chapter 2, for example, appears in the STAMP analysis of that accident at the appropriate levels of the control structure where they operated. The use of STAMP helps not only to identify the factors but also to understand the relationships among them.

While STAMP models will probably not be useful in law suits as they do not assign blame for the accident to a specific person or group, they do provide more help in understanding accidents by forcing examination of each part of the socio-technical system to see how it contributed to the loss—and there will usually be contributions at each level. Such understanding should help in learning how to engineer safer systems, including the technical, managerial, organizational, and regulatory aspects.

To accomplish this goal, a framework for classifying the factors that lead to accidents was derived from the basic underlying conceptual accident model (see figure 4.8). This classification can be used in identifying the factors involved in a particular accident and in understanding their role in the process leading to the loss. The accident investigation after the Black Hawk shootdown (analyzed in detail in the next chapter) identified 130 different factors involved in the accident. In the end, only the AWACS senior director was court-martialed, and he was acquitted. The more one knows about an accident process, the more difficult it is to find one person or part of the system responsible, but the easier it is to find effective ways to prevent similar occurrences in the future.

STAMP is useful not only in analyzing accidents that have occurred but in developing new and potentially more effective system engineering methodologies to prevent accidents. Hazard analysis can be thought of as investigating an accident before it occurs. Traditional hazard analysis techniques, such as fault tree analysis and various types of failure analysis techniques, do not work well for very complex systems, for software errors, human errors, and system design errors. Nor do they usually include organizational and management flaws. The problem is that these hazard analysis techniques are limited by a focus on failure events and the role of component failures in accidents; they do not account for component interaction accidents, the complex roles that software and humans are assuming in high-tech systems, the organizational factors in accidents, and the indirect relationships between events and actions required to understand why accidents occur.

STAMP provides a direction to take in creating these new hazard analysis and prevention techniques. Because in a system accident model everything starts from constraints, the new approach focuses on identifying the constraints required to maintain safety; identifying the flaws in the control structure that can lead to an accident (inadequate enforcement of the safety constraints); and then designing a control structure, physical system and operating conditions that enforces the constraints.

Such hazard analysis techniques augment the typical failure-based design focus and encourage a wider variety of risk reduction measures than simply adding redundancy and overdesign to deal with component failures. The new techniques also provide a way to implement *safety-guided design* so that safety analysis guides the design generation rather than waiting until a design is complete to discover it is unsafe. Part III describes ways to use techniques based on STAMP to prevent accidents through system design, including design of the operating conditions and the safety management control structure.

STAMP can also be used to improve performance analysis. Performance monitoring of complex systems has created some dilemmas. Computers allow the collection of massive amounts of data, but analyzing that data to determine whether the system is moving toward the boundaries of safe behavior is difficult. The use of an accident model based on system theory and the basic concept of safety constraints may provide directions for identifying appropriate safety metrics and leading indicators; determining whether control over the safety constraints is adequate; evaluating the assumptions about the technical failures and potential design errors, organizational structure, and human behavior underlying the hazard analysis; detecting errors in the operational and environmental assumptions underlying the design and the organizational culture; and identifying any maladaptive changes over time that could increase risk of accidents to unacceptable levels.

Finally, STAMP points the way to very different approaches to risk assessment. Currently, risk assessment is firmly rooted in the probabilistic analysis of failure events. Attempts to extend current PRA techniques to software and other new technology, to management, and to cognitively complex human control activities have been disappointing. This way forward may lead to a dead end. Significant progress in risk assessment for complex systems will require innovative approaches starting from a completely different theoretical foundation.

5   A Friendly Fire Accident

The goal of STAMP is to assist in understanding why accidents occur and to use that understanding to create new and better ways to prevent losses. This chapter and several of the appendices provide examples of how STAMP can be used to analyze and understand accident causation. The particular examples were selected to demonstrate the applicability of STAMP to very different types of systems and industries. A process, called CAST (Causal Analysis based on STAMP) is described in chapter 11 to assist in performing the analysis.

This chapter delves into the causation of the loss of a U.S. Army Black Hawk helicopter and all its occupants from friendly fire by a U.S. Air Force F-15 over northern Iraq in 1994. This example was chosen because the controversy and multiple viewpoints and books about the shootdown provide the information necessary to create most of the STAMP analysis. Accident reports often leave out important causal information (as did the official accident report in this case). Because of the nature of the accident, most of the focus is on operations. Appendix B presents an example of an accident where engineering development plays an important role. Social issues involving public health are the focus of the accident analysis in appendix C.

5.1 Background

After the Persian Gulf War, Operation Provide Comfort (OPC) was created as a multinational humanitarian effort to relieve the suffering of hundreds of thousands of Kurdish refugees who fled into the hills of northern Iraq during the war. The goal of the military efforts was to provide a safe haven for the resettlement of the refugees and to ensure the security of relief workers assisting them. The formal mission statement for OPC read: "To deter Iraqi behavior that may upset peace and order in northern Iraq."

In addition to operations on the ground, a major component of OPC's mission was to occupy the airspace over northern Iraq. To accomplish this task, a no-fly zone

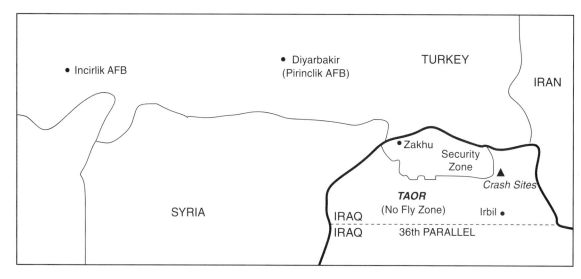

Figure 5.1
The no-fly zone and relevant surrounding locations.

(also called the TAOR or Tactical Area of Responsibility) was established that included all airspace within Iraq north of the 36th Parallel (see figure 5.1). Air operations were led by the Air Force to prohibit Iraqi aircraft from entering the no-fly zone while ground operations were organized by the Army to provide humanitarian assistance to the Kurds and other ethnic groups in the area.

U.S., Turkish, British, and French fighter and support aircraft patrolled the no-fly zone daily to prevent Iraqi warplanes from threatening the relief efforts. The mission of the Army helicopters was to support the ground efforts; the Army used them primarily for troop movement, resupply, and medical evacuation.

On April 15, 1994, after nearly three years of daily operations over the TAOR (Tactical Area of Responsibility), two U.S. Air Force F-15's patrolling the area shot down two U.S. Army Black Hawk helicopters, mistaking them for Iraqi Hind helicopters. The Black Hawks were carrying twenty-six people, fifteen U.S. citizens and eleven others, among them British, French, and Turkish military officers as well as Kurdish citizens. All were killed in one of the worst air-to-air friendly fire accidents involving U.S. aircraft in military history.

All the aircraft involved were flying in clear weather with excellent visibility, an AWACS (Airborne Warning and Control System) aircraft was providing surveillance and control for the aircraft in the area, and all the aircraft were equipped with electronic identification and communication equipment (apparently working properly) and flown by decorated and highly experienced pilots.

The hazard being controlled was mistaking a "friendly" (coalition) aircraft for a threat and shooting at it. This hazard, informally called *friendly fire*, was well known, and a control structure was established to prevent it. Appropriate constraints were established and enforced at each level, from the Joint Chiefs of Staff down to the aircraft themselves. Understanding why this accident occurred requires understanding why the control structure in place was ineffective in preventing the loss. Preventing future accidents involving the same control flaws requires making appropriate changes to the control structure, including establishing monitoring and feedback loops to detect when the controls are becoming ineffective and the system is migrating toward an accident, that is, moving toward a state of increased risk. The more comprehensive the model and factors identified, the larger the class of accidents that can be prevented.

For this STAMP example, information about the accident and the control structure was obtained from the original accident report [5], a GAO (Government Accountability Office) report on the accident investigation process and results [200], and two books on the shootdown—one originally a Ph.D. dissertation by Scott Snook [191] and one by Joan Piper, the mother of one of the victims [159]. Because of the extensive existing analysis, much of the control structure (shown in figure 5.3) can be reconstructed from these sources. A large number of acronyms are used in this chapter. They are defined in figure 5.2.

5.2   The Hierarchical Safety Control Structure to Prevent Friendly Fire Accidents

National Command Authority and Commander-in-Chief Europe

When the National Command Authority (the President and Secretary of Defense) directed the military to conduct Operation Provide Comfort, the U.S. Commander in Chief Europe (USCINCEUR) directed the creation of Combined Task Force (CTF) Provide Comfort.

A series of orders and plans established the general command and control structure of the CTF. These orders and plans also transmitted sufficient authority and guidance to subordinate component commands and operational units so that they could then develop the local procedures that were necessary to bridge the gap between general mission orders and specific subunit operations.

At the top of the control structure, the National Command Authority (the President and Secretary of Defense, who operate through the Joint Chiefs of Staff) provided guidelines for establishing Rules of Engagement (ROE). ROE govern the actions allowed by U.S. military forces to protect themselves and other personnel and property against attack or hostile incursion and specify a strict sequence of procedures to be followed prior to any coalition aircraft firing its weapons. They are

AAI	Air to Air Interrogation (used with IFF)
ACE	Airborne Command Element (the commander's representative in the AWACS)
ACO	Airspace Control Order (Guidance for all local air operations in OPC)
AFB	Air Force Base
AI	Airborne Intercept (a type of radar on fighter aircraft)
ARF	Aircraft Read Files (supplement to the ACO including the ROE)
ASO	Air Surveillance Officer (one of the positions in the AWACS)
ATO	Air Tasking Order (specific mission guidance for the day)
AWACS	Airborne Warning and Control System (a military air traffic control system in the sky)
BH	Black Hawk
BSD	Battle Staff Directive (late scheduling changes not making it into the ATO)
CTF	Combined Task Force
CFAC	Combined Forces Air Component (tactical control of all OPC aircraft operating in TAOR and operational control of AF aircraft)
DO	Director of Operations
GAO	U.S. Government Accountability Office
HQ-II	Have Quick (frequency hopping) radios
HUD	Heads Up Display
IFF	Identiification Friend or Foe
JOIC	Joint Operations and Intelligence Center
JSOC	Joint Special Operations Component (search and rescue operations inside Iraq)
JTIDS	Joint Tactical Information Distribution Center (provides ground with picture of airspace occupants)
MCC	Military Coordination Center (operational control of the Black Hawk helicopters)
MD	Mission Director (runs the mission from the ground)
Min Comm	Minimal Communications
NCA	National Command Authority (the President and the Secretary of Defense)
NFZ	No-Fly Zone
OPC	Operation Provide Comfort (multi-nation effort to protect Kurdish refugees)
ROE	Rules of Engagement (rules governing actions allowed by the U.S. military forces)
SD	Senior Director (one of the positions in the AWACS)
SITREP	Situation Report
SPINS	Mission-related Special Instructions
TACSAT	Tactical Satellite radios (used by Army helicopter pilots to communicate with MCC
TAOR	Tactical Area of Responsibility (another name for the No-Fly Zone)
USCINCEUR	U.S. Commander in Chief, Europe
VID	Visual Identification
WD	Weapons Director (a position in the AWACS)

Figure 5.2
Acronyms used in this chapter.

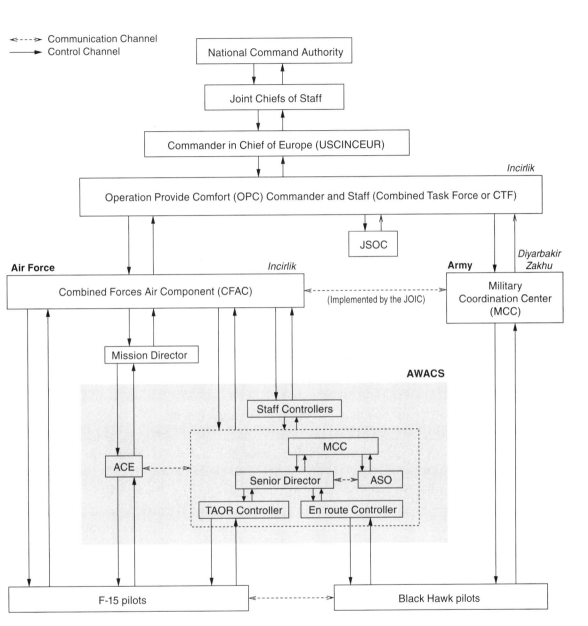

Figure 5.3
Control structure in the Iraqi no-fly zone.

based on legal, political, and military considerations and are intended to provide for adequate self-defense to ensure that military activities are consistent with current national objectives and that appropriate controls are placed on combat activities. Commanders establish ROE for their areas of responsibility that are consistent with the Joint Chiefs of Staff guidelines, modifying them for special operations and for changing conditions.

Because the ROE dictate how hostile aircraft or military threats are treated, they play an important role in any friendly fire accidents. The ROE in force for OPC were the peacetime ROE for the United States European Command with OPC modifications approved by the National Command Authority. These conservative ROE required a strict sequence of procedures to be followed prior to any coalition aircraft firing its weapons. The less aggressive peacetime rules of engagement were used even though the area had been designated a combat zone because of the number of countries involved in the joint task force. The goal of the ROE was to slow down any military confrontation in order to prevent the type of friendly fire accidents that had been common during Operation Desert Storm. Understanding the reasons for the shootdown of the Black Hawk helicopters requires understanding why the ROE did not provide an effective control to prevent friendly fire accidents.

Three System-Level Safety Constraints Related to This Accident:

1. The NCA and UNCINCEUR must establish a command and control structure that provides the ability to prevent friendly fire accidents.

2. The guidelines for ROE generated by the Joint Chiefs of Staff (with tailoring to suit specific operational conditions) must be capable of preventing friendly fire accidents in all types of situations.

3. The European Commander-in-Chief must review and monitor operational plans generated by the Combined Task Force, ensure they are updated as the mission changes, and provide the personnel required to carry out the plans.

Controls: The controls in place included the ROE guidelines, the operational orders, and review procedures for the controls (e.g., the actual ROE and Operational Plans) generated at the control levels below.

Combined Task Force (CTF)

The components of the Combined Task Force (CTF) organization relevant to the accident (and to preventing friendly fire) were a Combined Task Force staff, a Combined Forces Air Component (CFAC), and an Army Military Coordination Center. The Air Force fighter aircraft were co-located with CTF Headquarters and CFAC at Incirlik Air Base in Turkey while the U.S. Army helicopters were located with the Army headquarters at Diyarbakir, also in Turkey (see figure 5.1).

The Combined Task Force had three components under it (figure 5.3):

1. The Military Coordination Center (MCC) monitored conditions in the security zone and had operational control of Eagle Flight helicopters (the Black Hawks), which provided general aviation support to the MCC and the CTF.

2. The Joint Special Operations Component (JSOC) was assigned primary responsibility to conduct search-and-rescue operations should any coalition aircraft go down inside Iraq.

3. The Combined Forces Air Component (CFAC) was tasked with exercising tactical control of all OPC aircraft operating in the Tactical Area of Responsibility (TAOR) and operational control over Air Force aircraft.[1] The CFAC commander exercised daily control of the OPC flight mission through a Director of Operations (CFAC/DO), as well as a ground-based Mission Director at the Combined Task Force (CTF) headquarters in Incirlik and an Airborne Command Element (ACE) aboard the AWACS.

Operational orders were generated at the European Command level of authority that defined the initial command and control structure and directed the CTF commanders to develop an operations plan to govern OPC. In response, the CTF commander created an operations plan in July 1991 delineating the command relationships and organizational responsibilities within the CTF. In September 1991, the U.S. Commander-in-Chief, Europe, modified the original organizational structure in response to the evolving mission in northern Iraq, directing an increase in the size of the Air Force and the withdrawal of a significant portion of the ground forces.

The CTF was ordered to provide a supporting plan to implement the changes necessary in their CTF operations plan. The Accident Investigation Board found that although an effort was begun in 1991 to revise the operations plan, no evidence could be found in 1994 to indicate that the plan was actually updated to reflect the change in command and control relationships and responsibilities. The critical element of the plan with respect to the shootdown was that the change in mission led to the departure of an individual key to the communication between the Air Force and Army, without his duties being assigned to someone else. This example of asynchronous evolution plays a role in the loss.

---

1. Tactical control involves a fairly limited scope of authority, that is, the detailed and usually local direction and control of movement and maneuvers necessary to accomplish the assigned mission. Operational control, on the other hand, involves a broader authority to command subordinate forces, assign tasks, designate objectives, and give the authoritative direction necessary to accomplish the mission.

Command-Level Safety Constraints Related to the Accident:

1. Rules of engagement and operational orders and plans must be established at the command level that prevent friendly fire accidents. The plans must include allocating responsibility and establishing and monitoring communication channels to allow for coordination of flights into the theater of action.

2. Compliance with the ROE and operational orders and plans must be monitored. Alterations must be made in response to changing conditions and changing mission.

Controls: The controls included the ROE and operational plans plus feedback mechanisms on their effectiveness and application.

CFAC and MCC

The two parts of the Combined Task Force involved in the accident were the Army Military Coordination Center (MCC) and the Air Force Combined Forces Air Component (CFAC).

The shootdown obviously involved a communication failure: the F-15 pilots did not know the U.S. Army Black Hawks were in the area or that they were targeting friendly aircraft. Problems in communication between the three services (Air Force, Army, and Navy) are legendary. Procedures had been established to attempt to eliminate these problems in Operation Provide Comfort.

The Military Coordination Center (MCC) coordinated land and U.S. helicopter missions that supported the Kurdish people. In addition to providing humanitarian relief and protection to the Kurds, another important function of the Army detachment was to establish an ongoing American presence in the Kurdish towns and villages by showing the U.S. flag. This U.S. Army function was supported by a helicopter detachment called Eagle Flight.

All CTF components, with the exception of the Army Military Coordination Center lived and operated out of Incirlik Air Base in Turkey. The MCC operated out of two locations. A forward headquarters was located in the small village of Zakhu (see figure 5.1), just inside Iraq. Approximately twenty people worked in Zakhu, including operations, communications, and security personnel, medics, translators, and coalition chiefs. Zakhu operations were supported by a small administrative contingent working out of Pirinclik Air Base in Diyarbakir, Turkey. Pirinclik is also where the Eagle Flight Platoon of UH-60 Black Hawk helicopters was located. Eagle Flight helicopters made numerous (usually daily) trips to Zakhu to support MCC operations.

The Combined Forces Air Component (CFAC) Commander was responsible for coordinating the employment of all air operations to accomplish the OPC mission. He was delegated operational control of the Airborne Warning and Control System

(AWACS), U.S. Air Force (USAF) airlift, and the fighter forces. He had tactical control of the U.S. Army, U.S. Navy, Turkish, French, and British fixed wing and helicopter aircraft. The splintering of control between the CFAC and MCC commanders, along with communication problems between them, were major contributors to the accident.

In a complex coordination problem of this sort, communication is critical. Communications were implemented through the Joint Operations and Intelligence Center (JOIC). The JOIC received, delivered, and transmitted communications up, down, and across the CTF control structure. No Army liaison officer was assigned to the JOIC, but one was available on request to provide liaison between the MCC helicopter detachment and the CTF staff.

To prevent friendly fire accidents, pilots need to know exactly what friendly aircraft are flying in the no-fly zone at all times as well as know and follow the ROE and other procedures for preventing such accidents. The higher levels of control delegated the authority and guidance to develop local procedures[2] to the CTF level and below. These local procedures included:

- *Airspace Control Order (ACO):* The ACO contains the authoritative guidance for all local air operations in OPC. It covers such things as standard altitudes and routes, air refueling procedures, recovery procedures, airspace deconfliction responsibilities, and jettison procedures. The deconfliction procedures were a way to prevent interactions between aircraft that might result in accidents. For the Iraqi TAOR, fighter aircraft, which usually operated at high altitudes, were to stay above 10,000 feet above ground level while helicopters, which normally conducted low-altitude operations, were to stay below 400 feet. All flight crews were responsible for reviewing and complying with the information contained in the ACO. The CFAC Director of Operations was responsible for publishing the guidance, including the Airspace Control Order, for conducting OPC missions.

- *Aircrew Read Files (ARFs):* The Aircraft Read Files supplement the ACOs and are also required reading by all flight crews. They contain the classified rules of engagement (ROE), changes to the ACO, and recent amplification of how local commanders want air missions executed.

- *Air Tasking Orders (ATOs):* While the ACO and ARFs contain general information that applies to all aircraft in OPC, specific mission guidance was published in the daily ATOs. They contained the daily flight schedule, radio frequencies to be used, IFF codes (used to identify an aircraft as friend or foe),

---

2. The term *procedures* as used in the military denote standard and detailed courses of action that describe how to perform a task.

and other late-breaking information necessary to fly on any given day. All aircraft are required to have a hard copy of the current ATO with Special Instructions (SPINS) on board before flying. Each morning around 11:30 (1130 hours, in military time), the mission planning cell (or Frag shop) publishes the ATO for the following day, and copies are distributed to all units by late afternoon.

- *Battle Staff Directives (BSDs):* Any late scheduling changes that do not make it onto the ATO are published in last-minute Battle Staff Directives, which are distributed separately and attached to all ATOs prior to any missions flying the next morning.

- *Daily Flowsheets:* Military pilots fly with a small clipboard attached to their knees. These kneeboards contain boiled-down reference information essential to have handy while flying a mission, including the daily flowsheet and radio frequencies. The flowsheets are graphical depictions of the chronological flow of aircraft scheduled into the no-fly zone for that day. Critical information is taken from the ATO, translated into timelines, and reduced on a copier to provide pilots with a handy in-flight reference.

- *Local Operating Procedures and Instructions, Standard Operating Procedures, Checklists, and so on:* In addition to written material, real-time guidance is provided to pilots after taking off via radio through an unbroken command chain that runs from the OPC Commanding General, through the CFAC, through the mission director, through an Airborne Command Element (ACE) on board the AWACS, and ultimately to pilots.

The CFAC commander of operations was responsible for ensuring that aircrews were informed of all unique aspects of the OPC mission, including the ROE, upon their arrival. He was also responsible for publishing the Aircrew Read File (ARF), the Airspace Control Order (ACO), the daily Air Tasking Order, and mission-related special instructions (SPINS).

Safety Constraints Related to the Accident:

1. Coordination and communication among all flights into the TAOR must be established. Procedures must be established for determining who should be and is in the TAOR at all times.

2. Procedures must be instituted and monitored to ensure that all aircraft in the TAOR are tracked and fighters are aware of the location of all friendly aircraft in the TAOR.

3. The ROE must be understood and followed by those at lower levels.

4. All aircraft must be able to communicate effectively in the TAOR.

Controls: The controls in place included the ACO, ARFs, flowsheets, intelligence and other briefings, training (on the ROE, on aircraft identification, etc.), AWACS procedures for identifying and tracking aircraft, established radio frequencies and radar signals for the no-fly zone, a chain of command (OPC Commander to Mission Director to ACE to pilots), disciplinary actions for those not following the written rules, and a group (the JOIE) responsible for ensuring effective communication occurred.

Mission Director and Airborne Command Element

The Airborne Command Element (ACE) flies in the AWACS and is the commander's representative in the air, armed with up-to-the-minute situational information to make time-critical decisions. The ACE monitors all air operations and is in direct contact with the Mission Director located in the ground command post. He must also interact with the AWACS crew to identify reported unidentified aircraft.

The ground-based Mission Director maintains constant communication links with both the ACE up in the AWACS and with the CFAC commander on the ground. The Mission Director must inform the OPC commander immediately if anything happens over the no-fly zone that might require a decision by the commander or his approval. Should the ACE run into any situation that would involve committing U.S. or coalition forces, the Mission Director will communicate with him to provide command guidance. The Mission Director is also responsible for making weather-related decisions, implementing safety procedures, scheduling aircraft, and ensuring that the ATO is executed correctly.

The ROE in place at the time of the shootdown stated that aircrews experiencing unusual circumstances were to pass details to the ACE or AWACS, who would provide guidance on the appropriate response [200]. Exceptions were possible, of course, in cases of imminent threat. Aircrews were directed to first contact the ACE and, if that individual was unavailable, to then contact the AWACS. The six unusual circumstances/occurrences to be reported, as defined in the ROE, included "any intercept run on an unidentified aircraft." As stated, the ROE was specifically designed to slow down a potential engagement to allow time for those in the chain of command to check things out.

Although the written guidance was clear, there was controversy with respect to how it was or should have been implemented and who had decision-making authority. Conflicting testimony during the investigation of the shootdown about responsibility may either reflect after-the-fact attempts to justify actions or may instead reflect real confusion on the part of everyone, including those in charge, as to where the responsibility lay—perhaps a little of both.

Safety Constraints Related to the Accident:

1. The ACE and MD must follow procedures specified and implied by the ROE.

2. The ACE must ensure that pilots follow the ROE.

3. The ACE must interact with the AWACS crew to identify reported unidentified aircraft.

Controls: Controls to enforce the safety constraints included the ROE to provide overall principles for decision-making and to slow down engagements in order to prevent individual error or erratic behavior, the ACE up in the AWACS to augment communication by getting up-to-the-minute information about the state of the TAOR airspace and communicating with the pilots and AWACS crews, and the Mission Director on the ground to provide a chain of command from the pilots to the CFAC commander for real-time decision making.

AWACS Controllers

The AWACS (Airborne Warning and Control Systems) acts as an air traffic control tower in the sky. The AWACS OPC mission was to:

1. Control aircraft en route to and from the no-fly zone

2. Coordinate air refueling (for the fighter aircraft and the AWACS itself)

3. Provide airborne threat warning and control for all OPC aircraft operating inside the no-fly zone

4. Provide surveillance, detection, and identification of all unknown aircraft

An AWACS is a modified Boeing 707, with a saucer-shaped radar dome on the top, equipped inside with powerful radars and radio equipment that scan the sky for aircraft. A computer takes raw data from the radar dome, processes it, and ultimately displays tactical information on fourteen color consoles arranged in rows of three throughout the rear of the aircraft. AWACS have the capability to track approximately one thousand enemy aircraft at once while directing one hundred friendly ones [159].

The AWACS carries a flight crew (pilot, copilot, navigator, and flight engineer) responsible for safe ground and flight operation of the AWACS aircraft and a mission crew that has overall responsibility for the AWACS command, control, surveillance, communications, and sensor systems.

The mission crew of approximately nineteen people are under the direction of a mission crew commander (MCC). The MCC has overall responsibility for the AWACS mission and the management, supervision, and training of the mission crew. The mission crew members were divided into three sections:

1. *Technicians:*   The technicians are responsible for operating, monitoring, and maintaining the physical equipment on the aircraft.

2. *Surveillance:*   The surveillance section is responsible for the detection, tracking, identification, height measurement, display, and recording of surveillance data. As unknown targets appear on the radarscopes, surveillance technicians follow a detailed procedure to identify the tracks. They are responsible for handling unidentified and non-OPC aircraft detected by the AWACS electronic systems. The section is supervised by the air surveillance officer, and the work is carried out by an advanced air surveillance technician and three air surveillance technicians.

3. *Weapons:*   The weapons controllers are supervised by the senior director (SD). This section is responsible for the control of all assigned aircraft and weapons systems in the TAOR. The SD and three weapons directors are together responsible for locating, identifying, tracking, and controlling all friendly aircraft flying in support of OPC. Each weapons director was assigned responsibility for a specific task:

   · The enroute controller controlled the flow of OPC aircraft to and from the TAOR. This person also conducted radio and IFF checks on friendly aircraft outside the TAOR.

   · The TAOR controller provided threat warning and tactical control for all OPC aircraft within the TAOR.

   · The tanker controller coordinated all air refueling operations (and played no part in the accident so is not mentioned further).

To facilitate communication and coordination, the SD's console was physically located in the "pit" right between the MCC and the ACE (Airborne Command Element). Through internal radio nets, the SD synchronized the work of the weapons section with that of the surveillance section. He also monitored and coordinated the actions of his weapons directors to meet the demands of both the ACE and MCC.

Because those who had designed the control structure recognized the potential for some distance to develop between the training of the AWACS crew members and the continually evolving practice in the no-fly zone (another example of asynchronous evolution of the safety control structure), they had instituted a control by creating staff or instructor personnel permanently stationed in Turkey. Their job was to help provide continuity for U.S. AWACS crews who rotated through OPC on temporary duty status, usually for thirty-day rotations. This *shadow crew* flew with each new AWACS crew on their first mission in the TAOR to alert them as to how things were *really* done in OPC. Their job was to answer any questions the new crew might have about local procedures, recent occurrences, or changes in policy or interpretation that had come about since the last time they had been in the theater. Because the accident occurred on the first day for a new AWACS crew, instructor or staff personnel were also on board.

In addition to all these people, a Turkish controller flew on all OPC missions to help the crew interface with local air traffic control systems.

The AWACS typically takes off from Incirlik AFB approximately two hours before the first air refueling and fighter aircraft. Once the AWACS is airborne, the systems of the AWACS are brought on line, and a Joint Tactical Information Distribution System (JTIDS[3]) link is established with a Turkish Sector Operations Center (radar site). After the JTIDS link is confirmed, the CFAC airborne command element (ACE) initiates the planned launch sequence for the rest of the force. Normally, within a one-hour period, tanker and fighter aircraft take off and proceed to the TAOR in a carefully orchestrated flow. Fighters may not cross the political border into Iraq without AWACS coverage.

Safety Constraints Related to the Accident:

1. The AWACS mission crew must identify and track all aircraft in the TAOR. Friendly aircraft must not be identified as a threat (hostile).

2. The AWACS mission crew must accurately inform fighters about the status of all tracked aircraft when queried.

3. The AWACS mission crew must alert aircraft in the TAOR to any coalition aircraft not appearing on the flowsheet (ATO).

4. The AWACS crew must not fail to warn fighters about any friendly aircraft the fighters are targeting.

5. The JTIDS must provide the ground with an accurate picture of the airspace and its occupants.

Controls: Controls included procedures for identifying and tracking aircraft, training (including simulator missions), briefings, staff controllers, and communication channels. The SD and ASO provided real-time oversight of the crew's activities.

Pilots

Fighter aircraft, flying in formations of two and four aircraft, must always have a clear line of command. In the two-aircraft formation involved in the accident, the

---

3. The Joint Tactical Information Distribution System acts as a central component of the mission command and control system, providing ground commanders with a real-time downlink of the current air picture from AWACS. This information is then integrated with data from other sources to provide commanders with a more complete picture of the situation.

lead pilot is completely in charge of the flight and the wingman takes all of his commands from the lead.

The ACO (Airspace Control Order) stipulates that fighter aircraft may not cross the political border into Iraq without AWACS coverage and no aircraft may enter the TAOR until fighters with airborne intercept (AI) radars have searched the TAOR for Iraqi aircraft. Once the AI radar-equipped aircraft have "sanitized" the no-fly zone, they establish an orbit and continue their search for Iraqi aircraft and provide air cover while other aircraft are in the area. When they detect non-OPC aircraft, they are to intercept, identify, and take appropriate action as prescribed by the rules of engagement (ROE) and specified in the ACO.

After the area is sanitized, additional fighters and tankers flow to and from the TAOR throughout the six- to eight-hour daily flight schedule. This flying window is randomly selected to avoid predictability.

Safety Constraints Related to the Accident:

1. Pilots must know and follow the rules of engagement established and communicated from the levels above.

2. Pilots must know who is in the no-fly zone at all times and whether they should be there or not, i.e., they must be able to accurately identify the status of all other aircraft in the no-fly zone at all times and must not misidentify a friendly aircraft as a threat.

3. Pilots of aircraft in the area must be able to hear radio communications.

4. Fixed-wing aircraft must fly above 10,000 feet and helicopters must remain below 400 feet.

Controls: Controls included the ACO, the ATO, flowsheets, radios, IFF, the ROE, training, the AWACS, procedures to keep fighters and helicopters from coming into contact (for example, they fly at different altitudes), and special tactical radio frequencies when operating in the TAOR. Flags were displayed prominently on all aircraft in order to identify their origin.

Communication: Communication is important in preventing friendly fire accidents. The U.S. Army Black Hawk helicopters carried a full array of standard avionics, radio, IFF, and radar equipment as well as communication equipment consisting of FM, UHF, and VHF radios. Each day the FM and UHF radios were keyed with classified codes to allow pilots to *talk secure* in encrypted mode. The ACO directed that special frequencies were to be used when flying inside the TAOR.

Due to the line-of-sight limitations of their radios, the high mountainous terrain in northern Iraq, and the fact that helicopters tried to fly at low altitudes to use the terrain to mask them from enemy air defense radars, all Black Hawk flights into the no-fly zone also carried tactical satellite radios (TACSATs). These TACSATS were used to communicate with MCC operations. The helicopters had to land to place the TACSATs in operation; they cannot be operated from inside a moving helicopter.

The F-15's were equipped with avionics, communications, and electronic equipment similar to that on the Black Hawks, except that the F-15's were equipped with HAVE QUICK II (HQ-II) frequency-hopping radios while the helicopters were not. HQ-II defeated most enemy attempts to jam transmissions by changing frequencies many times per second. Although the F-15 pilots preferred to use the more advanced HQ technology, the F-15 radios were capable of communicating in a clear, non-HQ-II mode. The ACO directed that F-15s use the non-HQ-II frequency when specified aircraft that were not HQ-II capable flew in the TAOR. One factor involved in the accident was that Black Hawk helicopters (UH-60s) were *not* on the list of non-HQ-II aircraft that must be contacted using a non-HQ-II mode.

Identification: Identification of aircraft was assisted by systems called AAI/IFF (electronic Air-to-Air Interrogation/Identification Friend or Foe). Each coalition aircraft was equipped with an IFF transponder. Friendly radars (located in the AWACS, a fighter aircraft, or a ground site) execute what is called a *parrot check* to determine if the target being reflected on their radar screens is friendly or hostile. The AAI component (the interrogator) sends a signal to an airborne aircraft to determine its identity, and the IFF component answers or *squawks back* with a secret code—a numerically identifying pulse that changes daily and must be uploaded into aircraft using secure equipment prior to takeoff. If the return signal is valid, it appears on the challenging aircraft's visual display (radarscope). A compatible code has to be loaded into the cryptographic system of both the challenging and the responding aircraft to produce a friendly response.

An F-15's AAI/IFF system can interrogate using four identification signals or modes. The different types of IFF signals provide a form of redundancy. Mode I is a general identification signal that permits selection of 32 codes. Two Mode I codes were designated for use in OPC at the time of the accident: one for inside the TAOR and the other for outside. Mode II is an aircraft-specific identification mode allowing the use of 4,096 possible codes. Mode III provides a nonsecure friendly identification of both military and civilian aircraft and was not used in the TAOR. Mode IV is secure and provides high-confidence identification of friendly targets. According to the ACO, the primary means of identifying friendly aircraft in the Iraqi no-fly zone were to be modes I and IV in the IFF interrogation process.

Physical identification is also important in preventing friendly fire accidents. The ROE require that the pilots perform a visual identification of the potential threat. To assist in this identification, the Black Hawks were marked with six two-by-three-foot American flags. An American flag was painted on each door, on both sponsons,[4] on the nose, and on the belly of each helicopter [159]. A flag had been added to the side of each sponson because the Black Hawks had been the target of small-arms ground fire several months before.

5.3   The Accident Analysis Using STAMP

With all these controls and this elaborate control structure to protect against friendly fire accidents, which was a well-known hazard, how could the shootdown occur on a clear day with all equipment operational? As the Chairman of the Joint Chiefs of Staff said after the accident:

> In place were not just one, but a series of safeguards—some human, some procedural, some technical—that were supposed to ensure an accident of this nature could never happen. Yet, quite clearly, these safeguards failed.[5]

Using STAMP to understand why this accident occurred and to learn how to prevent such losses in the future requires determining why these safeguards were not successful in preventing the friendly fire. Various explanations for the accident have been posited. Making sense out of these conflicting explanations and understanding the accident process involved, including not only failures of individual system components but the unsafe interactions and miscommunications between components, requires understanding the role played in this process by each of the elements of the safety control structure in place at the time.

The next section contains a description of the proximate events involved in the loss. Then the STAMP analysis providing an explanation of why these events occurred is presented.

5.3.1   Proximate Events

Figure 5.4, taken from the official Accident Investigation Board Report, shows a timeline of the actions of each of the main actors in the proximate events—the AWACS, the F-15s, and the Black Hawks. It may also be helpful to refer back to figure 5.1, which contains a map of the area showing the relative locations of the important activities.

After receiving a briefing on the day's mission, the AWACS took off from Incirlik Air Base. When they arrived on station and started to track aircraft, the AWACS surveillance section noticed unidentified radar returns (from the Black Hawks). A "friendly general" track symbol was assigned to the aircraft and labeled as *H*,

---

4. Sponsons are auxiliary fuel tanks.

5. John Shalikashvili, chairman of the Joint Chief of Staff, from a cover letter to the twenty-one-volume report of the Aircraft Accident Investigation Board, 1994a, page 1.

TIME	AWACS	F-15s	BLACKHAWKS
0436	AWACS departs Incirlik AB		
0522			Black Hawks depart Diyarbakir
0545	AWACS declares "on Station." Surveillance section begins tracking a/c		
0616	"H" character programmed to appear on senior director's scope whenever Eagle Flight's IFF Mode 1, Code 42 is detected.		
0621	En route controller answers Black Hawks. Track annotated "EE01" for Eagle Flight		Black Hawks call AWACS on en route frequency at the entrance to TAOR
0624	Black Hawks' radar and IFF returns fade		Black Hawks land at Zakhu
0635		F-15s depart Incirlik AB	
0636	En route controller interrogates F-15s IFF Mode IV		
0654	AWACS receives Black Hawks' radio call. En route controller reinitiates EE01 symbol to resume tracking.		Black Hawks call AWACS to report en route from "Whiskey" to "Lima"
0655	"H" begins to be regularly displayed on SD's radar scope		
0705		F-15s check in with AWACS on en route frequency	
0711	Black Hawk's radar and IFF contacts fade; "H" ceases to be displayed on SD's scope; computer symbol continues to move at last known speed and direction.		Black hawks enter mountainous terrain
0713	ASO places arrow on SD scope in vicinity of Black Hawks' last known position		
0714	Arrow drops off SD's display		
0715	ACE replies to F-15s "...negative words." AWACS radar adjusted to low velocity detection settings	F-15s check in with ACE	
0720		F-15s enter TAOR and call AWACS	
0721	"EE01" symbol dropped by AWACS		
0722	TAOR controller responds "Clean there"	F-15 lead reports radar contact at 40 NMs	
0723	Intermittent IFF response appears in vicinity of F–15's reported radar contact		
0724	"H" symbol reappears on SD's scope		
0725	Black Hawk IFF response becomes more frequent. TAOR controller responds to F-15s with "Hits there."	F-15 lead calls "Contact" (radar return approx. 20 NMs)	
0726	Black Hawk IFF response continuous; radar returns intermittent		
0727	Enroute controller initiates "Unknown, Pending, Unevaluated symbol in vicinity of Black hawks IFF/radar returns; attempts IFF interrogation		
0728	Black Hawk IFF and radar returns fade	F-15 lead "visual" with a helicopter at 5 NM	
0728		F-15 lead conducts ID pass; calls "Tally two Hinds?"	
0728		F-15 wing conducts ID pass; calls "Tally Two."	
0729		F-15 lead instructs wing to "Arm hot"; calls AWACS and says "engaged"	
0730		F-15 pilots fire at helicopters helicopter	Black Hawks hit by missiles

Figure 5.4
The proximate chronological events leading to the accident.

denoting a helicopter. The Black Hawks (Eagle Flight) later entered the TAOR (no-fly zone) through Gate 1, checked in with the AWACS controllers who annotated the track with the identifier *EE01*, and flew to Zakhu. The Black Hawk pilots did not change their IFF (Identify Friend or Foe) Mode I code: The code for all friendly fixed-wing aircraft flying in Turkey on that day was 42, and the code for the TAOR was 52. They also remained on the enroute radio frequency instead of changing to the frequency to be used in the TAOR. When the helicopters landed at Zakhu, their radar and IFF (Identify Friend or Foe) returns on the AWACS radarscopes faded. Thirty minutes later, Eagle Flight reported their departure from Zakhu to the AWACS and said they were enroute from *Whiskey* (code name for Zakhu) to *Lima* (code name for Irbil, a town deep in the TAOR). The enroute controller reinitiated tracking of the helicopters.

Two F-15s were tasked that day to be the first aircraft in the TAOR and to *sanitize* it (check for hostile aircraft) before other coalition aircraft entered the area. The F-15s reached their final checkpoint before entering the TAOR approximately an hour after the helicopters had entered. They turned on all combat systems, switched their IFF Mode I code from 42 to 52, and switched to the TAOR radio frequency. They reported their entry into the TAOR to the AWACS.

At this point, the Black Hawks' radar and IFF contacts faded as the helicopters entered mountainous terrain. The AWACS computer continued to move the helicopter tracks on the radar display at the last known speed and direction, but the identifying *H* symbol (for helicopter) on the track was no longer displayed. The ASO placed an "attention arrow" (used to point out an area of interest) on the SD's scope at the point of the Black Hawk's last known location. This large arrow is accompanied by a blinking alert light on the SD's console. The SD did not acknowledge the arrow and after sixty seconds, both the arrow and the light were automatically dropped. The ASO then adjusted the AWACS radar to detect slow-moving objects.

Before entering the TAOR, the lead F-15 pilot checked in with the ACE and was told there were no relevant changes from previously briefed information ("negative words"). Five minutes later, the F-15's entered the TAOR, and the lead pilot reported their arrival to the TAOR controller. One minute later, the enroute controller finally dropped the symbol for the helicopters from the scope, the last remaining visual reminder that there were helicopters inside the TAOR.

Two minutes after entering the TAOR, the lead F-15 picked up hits on its instruments indicating that it was getting radar returns from a low and slow-flying aircraft. The lead F-15 pilot alerted his wingman and then locked onto the contact and used the F-15's air-to-air interrogator to query the target's IFF code. If it was a coalition aircraft, it should be squawking Mode I, code 52. The scope showed it was not. He reported the radar hits to the controllers in the AWACS, and the TAOR controller told him they had no radar contacts in that location ("clean there"). The wing pilot replied to the lead pilot's alert, noting that his radar also showed the target.

The lead F-15 pilot then switched the interrogation to the second mode (Mode IV) that all coalition aircraft should be squawking. For the first second it showed the right symbol, but for the rest of the interrogation (4 to 5 seconds) it said the target was not squawking Mode IV. The lead F-15 pilot then made a second contact call to the AWACS over the main radio, repeating the location, altitude, and heading of his target. This time the AWACS enroute controller responded that he had radar returns on his scope at the spot ("hits there") but did not indicate that these returns might be from a friendly aircraft. At this point, the Black Hawk IFF response was continuous but the radar returns were intermittent. The enroute controller placed an "unknown, pending, unevaluated" track symbol in the area of the helicopter's radar and IFF returns and attempted to make an IFF identification.

The lead F-15 pilot, after making a second check of Modes I and IV and again receiving no response, executed a visual identification pass to confirm that the target was hostile—the next step required in the rules of engagement. He saw what he thought were Iraqi helicopters. He pulled out his "goody book" with aircraft pictures in it, checked the silhouettes, and identified the helicopters as Hinds, a type of Russian aircraft flown by the Iraqis ("Tally two Hinds"). The F-15 wing pilot also reported seeing two helicopters ("Tally two"), but never confirmed that he had identified them as Hinds or as Iraqi aircraft.

The lead F-15 pilot called the AWACS and said they were engaging enemy aircraft ("Tiger Two[6] has tallied two Hinds, engaged"), cleared his wingman to shoot ("Arm hot"), and armed his missiles. He then did one final Mode I check, received a negative response, and pressed the button that released the missiles. The wingman fired at the other helicopter, and both were destroyed.

This description represents the chain of events, but it does not explain "why" the accident occurred except at the most superficial level and provides few clues as to how to redesign the system to prevent future occurrences. Just looking at these basic events surrounding the accident, it appears that mistakes verging on gross negligence were involved—undisciplined pilots shot down friendly aircraft in clear skies, and the AWACS crew and others who were supposed to provide assistance simply sat and watched without telling the F-15 pilots that the helicopters were there. An analysis using STAMP, as will be seen, provides a very different level of understanding. In the following analysis, the goal is to understand why the controls in place did not prevent the accident and to identify the changes necessary to prevent similar accidents in the future. A related type of hazard analysis can be used during system

---

6. Tiger One was the code name for the F-15 lead pilot, while Tiger Two denoted the wing pilot.

design and development (see chapters 8 and 9) to prevent such occurrences in the first place.

In the following analysis, the basic failures and dysfunctional interactions leading to the loss at the physical level are identified first. Then each level of the hierarchical safety control structure is considered in turn, starting from the bottom.

At each level, the context in which the behaviors took place is considered. The context for each level includes the hazards, the safety requirements and constraints, the controls in place to prevent the hazard, and aspects of the environment or situation relevant to understanding the control flaws, including the people involved, their assigned tasks and responsibilities, and any relevant environmental behavior-shaping factors. Following a description of the context, the dysfunctional interactions and failures at that level are described, along with the accident factors (see figure 4.8) that were involved.

5.3.2 Physical Process Failures and Dysfunctional Interactions

The first step in the analysis is to understand the physical failures and dysfunctional interactions within the physical process that were related to the accident. Figure 5.5 shows this information.

All the physical components worked exactly as intended, except perhaps for the IFF system. The fact that the Mode IV IFF gave an intermittent response has never been completely explained. Even after extensive equipment teardowns and reenactments with the same F-15s and different Black Hawks, no one has been able to explain why the F-15 IFF interrogator did not receive a Mode IV response [200]. The Accident Investigation Board report states: "The reason for the unsuccessful

Physical Process

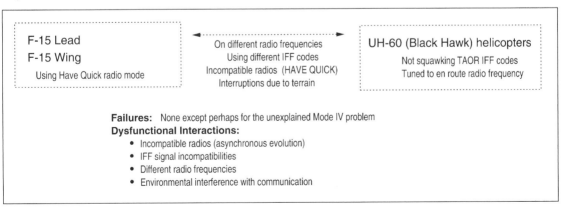

Figure 5.5
The physical level of the accident process.

Mode IV interrogation attempts cannot be established, but was probably attributable to one or more of the following factors: incorrect selection of interrogation modes, faulty air-to-air interrogators, incorrectly loaded IFF transponder codes, garbling of electronic responses, and intermittent loss of line-of-sight radar contact."[7]

There were several dysfunctional interactions and communication inadequacies among the correctly operating aircraft equipment. The most obvious unsafe interaction was the release of two missiles in the direction of two friendly aircraft, but there were also four obstacles to the type of fighter–helicopter communications that might have prevented that release.

1. The Black Hawks and F-15s were on different radio frequencies and thus the pilots could not speak to each other or hear the transmissions between others involved in the incident, the most critical of which were the radio transmissions between the two F-15 pilots and between the lead F-15 pilot and personnel onboard the AWACS. The Black Hawks, according to the Aircraft Control Order, should have been communicating on the TAOR frequency. Stopping here and looking only at this level, it appears that the Black Hawk pilots were at fault in not changing to the TAOR frequency, but an examination of the higher levels of control points to a different conclusion.

2. Even if they had been on the same frequency, the Air Force fighter aircraft were equipped with HAVE QUICK II (HQ-II) radios, while the Army helicopters were not. The only way the F-15 and Black Hawk pilots could have communicated would have been if the F-15 pilots switched to non-HQ mode. The procedures the pilots were given to follow did not tell them to do so. In fact, with respect to the two helicopters that were shot down, one contained an outdated version called HQ-I, which was not compatible with HQ-II. The other *was* equipped with HQ-II, but because not all of the Army helicopters supported HQ-II, CFAC refused to provide Army helicopter operations with the necessary cryptographic support required to synchronize their radios with the other OPC components.

   If the objective of the accident analysis is to assign blame, then the different radio frequencies could be considered irrelevant because the differing technology meant they could not have communicated even if they had been on the same frequency. If the objective, however, is to learn enough to prevent future accidents, then the different radio frequencies are relevant.

---

7. The commander of the U.S. Army in Europe objected to this sentence. He argued that nothing in the board report supported the possibility that the codes had been loaded improperly and that it was clear the Army crews were not at fault in this matter. The U.S. Commander in Chief, Europe, agreed with his view. Although the language in the opinion was not changed, the former said his concerns were addressed because the complaint had been included as an attachment to the board report.

3. The Black Hawks were not squawking the required IFF Mode I code for those flying within the TAOR. The GAO report states that Black Hawk pilots told them they routinely used the same Mode I code for outside the TAOR while operating within the TAOR and no one had advised them that it was incorrect to do so. But, again, the wrong Mode I code is only part of the story.

The Accident Investigation Board report concluded that the use of the incorrect Mode I IFF code by the Black Hawks was responsible for the F-15 pilots' failure to receive a Mode I response when they interrogated the helicopters. However, an Air Force special task force concluded that based on the descriptions of the system settings that the pilots testified they had used on the interrogation attempts, the F-15s should have received and displayed any Mode I or II response *regardless of the code* [200]. The AWACS was receiving friendly Mode I and II returns from the helicopters at the same time that the F-15s received no response. The GAO report concluded that the helicopters' use of the wrong Mode I code should not have prevented the F-15s from receiving a response. Confusing the situation even further, the GAO report cites the Accident Board president as telling the GAO investigators that because of the difference between the lead F-15 pilot's statement on the day of the incident and his testimony to the investigation board, it was difficult to determine the number of times the lead pilot had interrogated the helicopters [200].

4. Communication was also impeded by physical line-of-sight restrictions. The Black Hawks were flying in narrow valleys among very high mountains that disrupted communication depending on line-of-sight transmissions.

One reason for these dysfunctional interactions lies in the *asynchronous evolution* of the Army and Air Force technology, leaving the different services with largely incompatible radios. Looking only at the event chain or at the failures and dysfunctional interactions in the technical process—a common stopping point in accident investigations—gives a very misleading picture of the reasons this accident occurred. Examining the higher levels of control is necessary to obtain the information necessary to prevent future occurrences.

After the shootdown, the following changes were made:

• Updated radios were placed on Black Hawk helicopters to enable communication with fighter aircraft. Until the time the conversion was complete, fighters were directed to remain on the TAOR clear frequencies for deconfliction with helicopters.

• Helicopter pilots were directed to monitor the common TAOR radio frequency and to squawk the TAOR IFF codes.

5.3.3 The Controllers of the Aircraft and Weapons

The pilots directly control the aircraft, including the activation of weapons (figure 5.6). The context in which their decisions and actions took place is first described, followed by the dysfunctional interactions at this level of the control structure. Then the inadequate control actions are outlined and the factors that led to them are described.

Context in Which Decisions and Actions Took Place

Safety Requirements and Constraints: The safety constraints that must be enforced at this level of the sociotechnical control structure were described earlier. The F-15 pilots must know who is in the TAOR and whether they should be there or not—that is, they must be able to identify accurately the status of all other aircraft in the TAOR at all times so that a friendly aircraft is not identified as a threat. They must also follow the rules of engagement (ROE), which specify the procedures to be executed before firing weapons at any targets. As noted earlier in this chapter, the OPC ROE were devised by the OPC commander, based on guidelines created by the Joint Chiefs of Staff, and were purposely conservative because of the many multinational participants in OPC and the potential for friendly fire accidents. The ROE were designed to slow down any military confrontation, but were unsuccessful in this case. An important part of understanding this accident process and preventing repetitions is understanding why this goal was not achieved.

Controls: As noted in the previous section, the controls at this level included the rules and procedures for operating in the TAOR (specified in the ACO), information provided about daily operations in the TAOR (specified in the Air Tasking Order or ATO), flowsheets, communication and identification channels (radios and IFF), training, AWACS oversight, and procedures to keep fighters and helicopters from coming into contact (for example, the F-15s fly at different altitudes). National flags were required to be displayed prominently on all aircraft in order to facilitate identification of their origin.

Roles and Responsibilities of the F-15 Pilots: When conducting combat missions, aerial tactics dictate that F-15s always fly in pairs with one pilot as the lead and one as the wingman. They fly and fight as a team, but the lead is always in charge. The mission that day was to conduct a thorough radar search of the area to ensure that the TAOR was clear of hostile aircraft (to *sanitize* the airspace) before the other aircraft entered. They were also tasked to protect the AWACS from any threats. The wing pilot was responsible for looking 20,000 feet and higher with his radar while the lead pilot was responsible for the area 25,000 feet and below. The lead pilot had final responsibility for the 5,000-foot overlap area.

Environmental and Behavior-Shaping Factors for the F-15 Pilots: The lead pilot that day was a captain with nine years' experience in the Air Force. He had flown

F-15 Lead Pilot

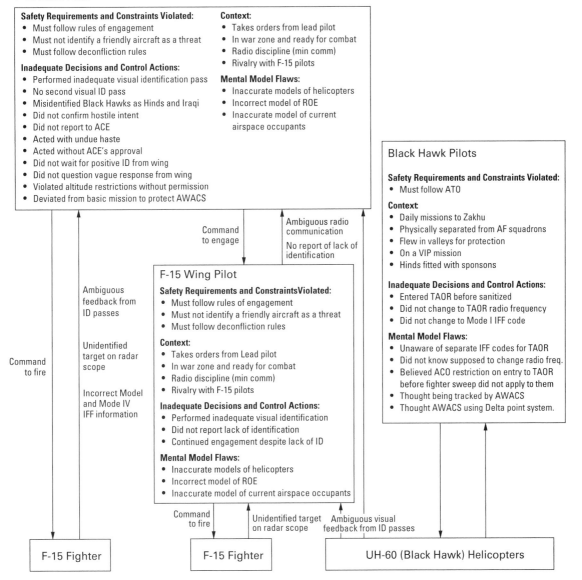

F-15 Lead Pilot

Safety Requirements and Constraints Violated:
- Must follow rules of engagement
- Must not identify a friendly aircraft as a threat
- Must follow deconfliction rules

Inadequate Decisions and Control Actions:
- Performed inadequate visual identification pass
- No second visual ID pass
- Misidentified Black Hawks as Hinds and Iraqi
- Did not confirm hostile intent
- Did not report to ACE
- Acted with undue haste
- Acted without ACE's approval
- Did not wait for positive ID from wing
- Did not question vague response from wing
- Violated altitude restrictions without permission
- Deviated from basic mission to protect AWACS

Context:
- Takes orders from lead pilot
- In war zone and ready for combat
- Radio discipline (min comm)
- Rivalry with F-15 pilots

Mental Model Flaws:
- Inaccurate models of helicopters
- Incorrect model of ROE
- Inaccurate model of current airspace occupants

Black Hawk Pilots

Safety Requirements and Constraints Violated:
- Must follow ATO

Context:
- Daily missions to Zakhu
- Physically separated from AF squadrons
- Flew in valleys for protection
- On a VIP mission
- Hinds fitted with sponsons

Inadequate Decisions and Control Actions:
- Entered TAOR before sanitized
- Did not change to TAOR radio frequency
- Did not change to Mode I IFF code

Mental Model Flaws:
- Unaware of separate IFF codes for TAOR
- Did not know supposed to change radio freq.
- Believed ACO restriction on entry to TAOR before fighter sweep did not apply to them
- Thought being tracked by AWACS
- Thought AWACS using Delta point system.

Command to engage

Ambiguous radio communication

No report of lack of identification

F-15 Wing Pilot

Safety Requirements and ConstraintsViolated:
- Must follow rules of engagement
- Must not identify a friendly aircraft as a threat
- Must follow deconfliction rules

Context:
- Takes orders from Lead pilot
- In war zone and ready for combat
- Radio discipline (min comm)
- Rivalry with F-15 pilots

Inadequate Decisions and Control Actions:
- Performed inadequate visual identification
- Did not report lack of identification
- Continued engagement despite lack of ID

Mental Model Flaws:
- Inaccurate models of helicopters
- Incorrect model of ROE
- Inaccurate model of current airspace occupants

Ambiguous feedback from ID passes

Unidentified target on radar scope

Incorrect Model and Mode IV IFF information

Command to fire

Command to fire

Unidentified target on radar scope

Ambiguous visual feedback from ID passes

F-15 Fighter

F-15 Fighter

UH-60 (Black Hawk) Helicopters

Figure 5.6
The analysis at the pilot level.

F-15s for over three years, including eleven combat missions over Bosnia and nineteen over northern Iraq protecting the no-fly zone. The mishap occurred on his sixth flight during his second tour flying in support of OPC.

The wing pilot was a lieutenant colonel and Commander of the 53rd Fighter Squadron at the time of the shootdown, and he was a highly experienced pilot. He had flown combat missions out of Incirlik during Desert Storm and had served in the initial group that set up OPC afterward. He was credited with the only confirmed kill of an enemy Hind helicopter during the Gulf War. That downing involved a *beyond visual range* shot, which means he never actually saw the helicopter.

F-15 pilots were rotated through every six to eight weeks. Serving in the no-fly zone was an unusual chance for peacetime pilots to have a potential for engaging in combat. The pilots were very aware they were going to be flying in unfriendly skies. They drew personal sidearms with live rounds, removed wedding bands and other personal items that could be used by potential captors, were supplied with *blood chits* offering substantial rewards for returning downed pilots, and were briefed about threats in the area. Every part of their preparation that morning drove home the fact that they could run into enemy aircraft: The pilots were making decisions in the context of being in a war zone and were ready for combat.

Another factor that might have influenced behavior, according to the GAO report, was rivalry between the F-15 and F-16 pilots engaged in Operation Provide Comfort (OPC). While such rivalry was normally perceived as healthy and leading to positive professional competition, at the time of the shootdown the rivalry had become more pronounced and intense. The Combined Task Force Commander attributed this atmosphere to the F-16 community's having executed the only fighter shootdown in OPC and all the shootdowns in Bosnia [200]. F-16 pilots are better trained and equipped to intercept low-flying helicopters. The F-15 pilots knew that F-16s would follow them into the TAOR that day. Any hesitation might have resulted in the F-16s getting another kill.

A final factor was a strong cultural norm of "radio discipline" (called *minimum communication* or *min comm*), which led to abbreviated phraseology in communication and a reluctance to clarify potential miscommunications. Fighter pilots are kept extremely busy in the cockpit; their cognitive capabilities are often stretched to the limit. As a result, any unnecessary interruptions on the radio are a significant distraction from important competing demands [191]. Hence, there was a great deal of pressure within the fighter community to minimize talking on the radio, which discouraged efforts to check accuracy and understanding.

Roles and Responsibilities of the Black Hawk Pilots: The Army helicopter pilots flew daily missions into the TAOR to visit Zakhu. On this particular day, a change of command had taken place at the US Army Command Center at Zakhu. The outgoing commander was to escort his replacement into the no-fly zone in order to introduce him to the two Kurdish leaders who controlled the area. The pilots were first scheduled to fly the routine leg into Zakhu, where they would pick up two Army colonels and carry other high-ranking VIPs representing the major players in OPC to the two Iraqi towns of Irbil and Salah ad Din. It was not uncommon for the Black Hawks to fly this far into the TAOR; they had done it frequently during the three preceding years of Operation Provide Comfort.

Environmental and Behavior-Shaping Factors for the Black Hawk Pilots: Inside Iraq, helicopters flew in terrain flight mode, that is, they hugged the ground, both to avoid midair collisions and to mask their presence from threatening ground-to-air Iraqi radars. There are three types of terrain flight: Pilots select the appropriate mode based on a wide range of tactical and mission-related variables. *Low-level* terrain flight is flown when enemy contact is not likely. *Contour* flying is closer to the ground than low level, and *nap-of-the-earth* flying is the lowest and slowest form of terrain flight, flown only when enemy contact is expected. Eagle Flight helicopters flew contour mode most of the time in northern Iraq. They liked to fly in the valleys and the low-level areas. The route they were taking the day of the shootdown was through a green valley between two steep, rugged mountains. The mountainous terrain provided them with protection from Iraqi air defenses during the one-hour flight to Irbil, but it also led to disruptions in communication.

Because of the distance and thus time required for the mission, the Black Hawks were fitted with *sponsons* or pontoon-shaped fuel tanks. The sponsons are mounted below the side doors, and each holds 230 gallons of extra fuel. The Black Hawks were painted with green camouflage, while the Iraqi Hinds' camouflage scheme was light brown and desert tan. To assist with identification, the Black Hawks were marked with three two-by-three-foot American flags—one on each door and one on the nose—and a fourth larger flag on the belly of the helicopter. In addition, two American flags had been painted on the side of each sponson.

Dysfunctional Interactions at This Level

Communication between the F-15 and Black Hawk pilots was obviously dysfunctional and related to the dysfunctional interactions in the physical process (incompatible radio frequencies, IFF codes, and anti-jamming technology) resulting in the ends of the communication channels not matching and information not being transmitted along the channel. Communication between the F-15 pilots was also hindered by the *minimum communication* policy that led to abbreviated messages and a reluctance to clarify potential miscommunications as described above as well as by the physical terrain.

Flawed or Inadequate Decisions and Control Actions

Both the Army helicopter pilots and the F-15 pilots executed inappropriate or inadequate control actions during their flights, beyond the obviously incorrect F-15 pilot commands to fire on two friendly aircraft.

Black Hawk Pilots:

- *The Army helicopters entered the TAOR before it had been sanitized by the Air Force.* The Air Control Order or ACO specified that a fighter sweep of the area must precede any entry of allied aircraft. However, because of the frequent trips of Eagle Flight helicopters to Zakhu, an official exception had been made to this policy for the Army helicopters. The Air Force fighter pilots had not been informed about this exception. Understanding this miscommunication requires looking at the higher levels of the control structure, particularly the communication structure at those levels.

- *The Army pilots did not change to the appropriate radio frequency to be used in the TAOR.* As noted earlier, however, even if they had been on the same frequency, they would have been unable to communicate with the F-15s because of the different anti-jamming technology of the radios.

- *The Army pilots did not change to the appropriate IFF Mode I signal for the TAOR.* Again, as noted above, the F-15s should still have been able to receive the Mode I response.

F-15 Lead Pilot: The accounts of and explanation for the unsafe control actions of the F-15 pilots differ greatly among those who have written about the accident. Analysis is complicated by the fact that any statements the pilots made after the accident were likely to have been influenced by the fact that they were being investigated on charges of negligent homicide—their stories changed significantly over time. Also, in the excitement of the moment, the lead pilot did not make the required radio call to his wingman requesting that he turn on the HUD[8] tape, and he also forgot to turn on his own tape. Therefore, evidence about certain aspects of what occurred and what was observed is limited to pilot testimony during the post-accident investigations and trials.

Complications also arise in determining whether the pilots followed the rules of engagement (ROE) specified for the no-fly zone, because the ROE are not public and the relevant section of the Accident Investigation Board Report is censored. Other sources of information about the accident, however, reference clear instances of Air Force pilot violations of the ROE.

---

8. Head-Up Display.

The following inadequate decisions and control actions can be identified for the lead F-15 pilot:

- *He did not perform a proper visual ID as required by the ROE and did not take a second pass to confirm the identification.*   F-15 pilots are not accustomed to flying close to the ground or to terrain. The lead pilot testified that because of concerns about being fired on from the ground and the danger associated with flying in a narrow valley surrounded by high mountains, he had remained high as long as possible and then dropped briefly for a visual identification that lasted between 3 and 4 seconds. He passed the helicopter on his left while flying more than 500 miles an hour and at a distance of about 1,000 feet off to the side and about 300 feet above the helicopter. He testified:

  > I was trying to keep my wing tips from hitting mountains and I accomplished two tasks simultaneously, making a call on the main radio and pulling out a guide that had the silhouettes of helicopters. I got only three quick interrupted glances of less than 1.25 seconds each. [159].

  The dark green Black Hawk camouflage blended into the green background of the valley, adding to the difficulty of the identification.

  The Accident Investigation Board used pilots flying F-15s and Black Hawks to recreate the circumstances under which the visual identification was made. The test pilots were unable to identify the Black Hawks, and they could not see any of the six American flags on each helicopter. The F-15 pilots could not have satisfied the ROE identification requirements using the type of visual identification passes they testified that they made.

- *He misidentified the helicopters as Iraqi Hinds.*   There were two basic incorrect decisions involved in this misidentification. The first was identifying the UH-60 (Black Hawk) helicopters as Russian Hinds, and the second was assuming that the Hinds were Iraqi. Both Syria and Turkey flew Hinds, and the helicopters could have belonged to one of the U.S. coalition partners. The Commander of the Operations Support Squadron, whose job was to run the weekly detachment squadron meetings, testified that as long as he had been in OPC, he had reiterated to the squadrons each week that they should be careful about misidentifying aircraft over the no-fly zone because there were so many nations and so many aircraft in the area and that any time F-15s or anyone else picked up a helicopter on radar, it was probably a U.S., Turkish, or United Nations helicopter:

  > Any time you intercept a helicopter as an unknown, there is always a question of procedures, equipment failure, and high terrain masking the line-of-sight radar. There

are numerous reasons why you would not be able to electronically identify a helicopter. Use discipline. It is better to miss a shot than be wrong. [159].

- *He did not confirm, as required by the ROE, that the helicopters had hostile intent before firing.* The ROE required that the pilot not only determine the type of aircraft and nationality, but to take into consideration the possibility the aircraft was lost, in distress, on a medical mission, or was possibly being flown by pilots who were defecting.

- *He violated the rules of engagement by not reporting to the Air Command Element (ACE).* According to the ROE, the pilot should have reported to the ACE (who is in his chain of command and physically located in the AWACS) that he had encountered an unidentified aircraft. He did not wait for the ACE to approve the release of the missiles.

- *He acted with undue and unnecessary haste that did not allow time for those above him in the control structure (who were responsible for controlling the engagement) to act.* The entire incident, from the first time the pilots received an indication about helicopters in the TAOR to shooting them down lasted only seven minutes. Pilots are allowed by the ROE to take action on their own in an emergency, so the question then becomes whether this situation was an emergency.

CFAC officials testified that there had been no need for haste. The slow-flying helicopters had traveled less than fourteen miles since the F-15s first picked them up on radar, they were not flying in a threatening manner, and they were flying southeast away from the Security Zone. The GAO report cites the Mission Director as stating that given the speed of the helicopters, the fighters had time to return to Turkish airspace, refuel, and still return and engage the helicopters before they could have crossed south of the 36th Parallel.

The helicopters also posed no threat to the F-15s or to their mission, which was to protect the AWACS and determine whether the area was clear. One expert later commented that even if they *had* been Iraqi Hinds, "A Hind is only a threat to an F-15 if the F-15 is parked almost stationary directly in front of it and says 'Kill me.' Other than that, it's probably not very vulnerable" [191].

Piper quotes Air Force Lt. Col. Tony Kern, a professor at the U.S. Air Force Academy, who wrote about this accident:

> Mistakes happen, but there was no rush to shoot these helicopters. The F-15s could have done multiple passes, or even followed the helicopters to their destination to determine their intentions. [159].

Any explanation behind the pilot's hasty action can only be the product of speculation. Snook attributes the fast reaction to the overlearned defensive responses taught to fighter pilots. Both Snook and the GAO report mention the rivalry with the F-16 pilots and a desire of the lead F-15 pilot to shoot down an enemy aircraft. F-16s would have entered the TAOR ten to fifteen minutes after the F-15s, potentially allowing the F-16 pilots to get credit for the downing of an enemy aircraft: F-16s are better trained and equipped to intercept low-flying helicopters. If the F-15 pilots had involved the chain of command, the pace would have slowed down, ruining the pilots' chance for a shootdown. In addition, Snook argues that this was a rare opportunity for peacetime pilots to engage in combat.

The goals and motivation behind any human action are unknowable (see section 2.7). Even in this case where the F-15 pilots survived the accident, there are many reasons to discount their own explanations, not the least of which is potential jail sentences. The explanations provided by the pilots right after the engagement differ significantly from their explanations a week later during the official investigations to determine whether they should be court-martialed. But in any case, there was no chance that such slow flying helicopters could have escaped two supersonic jet fighters in the open terrain of northern Iraq nor were they ever a serious threat to the F-15s. This situation, therefore, was not an emergency.

- *He did not wait for a positive ID from the wing pilot before firing on the helicopters and did not question the vague response when he got it:* When the lead pilot called out that he had visually identified two Iraqi helicopters, he asked the wing pilot to confirm the identification. The wingman called out "Tally Two" on his radio, which the lead pilot took as confirmation, but which the wing pilot later testified only meant he saw two helicopters but not necessarily Iraqi Hinds. The lead pilot did not wait for a positive identification from the wingman before starting the engagement.

- *He violated altitude restrictions without permission:* According to Piper, the commander of the OPC testified at one of the hearings,

> I regularly, routinely imposed altitude limitations in northern Iraq. On the fourteenth of April, the restrictions were a minimum of ten thousand feet for fixed-wing aircraft. This information was in each squadron's Aircrew Read File. Any exceptions had to have my approval. [159]

None of the other accident reports, including the official one, mentions this erroneous action on the part of the pilots. Because this control flaw was never investigated, it is not possible to determine whether the action resulted from a "reference channel" problem (i.e., the pilots did not know about the altitude restriction) or an "actuator" error (i.e., the pilots knew about it but chose to ignore it for an unknown reason.)

- *He deviated from the basic mission to protect the AWACS, leaving the AWACS open to attack:* The helicopter could have been a diversionary ploy. The mission of the first flight into the TAOR was to make sure it was safe for the AWACS and other aircraft to enter the restricted operating zone. Piper emphasizes that that was the only purpose of their mission [159]. Piper, who again is the only one who mentions it, cites testimony of the commander of OPC during one of the hearings when asked whether the F-15s exposed the AWACS to other air threats when they attacked and shot down the helicopters. The commander replied:

> Yes, when the F-15s went down to investigate the helicopters, made numerous passes, engaged the helicopters and then made more passes to visually reconnaissance the area, AWACS was potentially exposed for that period of time. [159]

Wing Pilot: The wing pilot, like the lead pilot, violated altitude restrictions and deviated from the basic mission. In addition:

- *He did not make a positive identification of the helicopters:* His visual identification was not even as close to the helicopters as the lead F-15 pilot, which was inadequate to recognize the helicopters, and the wing pilot's ID lasted only between two and three seconds. According to a *Washington Post* article, he told investigators that he never clearly saw the helicopters before reporting "Tally Two." In a transcript of one of his interviews with investigators, he said: "I did not identify them as friendly; I did not identify them as hostile. I expected to see Hinds based on the call my flight leader had made. I didn't see anything that disputed that."

  Although the wing had originally testified he could not identify the helicopters as Hinds, he reversed his statement between April and six months later when he testified at the hearing on whether to court-martial him that "I could identify them as Hinds" [159]. There is no way to determine which of these contradictory statements is true.

  Explanations for continuing the engagement without an identification could range from an inadequate mental model of the ROE, following the orders of the lead pilot and assuming that his identification had been proper, the strong influence on what one sees by what one expects to see, wanting the helicopters to be hostile, and any combination of these.

- *He did not tell the lead pilot that he had not identified the helicopters:* In the hearings to place blame for the shootdown, the lead pilot testified that he had radioed the wing pilot and said, "Tiger One has tallied two Hinds, confirm." Both pilots agree to this point, but then the testimony becomes contradictory.

The hearing in the fall of 1994 on whether the wing pilot should be charged with twenty-six counts of negligent homicide rested on the very narrow question of whether the lead pilot had called the AWACS announcing the engagement before or after the wing pilot responded to the lead pilot's directive to confirm whether the helicopters were Iraqi Hinds. The lead pilot testified that he had identified the helicopters as Hinds and then asked the wing to confirm the identification. When the wing responded with "Tally Two," the lead believed this response signaled confirmation of the identification. The lead then radioed the AWACS and reported, "Tiger Two has tallied two Hinds, engaged." The wing pilot, on the other hand, testified that the lead had called the AWACS with the "engaged" message before he (the wing pilot) had made his "Tally Two" radio call to the lead. He said his "Tally Two" call was in response to the "engaged" call, not the "confirm" call and simply meant that he had both target aircraft in sight. He argued that once the engaged call had been made, he correctly concluded that an identification was no longer needed.

The fall 1994 hearing conclusion about which of these scenarios actually occurred is different than the conclusions in the official Air Force accident report and that of the hearing officer in another hearing. Again, it is not possible nor necessary to determine blame here or to determine exactly which scenario is correct to conclude that the communications were ambiguous. The minimum communication policy was a factor here as was probably the excitement of a potential combat engagement. Snook suggests that the expectations of what the pilots expected to hear resulted in a filtering of the inputs. Such filtering is a well-known problem in airline pilots' communications with controllers. The use of well-established phraseology is meant to reduce it. But the calls by the wing pilot were nonstandard. In fact, Piper notes that in pilot training bases and programs that train pilots to fly fighter aircraft since the shootdown, these radio calls are used as examples of "the poorest radio communications possibly ever given by pilots during a combat intercept" [159].

- *He continued the engagement despite the lack of an adequate identification:* Explanations for continuing the engagement without an identification could range from an inadequate mental model of the ROE, following the orders of the lead pilot and assuming that the lead pilot's identification had been proper, wanting the helicopters to be hostile, and any combination of these. With only his contradictory testimony, it is not possible to determine the reason.

Some Reasons for the Flawed Control Actions and Dysfunctional Interactions

The accident factors shown in figure 4.8 can be used to provide an explanation for the flawed control actions. These factors here are divided into incorrect control

algorithms, inaccurate mental models, poor coordination among multiple controllers, and inadequate feedback from the controlled process.

Incorrect Control Algorithms: The Black Hawk pilots correctly followed the procedures they had been given (see the discussion of the CFAC–MCC level later).
These procedures were unsafe and were changed after the accident.

The F-15 pilots apparently did not execute their control algorithms (the procedures required by the rules of engagement) correctly, although the secrecy involved
in the ROE make this conclusion difficult to prove. After the accident, the ROE
were changed, but the exact changes made are not public.

Inaccurate Mental Models of the F-15 Pilots: There were many inconsistencies
between the mental models of the Air Force pilots and the actual process state. First,
they had an ineffective model of what a Black Hawk helicopter looked like. There
are several explanations for this, including poor visual recognition training and the
fact that Black Hawks with sponsons attached resemble Hinds. None of the pictures
of Black Hawks on which the F-15 pilots had been trained had these wing-mounted
fuel tanks. Additional factors include the speeds at which the F-15 pilots do their
visual identification (VID) passes and the angle at which the pilots passed over
their targets.

Both F-15 pilots received only limited visual recognition training in the previous
four months, partly due to the disruption of normal training caused by their wing's
physical relocation from one base to another in Germany. But the training was
probably inadequate even if it had been completed. Because the primary mission
of F-15s is air-to-air combat against other fast-moving aircraft, most of the operational training is focused on their most dangerous and likely threats—other high-
altitude fighters. In the last training before the accident, only five percent of the
slides depicted helicopters. None of the F-15 intelligence briefings or training ever
covered the camouflage scheme of Iraqi helicopters, which was light brown and
desert tan (in contrast to the forest green camouflage of the Black Hawks).

Pilots are taught to recognize many different kinds of aircraft at high speeds using
"beer shots," which are blurry pictures that resemble how the pilot might see those
aircraft while in flight. The Air Force pilots, however, received very little training in
the recognition of Army helicopters, which they rarely encountered because of the
different altitudes at which they flew. All the helicopter photos they did see during
training, which were provided by the Army, were taken from the ground—a perspective from which it was common for Army personnel to view them but not useful
for a fighter pilot in flight above them. None of the photographs were taken from
the above aft quadrant—the position from which most fighters would view a helicopter. Air Force visual recognition training and procedures were changed after
this accident.

The F-15 pilots also had an inaccurate model of the current airspace occupants, based on the information they had received about who would be in the airspace that day and when. They assumed and had been told in multiple ways that they would be the first coalition aircraft in the TAOR:

- The AGO specified that no coalition aircraft (fixed or rotary wing) was allowed to enter the TAOR before it was sanitized by a fighter sweep.

- The daily ATO and ARF included a list of all flights scheduled to be in the TAOR that day. The ATO listed the Army Black Hawk flights only in terms of their call signs, aircraft numbers, type of mission (transport), and general route (from Diyarbakir to the TAOR and back to Diyarbakir). All departure times were listed "as required" and no helicopters were mentioned on the daily flowsheet. Pilots fly with the flowsheet on kneeboards as a primary reference during the mission. The F-15s were listed as the very first mission into the TAOR; all other aircraft were scheduled to follow them.

- During preflight briefings that morning, the ATO and flowsheet were reviewed in detail. No mention was made of any Army helicopter flights not appearing on the flowsheet.

- The Battle Sheet Directive (a handwritten sheet containing last-minute changes to information published in the ATO and the ARF) handed to them before going to their aircraft contained no information about Black Hawk flights.

- In a radio call to the ground-based Mission Director just after engine start, the lead F-15 pilot was told that no new information had been received since the ATO was published.

- Right before entering the TAOR, the lead pilot checked in again, this time with the ACE in the AWACS. Again, he was not told about any Army helicopters in the area.

- At 1020, the lead pilot reported that they were on station. Usually at this time, the AWACS will give them a "picture" of any aircraft in the area. No information was provided to the F-15 pilots at this time, although the Black Hawks had already checked in with the AWACS on three separate occasions.

- The AWACS continued not to inform the pilots about Army helicopters during the encounter. The lead F-15 pilot twice reported unsuccessful attempts to identify radar contacts they were receiving, but in response they were not informed about the presence of Black Hawks in the area. After the first report, the TAOR controller responded with "Clean there," meaning he did not have a radar hit in that location. Three minutes later, after the second call, the TAOR controller replied, "Hits there." If the radar signal had been identified as a friendly aircraft, the controller would have responded, "Paint there."

- The IFF transponders on the F-15s did not identify the signals as from a friendly aircraft, as discussed earlier.

Various complex analyses have been proposed to explain why the F-15 pilots' mental models of the airspace occupants were incorrect and not open to reexamination once they received conflicting input. But a possible simple explanation is that they believed what they were told. It is well known in cognitive psychology that mental models are slow to change, particularly in the face of ambiguous evidence like that provided in this case. When operators receive input about the state of the system being controlled, they will first try to fit that information into their current mental model and will find reasons to exclude information that does not fit. Because operators are continually testing their mental models against reality (see figure 2.9), the longer a model has been held and the more different sources of information that led to that incorrect model, the more resistant the models will be to change due to conflicting information, particularly ambiguous information. The pilots had been told repeatedly and by almost everyone involved that there were no friendly helicopters in the TAOR at that time.

The F-15 pilots also may have had a misunderstanding about (incorrect model of) the ROE and the procedures required when they detected an unidentified aircraft. The accident report says that the ROE were reduced in briefings and in individual crew members' understandings to a simplified form. This simplification led to some pilots not being aware of specific considerations required prior to engagement, including identification difficulties, the need to give defectors safe conduct, and the possibility of an aircraft being in distress and the crew being unaware of their position. On the other hand, there had been an incident the week before and the F-15 pilots had been issued an oral directive reemphasizing the requirement for fighter pilots to report to the ACE. That directive was the result of an incident on April 7 in which F-15 pilots had initially ignored directions from the ACE to "knock off" or stop an intercept with an Iraqi aircraft. The ACE overheard the pilots preparing to engage the aircraft and contacted them, telling them to stop the engagement because he had determined that the hostile aircraft was outside the no-fly zone and because he was leery of a "bait and trap" situation.[9] The GAO report stated that CFAC officials told the GAO that the F-15 community was "very upset" about the intervention of the ACE during the knock-off incident

---

9. According to the GAO report, in such a strategy, a fighter aircraft is lured into an area by one or more enemy targets and then attacked by other fighter aircraft or surface-to-air missiles.

and felt he had interfered with the carrying out of the F-15 pilots' duties [200]. As discussed in chapter 2, there is no way to determine the motivation behind an individual's actions. Accident analysts can only present the alternative explanations.

Additional reasons for the lead pilot's incorrect mental model stem from ambiguous or missing feedback from the F-15 wing pilot, dysfunctional communication with the Black Hawks, and inadequate information provided over the reference channels from the AWACS and CFAC operations.

Inaccurate Mental Models of the Black Hawk Pilots: The Black Hawk control actions can also be linked to inaccurate mental models, that is, they were unaware there were separate IFF codes for flying inside and outside the TAOR and that they were supposed to change radio frequencies inside the TAOR. As will be seen later, they were actually told not to change frequencies. They had also been told that the AGO restriction on the entry of allied aircraft into the TAOR before the fighter sweep did not apply to them—an official exception had been made for helicopters. They understood that helicopters were allowed inside the TAOR without AWACS coverage as long as they stayed inside the security zone. In practice, the Black Hawk pilots frequently entered the TAOR prior to AWACS and fighter support without incident or comment, and therefore it became accepted practice.

In addition, because their radios were unable to pick up the HAVE QUICK communications between the F-15 pilots and between the F-15s and the AWACS, the Black Hawk pilots' mental models of the situation were incomplete. According to Snook, Black Hawk pilots testified during the investigation,

> We were not integrated into the entire system. We were not aware of what was going on with the F-15s and the sweep and the refuelers and the recon missions and AWACS. We had no idea who was where and when they were there. [191]

Coordination among Multiple Controllers: At this level, each component (aircraft) had a single controller and thus coordination problems did not occur. They were rife, however, at the higher control levels.

Feedback from the Controlled Process: The F-15 pilots received ambiguous information from their visual identification pass. At the speeds and altitudes they were traveling, it is unlikely that they would have detected the unique Black Hawk markings that identified them as friendly. The mountainous terrain in which they were flying limited their ability to perform an adequate identification pass and the green helicopter camouflage added to the difficulty. The feedback from the wingman to the lead F-15 pilot was also ambiguous and was most likely misinterpreted by the lead pilot. Both pilots apparently received incorrect IFF feedback.

Changes after the Accident

After the accident, Black Hawk pilots were:

- Required to strictly adhere to their ATO published routing and timing.
- Not allowed to operate in the TAOR unless under positive control of AWACS. Without AWACS coverage, only administrative helicopter flights between Diyarbakir and Zakhu were allowed, provided they were listed on the ATO.
- Required to monitor the common TAOR radio frequency.
- Required to confirm radio contact with AWACS at least every twenty minutes unless they were on the ground.
- Required to inform AWACS upon landing. They must make mandatory radio calls at each enroute point.
- If radio contact could not be established, required to climb to line-of-sight with AWACS until contact is reestablished.
- Prior to landing in the TAOR (including Zakhu), required to inform the AWACS of anticipated delays on the ground that would preclude taking off at the scheduled time.
- Immediately after takeoff, required to contact the AWACS and reconfirm IFF Modes I, II, and IV are operating. If they have either a negative radio check with AWACS or an inoperative Mode IV, they cannot proceed into the TAOR.

All fighter pilots were:

- Required to check in with the AWACS when entering the low-altitude environment and remain on the TAOR clear frequencies for deconfliction with helicopters.
- Required to make contact with AWACS using UHF, HAVE QUICK, or UHF clear radio frequencies and confirm IFF Modes I, II, and IV before entering the TAOR. If there was either a negative radio contact with AWACS or an inoperative Mode IV, they could not enter the TAOR.

Finally, white recognition strips were painted on the Black Hawk rotor blades to enhance their identification from the air.

5.3.4   The ACE and Mission Director

Context in Which Decisions and Actions Took Place

Safety Requirements and Constraints: The ACE and mission director must follow the procedures specified and implied by the ROE, the ACE must ensure that pilots follow the ROE, and the ACE must interact with the AWACS crew to identify reported unidentified aircraft (see figure 5.7).

Controls: The controls include the ROE to slow down the engagement and a chain of command to prevent individual error or erratic behavior.

Roles and Responsibilities: The ACE was responsible for controlling combat operations and for ensuring that the ROE were enforced. He flew in the AWACS so he could get up-to-the-minute information about the state of the TAOR airspace.

The ACE was always a highly experienced person with fighter experience. That day, the ACE was a major with nineteen years in the Air Force. He had perhaps more combat experience than anyone else in the Air Force under forty. He had logged 2,000 total hours of flight time and flown 125 combat missions, including 27 in the Gulf War, during which time he earned the Distinguished Flying Cross and two air medals for heroism. At the time of the accident, he had worked for four months as an ACE and flown approximately fifteen to twenty missions on the AWACS [191].

The Mission Director on the ground provided a chain of command for real-time decision making from the pilots to the CFAC commander. On the day of the accident, the Mission Director was a lieutenant colonel with more than eighteen years in the Air Force. He had logged more than 1,000 hours in the F-4 in Europe and an additional 100 hours worldwide in the F-15 [191].

Environmental and Behavior-Shaping Factors: No pertinent factors were identified in the reports and books on the accident.

Dysfunctional Interactions at This Level
The ACE was supposed to get information about unidentified or enemy aircraft from the AWACS mission crew, but in this instance they did not provide it.

Flawed or Inadequate Decisions and Control Actions
The ACE did not provide any control commands to the F-15s with respect to following the ROE or engaging and firing on the U.S. helicopters.

Reasons for Flawed Control Actions and Dysfunctional Interactions

Incorrect Control Algorithms: The control algorithms should theoretically have been effective, but they were never executed.

Inaccurate Mental Models: CFAC, and thus the Mission Director and ACE, exercised ultimate tactical control of the helicopters, but they shared the common view with the AWACS crew that helicopter activities were not an integral part of OPC air operations. In testimony after the accident, the ACE commented, "The way I understand it, only as a courtesy does the AWACS track Eagle Flight."

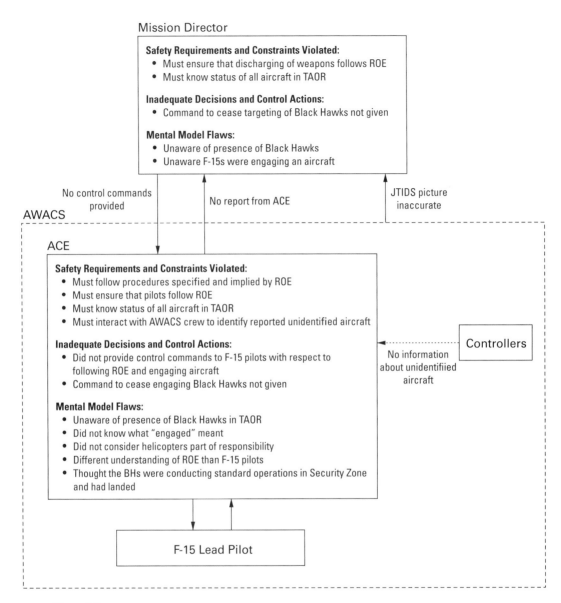

Figure 5.7
Analysis for the ACE and mission director.

The Mission Director and ACE also did not have the information necessary to exercise their responsibility. The ACE had an inaccurate model of where the Black Hawks were located in the airspace. He testified that he presumed the Black Hawks were conducting standard operations in the Security Zone and had landed [159]. He also testified that, although he had a radarscope, he had no knowledge of AWACS radar symbology: "I have no idea what those little blips mean." The Mission Director, on the ground, was dependent on the information about the current air-space state sent down from the AWACS via JTIDS (the Joint Tactical Information Distribution System).

The ACE testified that he assumed the F-15 pilots would ask him for guidance in any situation involving a potentially hostile aircraft, as required by the ROE. The ACE's and F-15 pilots' mental models of the ROE clearly did not match with respect to who had the authority to initiate the engagement of unidentified aircraft. The rules of engagement stated that the ACE was responsible, but some pilots believed they had authority when an imminent threat was involved. Because of security concerns, the actual ROE used were not disclosed during the accident investigation, but, as argued earlier, the slow, low-flying Black Hawks posed no serious threat to an F-15.

Although the F-15 pilot never contacted the ACE about the engagement, the ACE did hear the call of the F-15 lead pilot to the TAOR controller. The ACE testified to the Accident Investigation Board that he did not intervene because he believed the F-15 pilots were not committed to anything at the visual identi-fication point, and he had no idea they were going to react so quickly. Since being assigned to OPC, he said the procedure had been that when the F-15s or other fighters were investigating aircraft, they would ask for feedback from the ACE. The ACE and AWACS crew would then try to rummage around and find out whose aircraft it was and identify it specifically. If they were unsuccessful, the ACE would then ask the pilots for a visual identification [159]. Thus, the ACE probably assumed that the F-15 pilots would not fire at the helicopters without reporting to him first, which they had not done yet. At this point, they had simply requested an identification by the AWACS traffic controller. According to his understanding of the ROE, the F-15 pilots would not fire without his approval unless there was an immediate threat, which there was not. The ACE testified that he expected to be queried by the F-15 pilots as to what their course of action should be.

The ACE also testified at one of the hearings:

> I really did not know what the radio call "engaged" meant until this morning. I did not think the pilots were going to pull the trigger and kill those guys. As a previous right seater in an F-111, I thought "engaged" meant the pilots were going down to do a visual intercept. [159]

Coordination among Multiple Controllers: Not applicable

Feedback from Controlled Process: The F-15 lead pilot did not follow the ROE and report the identified aircraft to the ACE and ask for guidance, although the ACE did learn about it from the questions the F-15 pilots posed to the controllers on the AWACS aircraft. The Mission Director got incorrect feedback about the state of the airspace from JTIDS.

Time Lags: An unusual time lag occurred where the lag was in the controller and not in one of the other parts of the control loop.[10] The F-15 pilots responded faster than the ACE (in the AWACS) and Mission Director (on the ground) could issue appropriate control instructions (as required by the ROE) with regard to the engagement.

Changes after the Accident
There were no changes after the accident, although roles were clarified.

5.3.5   The AWACS Operators
This level of the control structure contains more examples of inconsistent mental models and asynchronous evolution. In addition, this control level provides interesting examples of the adaptation over time of specified procedures to accepted practice and of coordination problems. There were multiple controllers with confused and overlapping responsibilities for enforcing different aspects of the safety requirements and constraints (figure 5.8). The overlaps and boundary areas in the controlled processes led to serious coordination problems among those responsible for controlling aircraft in the TAOR.

Context in Which Decisions and Actions Took Place

Safety Requirements and Constraints: The general safety constraint involved in the accident at this level was to prevent misidentification of aircraft by the pilots and any friendly fire that might result. More specific requirements and constraints are shown in figure 5.8.

Controls: Controls included procedures for identifying and tracking aircraft, training (including simulator missions), briefings, staff controllers, and communication channels. The senior director and surveillance officer (ASO) provided real-time oversight of the crew's activities, while the mission crew commander (MCC) coordinated all the activities aboard the AWACS aircraft.

---

10. A similar type of time lag led to the loss of an F-18 when a mechanical failure resulted in inputs arriving at the computer interface faster than the computer was able to process them.

AWACS Mission Crew

Safety Requirements and Constraints Violated:
- Must identify and track all aircraft in TAOR
- Friendly aircraft must not be misidentified as hostile
- Must accurately inform fighters about status of all aircraft when queried
- Must alert fighters of any aircraft not appearing on flowsheet
- Must not fail to warn fighters about any friendly aircraft they are targeting
- Must provide ground with accurate picture of airspace and its occupants (through JTIDS)

Dysfunctional Interactions:
- Control of aircraft not handed off from enroute to TAOR controller
- Interactions between ASO and senior WD with respect to tracking the flight of the helicopters on the radarscope

Inadequate Decisions and Control Actions:
- Enroute controller did not tell BH pilots to change to TAOR frequency
- Enroute controller did not hand off control of BHs to TAOR controller
- Enroute controller did not monitor course of BHs while in TAOR
- Enroute controller did not use Delta point system to determine BH flight plan
- TAOR controller did not monitor course of helicopters in TAOR
- Nobody alerted F-15 pilots before they fired that the helicopters they were targeting were friendly
- Nobody warned pilots that friendly aircraft were in the area
- Did not try to stop the engagement
- Nobody told BH pilots that squawking wrong IFF code
- MCC did not relay information that was not on ATO about helicopters during morning briefing
- Shadow crew was not monitoring activities

Coordination Flaws:
- Confusion over who was tracking helicopters
- Confusion over responsibilities of surveillance and weapon directors
- No one assigned responsibility for monitoring helicopter traffic in NFZ
- Confusion over who had authority to initiate engagement

Context:
- Min Comm
- Poor morale, inadequate training, overworked
- Low activity at time of accident
- Terminal failure led to changed seating arrangement
- Airspace violations were rare

Mental Model Flaws:
- Did not think helicopters were an integral part of OPC air operations
- Inaccurate models of airspace occupants and where they were
- Thought helicopters only going to Zakhu

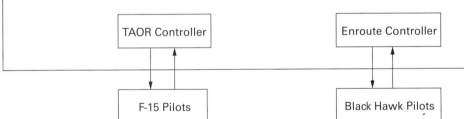

Figure 5.8
The analysis at the AWACS control level.

The Delta Point system, used since the inception of OPC, provided standard code names for real locations. These code names were used to prevent the enemy, who might be listening to radio transmissions, from knowing the helicopters' flight plans.

Roles and Responsibilities: The AWACS crew were responsible for identifying, tracking, and controlling all aircraft enroute to and from the TAOR; for coordinating air refueling; for providing airborne threat warning and control in the TAOR; and for providing surveillance, detection and identification of all unknown aircraft. Individual responsibilities are described in section 5.2.

The staff weapons director (instructor) was permanently assigned to Incirlik. He did all incoming briefings for new AWACS crews rotating into Incirlik and accompanied them on their first mission in the TAOR. The OPC leadership recognized the potential for some distance to develop between stateside spin-up training and continuously evolving practice in the TAOR. Therefore, as mentioned earlier, permanent *staff* or *instructor* personnel flew with each new AWACS crew on their maiden flight in Turkey. Two of these staff controllers were on the AWACS the day of the accident to answer any questions that the new crew might have about local procedures and, as described earlier, to inform them about adaptation of accepted practice from specified procedures.

The SD had worked as an AWACS controller for five years. This was his fourth deployment to OPC, his second as an SD, and his sixtieth mission over the Iraqi TAOR [159]. He worked as a SD more than two hundred days a year and had logged more than 2,383 hours flying time [191].

The enroute controller, who was responsible for aircraft outside the TAOR, was a first lieutenant with four years in the Air Force. He had finished AWACS training two years earlier (May 1992) and had served in the Iraqi TAOR previously [191].

The TAOR controller, who was responsible for controlling all air traffic flying within the TAOR, was a second lieutenant with more than nine years of service in the Air Force, but he had just finished controller's school and had had no previous deployments outside the continental United States. In fact, he had become mission ready only two months prior to the incident. This tour was his first in OPC and his first time as a TAOR controller. He had only controlled as a mission-ready weapons director on three previous training flights [191] and never in the role of TAOR controller. AWACS guidance at the time suggested that the most inexperienced controller be placed in the TAOR position: None of the reports on the accident provided the reasoning behind this practice.

The air surveillance officer (ASO) was a captain at the time of the shootdown. She had been mission-ready since October 1992 and was rated as an instructor ASO. Because the crew's originally assigned ASO was upgrading and could not make it to Turkey on time, she volunteered to fill in for him. She had already served for five and a half weeks in OPC at the time of the accident and was completing her third assignment to OPC. She worked as an ASO approximately two hundred days a year [191].

Environmental and Behavior-Shaping Factors: At the time of the shootdown, shrinking defense budgets were leading to base closings and cuts in the size of the military. At the same time, a changing political climate, brought about by the fall of the Soviet Union, demanded significant U.S. military involvement in a series of operations. The military (including the AWACS crews) were working at a greater pace than they had ever experienced due to budget cuts, early retirements, force outs, slowed promotions, deferred maintenance, and delayed fielding of new equipment. All of these factors contributed to poor morale, inadequate training, and high personnel turnover.

AWACS crews are stationed and trained at Tinker Air Force Base in Oklahoma and then deployed to locations around the world for rotations lasting approximately thirty days. Although all but one of the AWACS controllers on the day of the accident had served previously in the Iraqi no-fly zone, this was their first day working together and, except for the surveillance officer, the first day of their current rotation. Due to last minute orders, the team got only minimal training, including one simulator session instead of the two full three-hour sessions required prior to deploying. In the only session they did have, some of the members of the team were missing—the ASO, ACE, and MCC were unable to attend—and one was later replaced: As noted, the ASO originally designated and trained to deploy with this crew was instead shipped off to a career school at the last minute, and another ASO, who was just completing a rotation in Turkey, filled in.

The one simulator session they did receive was less than effective, partly because the computer tape provided by Boeing to drive the exercise was not current (another instance of asynchronous evolution). For example, the maps were out of date, and the rules of engagement used were different and much more restrictive than those currently in force in OPC. No Mode I codes were listed. The list of friendly participants in OPC did not include UH-60s (Black Hawks) and so on. The second simulation session was canceled because of a wing exercise.

Because the TAOR area had not yet been sanitized, it was a period of low activity: At the time, there were still only four aircraft over the no-fly zone—the two F-15s and the two Black Hawks. AWACS crews are trained and equipped to track literally hundreds of enemy and friendly aircraft during a high-intensity conflict. Many accidents occur during periods of low activity when vigilance is reduced compared to periods of higher activity.

The MCC sits with the other two key supervisors (SD and ACE) toward the front of the aircraft in a three-seat arrangement named the "Pit," where each has his own radarscope. The SD is seated to the MCC's left. Surveillance is seated in the rear.

Violations of the no-fly zone had been rare and threats few during the past three years, so that day's flight was expected to be an average one, and the supervisors in the Pit anticipated just another routine mission [159].

During the initial orbit of the AWACS, the technicians determined that one of the radar consoles was not operating. According to Snook, this type of problem was not uncommon, and the AWACS is therefore designed with extra crew positions. When the enroute controller realized his assigned console was not working properly, he moved from his normal position between the TAOR and tanker controllers, to a spare seat directly behind the senior director. This position kept him out of the view of his supervisor and also eliminated physical contact with the TAOR controller.

Dysfunctional Interactions among the Controllers

According to the formal procedures, control of aircraft was supposed to be handed off from the enroute controller to the TAOR controller when the aircraft entered the TAOR. This handoff did not occur for the Black Hawks, and the TAOR controller was not made aware of the Black Hawks' flight within the TAOR. Snook explains this communication error as resulting from the radar console failure, which interfered with communication between the TAOR and enroute controllers. But this explanation does not gibe with the fact that the *normal* procedure of the enroute controller was to continue to control helicopters without handing them off to the TAOR controller, even when the enroute and TAOR controllers were seated in their usual places next to each other. There may usually have been more informal interaction about aircraft in the area when they were seated next to each other, but there is no guarantee that such interaction would have occurred even with a different seating arrangement. Note that the helicopters had been dropped from the radar screens and the enroute controller had an incorrect mental model of where they were: He thought they were close to the boundary of the TAOR and was unaware they had gone deep within it. The enroute controller, therefore, could not have told the TAOR controller about the true location of the Black Hawks even if they had been sitting next to each other.

The interaction between the surveillance officer and the senior weapons director with respect to tracking the helicopter flight on the radar screen involved many dysfunctional interactions. For example, the surveillance officer put an attention arrow on the senior director's radarscope in an attempt to query him about the lost helicopter symbol that was floating, at one point, unattached to any track. The senior director did not respond to the attention arrow, and it automatically dropped off the screen after sixty seconds. The helicopter symbol (*H*) dropped off the radar screen when the radar and IFF returns from the Black Hawks faded and did not return until just before the engagement, removing any visual reminder to the AWACS crew that there were Black Hawks inside the TAOR. The accident investigation did not include an analysis of the design of the AWACS human–computer interface or how it might have contributed to the accident, although such an analysis is important in fully understanding why it made sense for the controllers to act the way they did.

During his court-martial for negligent homicide, the senior director argued that his radarscope did not identify the helicopters as friendly and that therefore he was not responsible. When asked why the Black Hawk identification was dropped from the radarscope, he gave two reasons. First, because it was no longer attached to any active signal, they assumed the helicopter had landed somewhere. Second, because the symbol displayed on their scopes was being relayed in real time through a JTIDS downlink to commanders on the ground, they were very concerned about sending out an inaccurate picture of the TAOR.

> Even if we suspended it, it would not be an accurate picture, because we wouldn't know for sure if that is where he landed. Or if he landed several minutes earlier, and where that would be. So, the most accurate thing for us to do at that time, was to drop the symbology [*sic*].

Flawed or Inadequate Decision Making and Control Actions

There were myriad inadequate control actions in this accident, involving each of the controllers in the AWACS. The AWACS crew work as a team so it is sometimes hard to trace incorrect decisions to one individual. While from each individual's stand-point the actions and decisions may have been correct, when put together as a whole the decisions were incorrect.

The enroute controller never told the Black Hawk pilots to change to the TAOR frequency that was being monitored by the TAOR controller and did not hand off control of the Black Hawks to the TAOR controller. The established practice of not handing off the helicopters had probably evolved over time as a more efficient way of handling traffic—another instance of asynchronous evolution. Because the heli-copters were usually only at the very border of the TAOR and spent very little time there, the overhead of handing them off twice within a short time period was con-sidered inefficient by the AWACS crews. As a result, the procedures used had changed over time to the more efficient procedure of keeping them under the control of the enroute controller. The AWACS crews were not provided with written guidance or training regarding the control of helicopters within the TAOR, and, in its absence, they adapted their normal practices for fixed-wing aircraft as best they could to apply them to helicopters.

In addition to not handing off the helicopters, the enroute controller did not monitor the course of the Black Hawks while they were in the TAOR (after leaving Zakhu), did not take note of the flight plan (from *Whiskey* to *Lima*), did not alert the F-15 pilots there were friendly helicopters in the area, did not alert the F-15 pilots before they fired that the helicopters they were targeting were friendly, and did not tell the Black Hawk pilots that they were on the wrong frequency and were squawking the wrong IFF Mode I code.

The TAOR controller did not monitor the course of the Black Hawks in the TAOR and did not alert the F-15 pilots before they fired that the helicopters they were targeting were friendly. None of the controllers warned the F-15 pilots at any time that there were friendly helicopters in the area nor did they try to stop the engagement. The accident investigation board found that because Army helicopter activities were not normally known at the time of the fighter pilots' daily briefings, normal procedures were for the AWACS crews to receive real-time information about their activities from the helicopter crews and to relay that information on to the other aircraft in the area. If this truly was established practice, it clearly did not occur on that day.

The controllers were supposed to be tracking the helicopters using the Delta Point system, and the Black Hawk pilots had reported to the enroute controller that they were traveling from *Whiskey* to *Lima*. The enroute controller testified, however, that he had no idea of the towns to which the code names *Whiskey* and *Lima* referred. After the shootdown, he went in search of the card defining the call signs and finally found it in the Surveillance Section [159]. Clearly, tracking helicopters using call signs was not a common practice or the charts would have been closer at hand. In fact, during the court-martial of the senior director, the defense was unable to locate any AWACS crewmember at Tinker AFB (where AWACS crews were stationed and trained) who could testify that he or she had *ever* used the Delta Point system [159] although clearly the Black Hawk pilots thought it was being used because they provided their flight plan using Delta Points.

None of the controllers in the AWACS told the Black Hawk helicopters that they were squawking the wrong IFF code for the TAOR. Snook cites testimony from the court-martial of the senior director that posits three related explanations for this lack of warning: (1) the minimum communication (min comm) policy, (2) a belief by the AWACS crew that the Black Hawks should know what they were doing, and (3) pilots not liking to be told what to do. None of these explanations provided during the trial is very satisfactory and appear to be after-the-fact ratio-nalizations for the controllers not doing their job when faced with possible court-martial and jail terms. Given that the controllers acknowledged that the Army helicopters never squawked the right codes and had not done so for months, there must have been other communication channels that could have been used besides real-time radio communication to remedy this situation, so the min comm policy is not an adequate explanation. Arguing that the pilots should know what they were doing is simply an abdication of responsibility, as is the argument that pilots did not like being told what to do. A different perspective, and one that likely applies to all the controllers, was provided by the staff weapons director, who testified, "For a helicopter, if he's going to Zakhu, I'm not that concerned about him going beyond that. So, I'm not really concerned about having an F-15 needing to identify this guy." [159]

The mission crew commander had provided the crew's morning briefing. He spent some time going over the activity flowsheet, which listed all the friendly aircraft flying in the OPC that day, their call signs, and the times they were scheduled to enter the TAOR. According to Piper (but nobody else mentions it), he failed to note the helicopters, even though their call signs and their IFF information had been written on the margin of his flowsheet.

The shadow crew always flew with new crews on their first day in OPC, but the task of these instructors does not seem to have been well defined. At the time of the shootdown, one was in the galley "taking a break," and the other went back to the crew rest area, read a book, and took a nap. The staff weapons director, who was asleep in the back of the AWACS, during the court-martial of the senior director testified that his purpose on the mission was to be the "answer man," just to answer any questions they might have. This was a period of very little activity in the area (only the two F-15s were supposed to be in the TAOR), and the shadow crew members may have thought their advice was not needed at that time.

When the staff weapons director went back to the rest area, the only symbol displayed on the scopes of the AWACS controllers was the one for the helicopters (*EE01*), which they thought were going to Zakhu only.

Because many of the dysfunctional actions of the crew *did* conform to the established practice (e.g., not handing off helicopters to the TAOR controller), it is unclear what different result might have occurred if the shadow crew had been in place. For example, the staff weapons director testified during the hearings and trial that he had seen helicopters out in the TAOR before, past Zakhu, but he really did not feel it was necessary to brief crews about the Delta Point system to determine a helicopter's destination [159].[11]

Reasons for the Flawed Control

Inadequate Control Algorithms: This level of the accident analysis provides an interesting example of the difference between prescribed procedures and established practice, the adaptation of procedures over time, and migration toward the boundaries of safe behavior. Because of the many helicopter missions that ran from Diyarbakir to Zakhu and back, the controllers testified that it did not seem worth

---

11. Even if the actions of the shadow crew did not contribute to this particular accident, we can take advantage of the accident investigation to perform a safety audit on the operation of the system and identify potential improvements.

handing them off and switching them over to the TAOR frequency for only a few minutes. Established practice (keeping the helicopters under the control of the enroute controller instead of handing them off to the TAOR controller) appeared to be safe until the day the helicopters' behavior differed from normal, that is, they stayed longer in the TAOR and ventured beyond a few miles inside the boundaries. Established practice no longer assured safety under these conditions. A complicating factor in the accident was the universal misunderstanding of each of the controllers' responsibilities with respect to tracking Army helicopters.

Snook suggests that the *min comm* norm contributed to the AWACS crew's general reluctance to enforce rules, contributed to AWACS not correcting Eagle Flight's improper Mode I code, and discouraged controllers from pushing helicopter pilots to the TAOR frequency when they entered Iraq because they were reluctant to say more than absolutely necessary.

According to Snook, there were also no explicit or written procedures regarding the control of helicopters. He states that radio contact with helicopters was lost frequently, but there were no procedures to follow when this occurred. In contrast, Piper claims the AWACS operations manual says:

> Helicopters are a high interest track and should be hard copied every five minutes in turkey and every two minutes in Iraq. These coordinates should be recorded in a special log book, because radar contact with helicopters is lost and the radar symbology [*sic*] can be suspended. [159].

There is no information in the publicly available parts of the accident report about any special logbook or whether such a procedure was normally followed.

Inaccurate and Inconsistent Mental Models: In general, the AWACS crew (and the ACE) shared the common view that helicopter activities were not an integral part of OPC air operations. There was also a misunderstanding about which provisions of the ATO applied to Army helicopter activities.

Most of the people involved in the control of the F-15s were unaware of the presence of the Black Hawks in the TAOR that day, the lone exception perhaps being the enroute controller who knew they were there but apparently thought they would stay at the boundaries of the TAOR and thus were far from their actual location deep within it. The TAOR controller testified that he had never talked to the Black Hawks: Following their two check-ins with the enroute controller, the helicopters had remained on the enroute frequency (as was the usual, accepted practice), even as they flew deep into the TAOR.

The enroute controller, who had been in contact with the Black Hawks, had an inaccurate model of where the helicopters were. When the Black Hawk pilots originally reported their takeoff from the Army Military Coordination Center at Zakhu, they contacted the enroute controller and said they were bound for *Lima*. The enroute controller did not know to what city the call sign *Lima* referred and did not try to look up this information. Other members of the crew also had inaccurate models of their responsibilities, as described in the next section. The Black Hawk pilots clearly thought the AWACS was tracking them and also thought the controllers were using the Delta Point system—otherwise helicopter pilots would not have provided the route names in that way.

The AWACS crews did not appear to have accurate models of the Black Hawks mission and role in OPC. Some of the flawed control actions seem to have resulted from a mental model that helicopters only went to Zakhu and therefore did not need to be tracked or to follow the standard TAOR procedures.

As with the pilots and their visual recognition training, the incorrect mental models may have been at least partially the result of the inadequate AWACS training the team received.

Coordination among Multiple Controllers: As mentioned earlier, coordination problems are pervasive in this accident due to overlapping control responsibilities and confusion about responsibilities in the boundary areas of the controlled process. Most notably, the helicopters usually operated close to the boundary of the TAOR, resulting in confusion over who was or should be controlling them.

The official accident report noted a significant amount of confusion within the AWACS mission crew regarding the tracking responsibilities for helicopters [5]. The mission crew commander testified that nobody was specifically assigned responsibility for monitoring helicopter traffic in the no-fly zone and that his crew believed the helicopters were not included in their orders [159]. The staff weapons director made a point of not knowing what the Black Hawks do: "It was some kind of a squirrely mission" [159]. During the court-martial of the senior director, the AWACS tanker controller testified that in the briefing the crew received upon arrival at Incirlik, the staff weapons director had said about helicopters flying in the no-fly zone, "They're there, but don't pay any attention to them." The enroute controller testified that the handoff procedures applied only to fighters. "We generally have no set procedures for any of the helicopters. . . . We never had any [verbal] guidance [or training] at all on helicopters" [159].

Coordination problems also existed between the activities of the surveillance personnel and the other controllers. During the investigation of the accident, the ASO testified that surveillance's responsibility was south of the 36th Parallel, and the other controllers were responsible for tracking and identifying all aircraft north of the 36th Parallel. The other controllers suggested that surveillance was responsible for tracking and identifying all unknown aircraft, regardless of location. In fact, Air Force regulations say that surveillance had tracking responsibility for unknown and unidentified tracks throughout the TAOR. It is not possible through the testimony alone, again because of the threat of court-martial, to piece out exactly what was the problem here, including simply a migration of normal operations from specified operations. At the least, it is clear that there was confusion about who was in control of what.

One possible explanation for the lack of coordination among controllers at this level of the hierarchical control structure is that, as suggested by Snook, this particular group had never trained together as a team [191]. But given the lack of procedures for handling helicopters and the confusion even by experienced controllers and the staff instructors about responsibilities for handling helicopters, Snook's explanation is not very convincing. A more plausible explanation is simply a lack of guidance and delineation of responsibilities by the management level above. And even if the roles of everyone in such a structure had been well defined originally, uncontrolled local adaptation to more efficient procedures and asynchronous evolution of the different parts of the control structure created dysfunctionalities as time passed. The helicopters and fixed wing aircraft had separate control structures that only joined fairly high up on the hierarchy and, as is described in the next section, there were communication problems between the components at the higher levels of the control hierarchy, particularly between the Army Military Coordination Center (MCC) and the Combined Forces Air Component (CFAC) headquarters.

Feedback from the Controlled Process: Signals to the AWACS from the Black Hawks were inconsistent due to line-of-sight limitations and the mountainous terrain in which the Black Hawks were flying. The helicopters used the terrain to mask themselves from air defense radars, but this terrain masking also caused the radar returns from the Black Hawks to the AWACS (and to the fighters) to fade at various times.

Time Lags: Important time lags contributed to the accident, such as the delay of radio reports from the Black Hawk helicopters due to radio signal transmission problems and their inability to use the TACSAT radios until they had landed. As with the ACE, the speed with which the F-15 pilots acted also provided the controllers with little time to evaluate the situation and respond appropriately.

Changes after the Accident
Many changes were instituted with respect to AWACS operations after the accident:

- Confirmation of a positive IFF Mode IV check was required for all OPC aircraft prior to their entry into the TAOR.

- The responsibilities for coordination of air operations were better defined.

- All AWACS aircrews went through a one-time retraining and recertification program, and every AWACS crewmember had to be recertified.

- A plan was produced to reduce the temporary duty of AWACS crews to 120 days a year. In the end, it was decreased from 166 to 135 days per year from January 1995 to July 1995. The Air Combat Command planned to increase the number of AWACS crews.

- AWACS control was required for all TAOR flights.

- In addition to normal responsibilities, AWACS controllers were required to specifically maintain radar surveillance of all TAOR airspace and to issue advisory/deconflicting assistance on all operations, including helicopters.

- The AWACS controllers were required to periodically broadcast friendly helicopter locations operating in the TAOR to all aircraft.

Although not mentioned anywhere in the available documentation on the accident, it seems reasonable that either the AWACS crews started to use the Delta Point system or the Black Hawk pilots were told not to use it and an alternative means for transmitting flight plans was mandated.

5.3.6 The Higher Levels of Control

Fully understanding the behavior at any level of the sociotechnical control structure requires understanding how and why the control at the next higher level allowed or contributed to the inadequate control at the current level. In this accident, many of the erroneous decisions and control actions at the lower levels can only be fully understood by examining this level of control.

Context in Which Decisions and Actions Took Place

Safety Requirements and Constraints Violated: There were many safety constraints violated at the higher levels of the control structure—the Military Coordination Center, Combined Forces Air Component, and CTF commander—and several people were investigated for potential court-martial and received official letters of reprimand. These safety constraints include: (1) procedures must be instituted that delegate appropriate responsibility, specify tasks, and provide effective training to all those responsible for tracking aircraft and conducting combat operations; (2) procedures must be consistent or at least complementary for everyone involved in TAOR airspace operations; (3) performance must be monitored (feedback channels established) to ensure that safety-critical activities are being carried out correctly and that local adaptations have not moved operations beyond safe limits; (4) equipment and procedures must be coordinated between the Air Force and Army to make sure that communication channels are effective and that asynchronous evolution has not occurred; (5) accurate information about scheduled flights must be provided to the pilots and the AWACS crews.

CFAC and MCC

Safety Requirements and Constraints Violated:
- Procedures must be instituted and monitored to ensure that all aircraft in the TAOR are tracked and fighters are aware of theiir location
- Coordination and communication among all flights into the TAOR must be established. Procedures must be established for determining who should be and who is in the TAOR at all times
- The ROE must be understood and followed by those at lower levels
- All aircraft must be able to communicate effectively in the TAOR

Context:
- CFAC operations were physically separated from MCC
- Air Force was able to operate with fixed and rigid schedules while Army mission required flexible scheduling

Dysfunctional Communication and Interactions:
- Did not receive timely detailed flight info on planned helicopte activities
- Information about flights not distributed to all those needing to know it
- Information channels primarily one way
- Mode I code change never got to MCC
- Two versions of ATO
- Helicopter flight plans distributed to F-16 pilots but not to F-15 pilots

Mental Model Flaws:
- Commander thought procedures were being followed, helicopters were being tracked, F-15 pilots were receiving helicopter flight schedules
- Thought Army and Air Force ATOs consistent

Inadequate Decisions and Control Actions:
- Black Hawks allowed to enter TAOR before fighter sweep but F-15 and AWACS crews not informed of this exception
- No requirement for helicopters to file detailed flight plans and follow them
- No procedures for dealing with last minute changes in helicopter flight plans
- F-15 pilots not told to use non-HQ mode for Army helicopters
- No procedures specified to pass SITREP into to CFAC
- Inadequate training on ROE provided to new rotators
- Inadequate discipline
- Inadequate pilot training provided on visual identification
- Inadequate simulator and spin-up training for AWACS crews
- Handoff procedures not established for helicopters. No explicit or writtten procedures, verbal guidance, or training provided regarding control of helicopters by AWACS
- Rules and procedures did not provide adequate control over unsafe F-15 pilot behavior
- Army pilots given wrong information about IFF codes
- Inadequate procedures specified for shadow crew

Inadequate Coordination
- Nobody thought they were responsible for coordinating helicopter flights

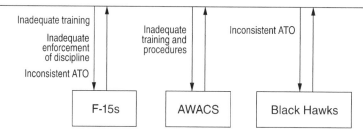

Inadequate training

Inadequate enforcement of discipline

Inconsistent ATO

Inadequate training and procedures

Inconsistent ATO

| F-15s | AWACS | Black Hawks |

Figure 5.9
Analysis at the CFAC-MCC level.

Controls: The controls in place included operational orders and plans to designate roles and responsibilities as well as a management structure, the ACO, coordination meetings and briefings, a chain of command (OPC commander to mission director to ACE to pilots), disciplinary actions for those not following the written rules, and a group (the Joint Operations and Intelligence Center or JOIC) responsible for ensuring effective communication occurred.

Roles and Responsibilities: The MCC had operational control over the Army helicopters while the CFAC had operational control over fixed-wing aircraft and tactical control over all aircraft in the TAOR. The Combined Task Force commander general (who was above both the CFAC and MCC) had ultimate responsibility for the coordination of fixed-wing aircraft flights with Army helicopters.

While specific responsibilities of individuals might be considered here in an official accident analysis, treating the CFAC and MCC as entities is sufficient for the purposes of this analysis.

Environmental and Behavior-Shaping Factors: The Air Force operated on a predictable, well-planned, and tightly executed schedule. Detailed mission packages were organized weeks and months in advance. Rigid schedules were published and executed in preplanned *packages*. In contrast, Army aviators had to react to constantly changing local demands, and they prided themselves on their flexibility [191]. Because of the nature of their missions, exact takeoff times and detailed flight plans for helicopters were virtually impossible to schedule in advance. They were even more difficult to execute with much rigor. The Black Hawks' flight plan contained their scheduled takeoff time, transit routes between Diyarbakir through Gate 1 to Zakhu, and their return time. Because the Army helicopter crews rarely knew exactly where they would be going within the TAOR until after they were briefed at the Military Coordination Center at Zakhu, most flight plans only indicated that Eagle Flight would be "operating in and around the TAOR."

The physical separation of the Army Eagle Flight pilots from the CFAC operations and Air Force pilots at Incirlik contributed to the communication difficulties that already existed between the services.

Dysfunctional Interactions among Controllers

Dysfunctional communication at this level of the control structure played a critical role in the accident. These communication flaws contributed to the coordination flaws at this level and at the lower levels.

A critical safety constraint to prevent friendly fire requires that the pilots of the fighter aircraft know who is in the no-fly zone and whether they are supposed to be there. However, neither the CTF staff nor the Combined Forces Air Component staff requested nor received timely, detailed flight information on planned

MCC helicopter activities in the TAOR. Consequently, the OPC daily Air Tasking Order was published with little detailed information regarding U.S. helicopter flight activities over northern Iraq.

According to the official accident report, specific information on routes of flight and times of MCC helicopter activity in the TAOR was normally available to the other OPC participants only when AWACS received it from the helicopter crews by radio and relayed the information on to the pilots [5]. While those at the higher levels of control may have thought this relaying of flight information was occurring, that does not seem to be the case given that the Delta point system (wherein the helicopter crews provided the AWACS controllers with their flight plan) was not used by the AWACS controllers: When the helicopters went beyond Zakhu, the AWACS controllers did not know their flight plans and therefore could not relay that information to the fighter pilots and other OPC participants.

The weekly flight schedules the MCC provided to the CFAC staff were not complete enough for planning purposes. While the Air Force could plan their missions in advance, the different type of Army helicopter missions had to be flexible to react to daily needs. The MCC daily mission requirements were generally based on the events of the previous day. A weekly flight schedule was developed and provided to the CTF staff, but a firm itinerary was usually not available until after the next day's ATO was published. The weekly schedule was briefed at the CTF staff meetings on Mondays, Wednesday, and Fridays, but the information was neither detailed nor firm enough for effective rotary-wing and fixed-wing aircraft coordination and scheduling purposes [5].

Each daily ATO was published showing several Black Hawk helicopter lines. Of these, two helicopter lines (two flights of two helicopters each) were listed with call signs (Eagle 01/02 and Eagle 03/04), mission numbers, IFF Mode II codes, and a route of flight described only as LLTC (the identifier for Diyarbakir) to TAOR to LLTC. No information regarding route or duration of flight time within the TAOR was given on the ATO. Information concerning takeoff time and entry time into the TAOR was listed as *A/R* (as required).

Every evening, the MCC at Zakhu provided a situation report (SITREP) to the JOIC (located at Incirlik), listing the helicopter flights for the following day. The SITREP did not contain complete flight details and arrived too late to be included in the next day's ATO. The MCC would call the JOIC the night prior to the scheduled mission to "activate" the ATO line. There were, however, no procedures in place to get the SITREP information from the JOIC to those needing to know it in CFAC.

After receiving the SITREP, a duty officer in the JOIC would send takeoff times and gate times (the times the helicopters would enter northern Iraq) to Turkish operations for approval. Meanwhile, an intelligence representative to the JOIC consolidated the MCC weekly schedule with the SITREP and used secure intelligence channels to pass this updated information to some of his counterparts in operational squadrons who had requested it. No procedures existed to pass this information from the JOIC to those in CFAC with tactical responsibility for the helicopters (through the ACE and Mission Director) [5]. Because CFAC normally determined who would fly when, the information channels were designed primarily for one-way communications outward and downward.

In the specific instance involved in the shootdown, the MCC weekly schedule was provided on April 8 to the JOIC and thence to the appropriate person in CFAC. That schedule showed a two-ship, MCC helicopter administrative flight scheduled for April 14. According to the official accident report, two days before (April 12) the MCC Commander had requested approval for an April 14 flight outside the Security Zone from Zakhu to the towns of Irbil and Salah ad Din. The OPC commanding general approved the written request on April 13, and the JOIC transmitted the approval to the MCC but apparently the information was not provided to those responsible for producing the ATO. The April 13 SITREP from MCC listed the flight as "mission support," but contained no other details. Note more information was available earlier than normal in this instance, and it could have been included in the ATO but the established communication channels and procedures did not exist to get it to the right places. The MCC weekly schedule update, received by the JOIC on the evening of April 13 along with the MCC SITREP, gave the destinations for the mission as Salah ad Din and Irbil. This information was not passed to CFAC.

Late in the afternoon on April 13, MCC contacted the JOIC duty officer and activated the ATO line for the mission. A takeoff time of 0520 and a gate time of 0625 were requested. No takeoff time or route of flight beyond Zakhu was specified. The April 13 SITREP, the weekly flying schedule update, and the ATO-line activation request were received by the JOIC too late to be briefed during the Wednesday (April 13) staff meetings. None of the information was passed to the CFAC scheduling shop (which was responsible for distributing last minute changes to the ATO through various sources such as the Battle Staff Directives, morning briefings, and so on), to the ground-based Mission Director, nor to the ACE on board the AWACS [5]. Note that this flight was not a routine food and medical supply run, but instead it carried sixteen high-ranking VIPs and required the personal attention and approval of the CTF Commander. Yet information about the flight was never communicated to the people who needed to know about it [191]. That is, the information went up from the MCC to the CTF staff, but not across from MCC to CFAC nor down from the CTF staff to CFAC (see figure 5.3).

A second example of a major dysfunctional communication involved the communication of the proper radio frequencies and IFF codes to be used in the TAOR.

About two years before the shootdown, someone in the CFAC staff decided to change the instructions pertaining to IFF modes and codes. According to Snook, no one recalled exactly how or why this change occurred. Before the change, all aircraft squawked a single Mode I code everywhere they flew. After the change, all aircraft were required to switch to a different Mode I code while flying in the no-fly zone. The change was communicated through the daily ATO. However, after the accident it was discovered that the Air Force's version of the ATO was not exactly the same as the one received electronically by the Army aviators—another instance of asynchronous evolution and lack of linkup between system components. For at least two years, there existed two versions of the daily ATO: one printed out directly by the Incirlik Frag Shop and distributed locally by messenger to all units at Incirlik Air Base, and a second one transmitted electronically through an Air Force communications center (the JOIC) to Army helicopter operations at Diyarbakir. The one received by the Army aviators was identical in all respects to the one distributed by the Frag Shop, *except* for the changed Mode I code information contained in the SPINS. The ATO that Eagle Flight received contained no mention of two Mode I codes [191].

What about the confusion about the proper radio frequency to be used by the Black Hawks in the TAOR? Piper notes that the Black Hawk pilots were told to use the enroute frequency while flying in the TAOR. The commander of OPC testified after the accident that the use by the Black Hawks of the enroute radio frequency rather than the TAOR frequency had been briefed to him as a *safety measure* because the Black Hawk helicopters were not equipped with HAVE QUICK technology. The ACO (Aircraft Control Order) required the F-15s to use non–HAVE QUICK mode when talking to specific types of aircraft (such as F-1s) that, like the Black Hawks, did not have the new technology. The list of non-HQ aircraft provided to the F-15 pilots, however, for some reason did not include UH-60s. Apparently the decision was made to have the Black Hawks use the enroute radio frequency but this decision was never communicated to those responsible for the F-15 procedures specified in the ACO. Note that a thorough investigation of the higher levels of control, as is required in a STAMP-based analysis, is necessary to explain properly the use of the enroute radio frequency by the Black Hawks. Of the various reports on the shootdown, only Piper notes the fact that an exception had been made for Army helicopters for safety reasons—the official accident report, Snook's detailed book on the accident, and the GAO report do not mention this fact! Piper found out about it from her attendance at the public hearings and trial. This omission of important information from the accident reports is an interesting example of how incomplete investigation of the higher levels of control can lead to incorrect causal analysis. In her book, Piper questions why the Accident Investigation Board, while producing twenty-one volumes of evidence, never asked the commander of OPC about the radio frequency and other problems found during the investigation.

Other official exceptions were made for the helicopter operations, such as allowing them in the Security Zone without AWACS coverage. Using STAMP, the accident can be understood as a dynamic process where the operations of the Army and Air Force adapted and diverged without effective communication and coordination.

Many of the dysfunctional communications and interactions stem from asynchronous evolution of the mission and the operations plan. In response to the evolving mission in northern Iraq, air assets were increased in September 1991 and a significant portion of the ground forces were withdrawn. Although the original organizational structure of the CTF was modified at this time, the operations plan was not. In particular, the position of the person who was in charge of communication and coordination between the MCC and CFAC was eliminated without establishing an alternative communication channel.

Unsafe asynchronous evolution of the safety control structure can be prevented by proper documentation of safety constraints, assumptions, and their controls during system design and checking before changes are made to determine if the constraints and assumptions are violated by the design. Unintentional changes and migration of behavior outside the boundaries of safety can be prevented by various means, including education, identifying and checking leading indicators, and targeted audits. Part III describes ways to prevent asynchronous evolution from leading to accidents.

Flawed or Inadequate Control Actions

There were many flawed or missing control actions at this level, including:

• *The Black Hawk pilots were allowed to enter the TAOR without AWACS coverage and the F-15 pilots and AWACS crews were not informed about this exception to the policy.* This control problem is an example of the problems of distributed decision making with other decision makers not being aware of the decisions of others (see the Zeebrugge example in figure 2.2).

Prior to September 1993, Eagle Flight helicopters flew any time required, before the fighter sweeps and without fighter coverage, if necessary. After September 1993, helicopter flights were restricted to the security zone if AWACS and fighter coverage were not on station. But for the mission on April 14, Eagle Flight requested and received permission to execute their flight outside the security zone. A CTF policy letter dated September 1993 implemented the following policy for UH-60 helicopter flights supporting the MCC: "All UH-60 flights into Iraq outside of the security zone require AWACS coverage." Helicopter flights had routinely been flown within the TAOR security zone without AWACS or fighter coverage and CTF personnel at various levels were aware of this. MCC personnel were aware of the requirement to have

AWACS coverage for flights outside the security zone and complied with that requirement. However, the F-15 pilots involved in the accident, relying on the written guidance in the ACO, believed that no OPC aircraft, fixed or rotary wing, were allowed to enter the TAOR prior to a fighter sweep [5].

At the same time, the Black Hawks also thought they were operating correctly. The Army Commander at Zakhu had called the Commander of Operations, Plans, and Policy for OPC the night before the shootdown and asked to be able to fly the mission without AWACS coverage. He was told that they must have AWACS coverage. From the view of the Black Hawks pilots (who had reported in to the AWACS during the flight and provided their flight plan and destinations) they were complying and were under AWACS control.

- *Helicopters were not required to file detailed flight plans and follow them.* Effective procedures were not established for communicating last minute changes or updates to the Army flight plans that had been filed.

- *F-15 pilots were not told to use non-HQ mode for helicopters.*

- *No procedures were specified to pass SITREP information to CFAC.* Helicopter flight plans were not distributed to CFAC and the F-15 pilots, but they were given to the F-16 squadrons. Why was one squadron informed, while another one, located right across the street, was not? F-15s are designed primarily for air superiority—high altitude aerial combat missions. F-16s, on the other hand, are all-purpose fighters. Unlike F-15s, which rarely flew low-level missions, it was common for F-16s to fly low-level missions where they might encounter the low-flying Army helicopters. As a result, to avoid low-altitude midair collisions, staff officers in F-16 squadrons requested details concerning helicopter operations from the JOIC, went to pick it up from the mail pickup point on the post, and passed it on to the pilots during their daily briefings; F-15 planners did not [191].

- *Inadequate training on the ROE was provided for new rotators.* Piper claims that OPC personnel did not receive consistent, comprehensive training to ensure they had a thorough understanding of the rules of engagement and that many of the aircrews new to OPC questioned the need for the less aggressive rules of engagement in what had been designated a combat zone [159]. Judging from these complaints (details can be found in [159]) and incidents involving F-15 pilots, it appears that the pilots did not fully understand the ROE purpose or need.

- *Inadequate training was provided to the F-15 pilots on visual identification.*

- *Inadequate simulator and spin-up training was provided to the AWACS crews.* Asynchronous evolution occurred between the changes in the training materials and the actual situation in the no-fly zone. In addition, there were no controls to ensure the required simulator sessions were provided and that all members of the crew participated.

- *Handoff procedures were never established for helicopters.*   In fact, no explicit or written procedures, verbal guidance, or training of any kind were provided to the AWACS crews regarding the control of helicopters within the TAOR [191]. The AWACS crews testified during the investigation that they lost contact with helicopters all the time, but there were no procedures to follow when that occurred.

- *Inadequate procedures were specified and enforced for how the shadow crew would instruct the new crews.*

- *The rules and procedures established for the operation did not provide adequate control over unsafe F-15 pilot behavior, adequate enforcement of discipline, or adequate handling of safety violations.*   The CFAC Assistant Director of Operations told the GAO investigators that there was very little F-15 oversight in OPC at the time of the shootdown. There had been so many flight discipline incidents leading to close calls that a group safety meeting had been held a week before the shootdown to discuss it. The flight discipline and safety issues included midair close calls, unsafe incidents when refueling, and unsafe takeoffs. The fixes (including the meeting) obviously were not effective. But the fact that there were a lot of close calls indicates serious safety problems existed and were not handled adequately.

  The CFAC Assistant Director of Operations also told the GAO that contentious issues involving F-15 actions had become common topics of discussion at Detachment Commander meetings. No F-15 pilots were on the CTF staff to communicate with the F-15 group about these problems. The OPC Commander testified that there was no tolerance for mistakes or unprofessional flying at OPC and that he had regularly sent people home for violation of the rules—the majority of those he sent home were F-15 pilots, suggesting that there were serious problems in discipline and attitude among this group [159].

- *The Army pilots were given the wrong information about the IFF codes and radio frequencies to use in the TAOR.*   As described above, this mismatch resulted from asynchronous evolution and lack of linkup (consistency) between process controls, that is, the two different ATOs. It provides yet another example of the danger involved in distributed decision making (again see figure 2.2).

Reasons for the Flawed Control

Ineffective Control Algorithms: Almost all of the control flaws at this level relate to the existence and use of ineffective control algorithms. Equipment and procedures were not coordinated between the Air Force and the Army to make sure that communication channels were effective and that asynchronous evolution had not occurred. The last CTF staff member who appears to have actively coordinated rotary-wing flying activities with the CFAC organization departed in January 1994. No representative of the MCC was specifically assigned to the CFAC for coordination purposes. Since December 1993, no MCC helicopter detachment representative had attended the CFAC weekly scheduling meetings. The Army liaison officer, attached to the MCC helicopter detachment at Zakhu and assigned to Incirlik AB, was new on station (he arrived in April 1994) and was not fully aware of the relationship of the MCC to the OPC mission [5].

Performance was not monitored to ensure that safety-critical activities were carried out correctly, that local adaptations had not moved operations beyond safe limits, and that information was being effectively transmitted and procedures followed. Effective controls were not established to prevent unsafe adaptations.

The feedback that was provided about the problems at the lower levels was ignored. For example, the Piper account of the accident includes a reference to helicopter pilots' testimony that six months before the shootdown, in October 1993, they had complained that the fighter aircraft were using their radar to lock onto the Black Hawks an unacceptable number of times. The Army helicopter pilots had argued there was an urgent need for the Black Hawk pilots to be able to communicate with the fixed-wing aircraft, but nothing was changed until after the accident, when new radios were installed in the Black Hawks.

Inaccurate Mental Models: The commander of the Combined Task Force thought that the appropriate control and coordination was occurring. This incorrect mental model was supported by the feedback he received flying as a regular passenger on board the Army helicopter flights, where it was his perception that the AWACS was monitoring their flight effectively. The Army helicopter pilots were using the Delta Point system to report their location and flight plans, and there was no indication from the AWACS that the messages were being ignored. The CTF Commander testified that he believed the Delta Point system was standard on all AWACS missions. When asked at the court-martial of the AWACS senior director whether the AWACS crew were tracking Army helicopters, the OPC Commander replied:

> Well, my experience from flying dozens of times on Eagle Flight, which that—for some eleven hundred and nine days prior to this event, that was—that was normal procedures for them to flight follow. So, I don't know that they had something written about it, but I know that it seemed very obvious and clear to me as a passenger on Eagle Flight numerous times that that was occurring. [159]

The commander was also an active F-16 pilot who attended the F-16 briefings. At these briefings he observed that Black Hawk times were part of the daily ATOs received by the F-16 pilots and assumed that all squadrons were receiving the same information. However, as noted, the head of the squadron with which the commander flew had gone out of his way to procure the Black Hawk flight information, while the F-15 squadron leader had not.

Many of those involved at this level were also under the impression that the ATOs provided to the F-15 pilots and to the Black Hawks pilots were consistent, that required information had been distributed to everyone, that official procedures were understood and being followed, and so on.

Coordination among Multiple Controllers: There were clearly problems with overlapping and boundary areas of control between the Army and the Air Force. Coordination problems between the services are legendary and were not handled adequately here. For example, two different versions of the ATO were provided to the Air Force and the Army pilots. The Air Force F-15s and the Army helicopters had separate control structures, with a common control point fairly high above the physical process. The problems were complicated by the differing importance of flexibility in flight plans between the two services. One symptom of the problem was that there was no requirement for helicopters to file detailed flight plans and follow them and no procedures established to deal with last minute changes. These deficiencies were also related to the shared control of helicopters by MCC and CFAC and complicated by the physical separation of the two headquarters.

During the accident investigation, a question was raised about whether the Combined Task Force Chief of Staff was responsible for the breakdown in staff communication. After reviewing the evidence, the hearing officer recommended that no adverse action be taken against the Chief of Staff because he (1) had focused his attention according to the CTF Commander's direction, (2) had neither specific direction nor specific reason to inquire into the transmission of info between his Director of Operations for Plans and Policy and the CFAC, (3) had been the most recent arrival and the only senior Army member of a predominantly Air Force staff and therefore generally unfamiliar with air operations, and (4) had relied on experienced colonels under whom the deficiencies had occurred [200]. This conclusion was obviously influenced by the goal of trying to establish blame. Ignoring the blame aspects, the conclusion gives the impression that nobody was in charge and everyone thought someone else was.

According to the official accident report, the contents of the ACO largely reflected the guidance given in the operations plan dated September 7, 1991. But that was the plan provided before the mission had changed. The accident report concludes that key CTF personnel at the time of the accident were either unaware of the existence of this particular plan or considered it too outdated to be applicable. The accident report states, "Most key personnel within the CFAC and CTF staff did not consider coordination of MCC helicopter activities to be part of their respective CFAC/CTF responsibilities" [5].

Because of the breakdown of clear guidance from the Combined Task Force staff to its component organizations (CFAC and MCC), they did not have a clear understanding of their respective responsibilities. Consequently, MCC helicopter activities were not fully integrated with other OPC air operations in the TAOR.

5.4 Conclusions from the Friendly Fire Example

When looking only at the proximate events and the behavior of the immediate participants in the accidental shootdown, the reasons for this accident appear to be gross mistakes by the technical system operators (the pilots and AWACS crew). In fact, a special Air Force task force composed of more than 120 people in six commands concluded that two breakdowns in individual performance contributed to the shootdown: (1) the AWACS mission crew did not provide the F-15 pilots an accurate picture of the situation and (2) the F-15 pilots misidentified the target. From the twenty-one-volume accident report produced by the Accident Investigation Board, Secretary of Defense William Perry summarized the "errors, omissions, and failures" in the "chain of events" leading to the loss as:

- The F-15 pilots misidentified the helicopters as Iraqi Hinds.
- The AWACS crew failed to intervene.
- The helicopters and their operations were not integrated into the Task Force running the no-fly zone operations.
- The Identity Friend or Foe (IFF) systems failed.

According to Snook, the military community has generally accepted these four "causes" as the explanation for the shootdown.

While there certainly were mistakes made at the pilot and AWACS levels, the use of the STAMP analysis paints a much more complete explanation of the role of the environment and other factors that influenced their behavior including: inconsistent, missing, or inaccurate information; incompatible technology; inadequate coordination; overlapping areas of control and confusion about who was responsible for what; a migration toward more efficient but less safe operational procedures over time without any controls and checks on the potential adaptations; inadequate training; and in general a control structure that did not enforce the safety constraints. Boiling down this very complex accident to four "causes" and assigning blame in this way inhibits learning from the events. The more complete STAMP analysis was possible only because individuals outside the military, some of whom were relatives of the victims, did not accept the simple analysis provided in the accident report and did their own uncovering of the facts.

STAMP views an accident as a dynamic process. In this case, Army and Air Force operations adapted and diverged without communication and coordination. OPC had operated incident-free for over three years at the time of the shootdown. During that time, local adaptations to compensate for inadequate control from above had managed to mask the ongoing problems until a situation occurred where local adaptations did not work. A lack of awareness at the highest levels of command of the severity of the coordination, communication, and other problems is a key factor in this accident.

Nearly all the types of causal factors identified in section 4.5 can be found in this accident. This fact is not an anomaly: Most accidents involve a large number of these factors. Concentrating on an event chain focuses attention on the proximate events associated with the accident and thus on the principle local actors, in this case, the pilots and the AWACS personnel. Treating an accident as a control problem using STAMP clearly identifies other organizational factors and actors and the role they played. Most important, without this broader view of the accident, only the symptoms of the organizational problems may be identified and eliminated without significantly reducing risk of a future accident caused by the same systemic factors but involving different symptoms at the lower technical and operational levels of the control structure.

More information on how to build multiple views of an accident using STAMP in order to aid understanding can be found in chapter 11. More examples of STAMP accident analyses can be found in the appendixes.

III USING STAMP

STAMP provides a new theoretical foundation for system safety on which new, more powerful techniques and tools for system safety can be constructed. Part III presents some practical methods for engineering safer systems. All the techniques described in part III have been used successfully on real systems. The surprise to those trying them has been how well they work on enormously complex systems and how economical they are to use. Improvements and even more applications of the theory to practice will undoubtedly be created in the future.

6 Engineering and Operating Safer Systems Using STAMP

Part III of this book is for those who want to build safer systems without incurring enormous and perhaps impractical financial, time, and performance costs. The belief that building and operating safer systems requires such penalties is widespread and arises from the way safety engineering is usually done today. It need not be the case. The use of top-down system safety engineering and safety-guided design based on STAMP can not only enhance the safety of these systems but also potentially reduce the costs associated with engineering for safety. This chapter provides an overview, while the chapters following it provide details about how to implement this cost-effective safety process.

6.1 Why Are Safety Efforts Sometimes Not Cost-Effective?

While there are certainly some very effective safety engineering programs, too many expend a large amount of resources with little return on the investment in terms of improved safety. To fix a problem, we first need to understand it. Why are safety efforts sometimes not cost-effective? There are five general answers to this question:

1. Safety efforts may be superficial, isolated, or misdirected.
2. Safety activities often start too late.
3. The techniques used are not appropriate for the systems we are building today and for new technology.
4. Efforts may be narrowly focused on the technical components.
5. Systems are usually assumed to be static throughout their lifetime.

Superficial, isolated, or misdirected safety engineering activities: Often, safety engineering consists of performing a lot of very costly and tedious activities of limited usefulness in improving safety in the final system design. Childs calls this "cosmetic system safety" [37]. Detailed hazard logs are created and analyses performed, but these have limited impact on the actual system design. Numbers are associated with unquantifiable properties. These numbers always seem to support whatever numerical requirement is the goal, and all involved feel as if they have done their jobs. The safety analyses provide the answer the customer or designer wants—that the system is safe—and everyone is happy. Haddon-Cave, in the 2009 Nimrod MR2 accident report, called such efforts *compliance only exercises* [78]. The results impact certification of the system or acceptance by management, but despite all the activity and large amounts of money spent, the safety of the system has been unaffected.

A variant of this problem is that safety activities may be isolated from the engineers and developers building the system. Too often, safety professionals are separated from engineering design and placed within a mission assurance organization. Safety cannot be assured without its already being part of the design; systems must be constructed to be safe from the beginning. Separating safety engineering from design engineering is almost guaranteed to make the effort and resources expended a poor investment. Safety engineering is effective when it participates in and provides input to the design process, not when it focuses on making arguments about the artifacts created after the major safety-related decisions have been made.

Sometimes the major focus of the safety engineering efforts is on creating a *safety case* that proves the completed design is safe, often by showing that a particular process was followed during development. Simply following a process does not mean that the process was effective, which is the basic limitation of many process assurance activities. In other cases the arguments go beyond the process, but they start from the assumption that the system is safe and then focus on showing the conclusion is true. Most of the effort is spent in seeking evidence that shows the system is safe while not looking for evidence that the system is *not* safe. The basic mindset is wrong, so the conclusions are biased.

One of the reasons System Safety has been so successful is that it takes the opposite approach: an attempt is made to show that the system is *unsafe* and to identify hazardous scenarios. By using this alternative perspective, paths to hazards are often identified that were missed by the engineers, who tend to focus on what they want to happen, not what they do *not* want to happen.

If safety-guided design, as defined in part III of this book, is used, the "safety case" is created along with the design. Developing the certification argument becomes trivial and consists primarily of simply gathering the documentation that has been created during the development process.

Safety efforts start too late: Unlike the examples of ineffective safety activities above, the safety efforts may involve potentially useful activities, but they may start too late. Frola and Miller claim that 70–80 percent of the most critical decisions related to the safety of the completed system are made during early concept development [70]. Unless the safety engineering effort impacts these decisions, it is unlikely to have much effect on safety. Too often, safety engineers are busy doing safety analyses, while the system engineers are in parallel making critical decisions about system design and concepts of operation that are not based on that hazard analysis. By the time the system engineers get the information generated by the safety engineers, it is too late to have a significant impact on design decisions.

Of course, engineers normally do try to consider safety early, but the information commonly available is only whether a particular function is safety-critical or not. They are told that the function they are designing can contribute to an accident, with perhaps some letter or numerical "score" of how critical it is, but not much else. Armed only with this very limited information, they have no choice but to focus safety design efforts on increasing the component's reliability by adding redundancy or safety margins. These features are often added without careful analysis of whether they are needed or will be effective for the specific hazards related to that system function. The design then becomes expensive to build and maintain without necessarily having the maximum possible (or sometimes any) impact on eliminating or reducing hazards. As argued earlier, redundancy and overdesign, such as building in safety margins, are effective primarily for purely electromechanical components and component failure accidents. They do not apply to software and miss component interaction accidents entirely. In some cases, such design techniques can even *contribute* to component interaction accidents when they add to the complexity of the design.

Most of our current safety engineering techniques start from detailed designs. So even if they are conscientiously applied, they are useful only in evaluating the safety of a completed design, not in guiding the decisions made early in the design creation process. One of the results of evaluating designs after they are created is that engineers are confronted with important safety concerns only after it is too late or too expensive to make significant changes. If and when the system and component design engineers get the results of the safety activities, often in the form of a critique of the design late in the development process, the safety concerns are frequently ignored or argued away because changing the design at that time is too costly. Design reviews then turn into contentious exercises where one side argues that the system has serious safety limitations while the other side argues that those limitations do not exist, they are not serious, or the safety analysis is wrong.

The problem is not a lack of concern by designers; it's simply that safety concerns about their design are raised at a time when major design changes are not possible — the design engineers have no other option than to defend the design they have. If they lose that argument, then they must try to patch the current design; starting over with a safer design is, in almost all cases, impractical. If the designers had the information necessary to factor safety into their early decision making, then the process of creating safer designs need cost no more and, in fact, will cost less due to two factors: (1) reduced rework after the decisions made are found to be flawed or to provide inadequate safety and (2) less unnecessary overdesign and unneeded protection.

The key to having a cost-effective safety effort is to embed it into a system engineering process starting from early concept development and then to design safety into the system as the design decisions are made. Costs are much less when safety is built into the system design from the beginning rather than added on or retrofitted later.

The techniques used are not appropriate for today's systems and new technology: The assumptions of the major safety engineering techniques currently used, almost all of which stem from decades past, do not match the assumptions underlying the technology and complexity of the systems being built today or the new emerging causes of accidents: They do not apply to human or software errors or flawed management decision making, and they certainly do not apply to weaknesses in the organizational structure or social infrastructure systems. These contributors to accidents do not "fail" in the same way assumed by the current safety analysis tools.

But with no other tools to use, safety engineers attempt to force square pegs into round holes, hoping this will be sufficient. As a result, nothing much is accomplished beyond expending time, money, and other resources. It's time we face up to the fact that new safety engineering techniques are needed to handle those aspects of systems that go beyond the analog hardware components and the relatively simple designs of the past for which the current techniques were invented. Chapter 8 describes a new hazard analysis technique based on STAMP, called STPA, but others are possible. The important thing is to confront these problems head on and not ignore them and waste our time misapplying or futilely trying to extend techniques that do not apply to today's systems.

The safety efforts are focused on the technical components of the system: Many safety engineering (and system engineering, for that matter) efforts focus on the technical system details. Little effort is made to consider the social, organizational, and human components of the system in the design process. Assumptions are made that operators will be trained to do the right things and that they will adapt to whatever design they are given. Sophisticated human factors and system analysis input is lacking, and when accidents inevitably result, they are blamed on the operators for not behaving the way the designers thought they would. To give just one example (although most accident reports contain such examples), one of the four causes, all of which cited pilot error, identified in the loss of the American Airlines B757 near Cali, Colombia (see chapter 2), was "Failure of the flight crew to revert to basic radio navigation when the FMS-assisted navigation became confusing and demanded an excessive workload in a critical phase of the flight." A more useful alternative statement of the cause might have been "An FMS system that confused the operators and demanded an excessive workload in a critical phase of flight."

Virtually all systems contain humans, but engineers are often not taught much about human factors and draw convenient boundaries around the technical components, focusing their attention inside these artificial boundaries. Human factors experts have complained about the resulting *technology-centered automation* [208], where the designers focus on technical issues and not on supporting operator tasks. The result is what has been called "clumsy" automation that increases the chance of human error [183, 22, 208]. One of the new assumptions for safety in chapter 2 is that operator "error" is a product of the environment in which it occurs.

A variant of the problem is common in systems using information technology. Many medical information systems, for example, have not been as successful as they might have been in increasing safety and have even led to new types of hazards and losses [104, 140]. Often, little effort is invested during development in considering the usability of the system by medical professionals or of the impact, not always positive, that the information system design will have on workflow and on the practice of medicine.

Automation is commonly assumed to be safer than manual systems because the hazards associated with the manual systems are eliminated. Inadequate consideration is given to whether new, and maybe even worse, hazards are introduced by the automated system and how to prevent or minimize these new hazards. The aviation industry has, for the most part, learned this lesson for cockpit and flight control design, where eliminating errors of commission simply created new errors of omission [181, 182] (see chapter 9), but most other industries are far behind in this respect.

Like other safety-related system properties that are ignored until too late, operators and human-factors experts often are not brought into the early design process or they work in isolation from the designers until changes are extremely expensive to make. Sometimes, human factors design is not considered until after an accident, and occasionally not even then, almost guaranteeing that more accidents will occur.

To provide cost-effective safety engineering, the system and safety analysis and design process needs to consider the humans in systems—including those that are not directly controlling the physical processes—not separately or after the fact but starting at concept development and continuing throughout the life cycle of the system.

Systems are assumed to be static throughout their lifetimes: It is rare for engineers to consider how the system will evolve and change over time. While designing for maintainability may be considered, unintended changes are often ignored. Change is a constant for all systems: physical equipment ages and degrades over its lifetime and may not be maintained properly; human behavior and priorities usually change over time; organizations change and evolve, which means the safety control structure itself will evolve. Change may also occur in the physical and social environment within which the system operates and with which it interacts. To be effective, controls need to be designed that will reduce the risk associated with all these types of changes. Not only are accidents expensive, but once again planning for system change can reduce the costs associated with the change itself. In addition, much of the effort in operations needs to be focused on managing and reacting to change.

6.2   The Role of System Engineering in Safety

As the systems we build and operate increase in size and complexity, the use of sophisticated system engineering approaches becomes more critical. Important system-level (emergent) properties, such as safety, must be built into the design of these systems; they cannot be effectively added on or simply measured afterward.

While system engineering was developed originally for technical systems, the approach is just as important and applicable to social systems or the social components of systems that are usually not thought of as "engineered." All systems are engineered in the sense that they are designed to achieve specific goals, namely to satisfy requirements and constraints. So ensuring hospital safety or pharmaceutical safety, for example, while not normally thought of as engineering problems, falls within the broad definition of engineering. The goal of the system engineering process is to create a system that satisfies the mission while maintaining the constraints on how the mission is achieved.

Engineering is a way of organizing that design process to achieve the most cost-effective results. Social systems may not have been "designed" in the sense of a purposeful design process but may have evolved over time. Any effort to change such systems in order to improve them, however, can be thought of as a redesign or reengineering process and can again benefit from a system engineering approach. When using STAMP as the underlying causality model, engineering or reengineering safer systems means designing (or redesigning) the safety-control structure and the controls designed into it to ensure the system operates safely, that is, without unacceptable losses. What is being controlled—chemical manufacturing processes, spacecraft or aircraft, public health, safety of the food supply, corporate fraud, risks in the financial system—is irrelevant in terms of the general process, although significant differences will exist in the types of controls applicable and the design of those controls. The process, however, is very similar to a regular system engineering process.

The problem is that most engineering and even many system engineering techniques were developed under conditions and assumptions that do not hold for complex social systems, as discussed in part I. But STAMP and new system-theoretic approaches to safety can point the way forward for both complex technical *and* social processes. The general engineering and reengineering process described in part III applies to all systems.

6.3   A System Safety Engineering Process

In STAMP, accidents and losses result from not enforcing safety constraints on behavior. Not only must the original system design incorporate appropriate constraints to ensure safe operations, but the safety constraints must continue to be enforced as changes and adaptations to the system design occur over time. This goal forms the basis for safe management, development, and operations.

There is no agreed upon best system engineering process and probably cannot be one—the process needs to match the specific problem and environment in which it is being used. What is described in part III of this book is how to integrate system safety into any reasonable system engineering process. Figure 6.1 shows the three major components of a cost-effective system safety process: management, development, and operations.

6.3.1   Management
Safety starts with management leadership and commitment. Without these, the efforts of others in the organization are almost doomed to failure. Leadership creates culture, which drives behavior.

Besides setting the culture through their own behavior, managers need to establish the organizational safety policy and create a safety control structure with appropriate responsibilities, accountability and authority, safety controls, and feedback channels. Management must also establish a safety management plan and ensure that a safety information system and continual learning and improvement processes are in place and effective.

Chapter 13 discusses management's role and responsibilities in safety.

6.3.2   Engineering Development
The key to having a cost-effective safety effort is to embed it into a system engineering process from the very beginning and to design safety into the system as the design decisions are made. All viewpoints and system components must be included

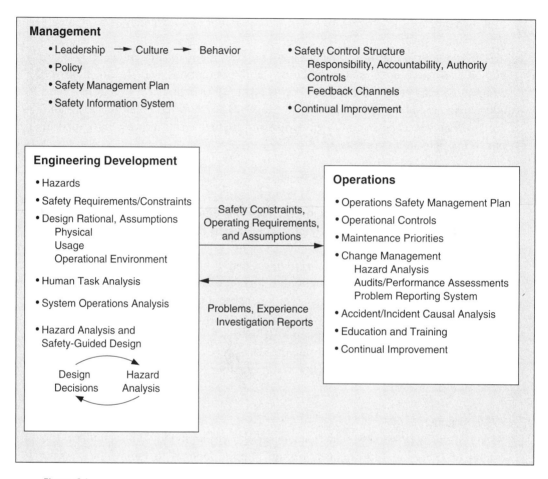

Figure 6.1
The components of a system safety engineering process based on STAMP.

in the process and information used and documented in a way that is accessible, understandable, and helpful.

System engineering starts with first determining the goals of the system. Potential hazards to be avoided are then identified. From the goals and system hazards, a set of system functional and safety requirements and constraints are identified that set the foundation for design, operations, and management. Chapter 7 describes how to establish these fundamentals.

To start safety engineering early enough to be cost-effective, safety must be considered from the early concept formation stages of development and continue throughout the life cycle of the system. Design decisions should be guided by safety considerations while at the same time taking other system requirements and constraints into account and resolving conflicts. The hazard analysis techniques used must not require a completed design and must include all the factors involved in accidents. Chapter 8 describes a new hazard analysis technique, based on the STAMP model of causation, that provides the information necessary to design safety into the system, and chapter 9 shows how to use it in a safety-guided design process. Chapter 9 also presents general principles for safe design including how to design systems and system components used by humans that do not contribute to human error.

Documentation is critical not only for communication in the design and development process but also because of inevitable changes over time. That documentation must include the rationale for the design decisions and traceability from high-level requirements and constraints down to detailed design features. After the original system development is finished, the information necessary to operate and maintain it safely must be passed in a usable form to operators and maintainers. Chapter 10 describes how to integrate safety considerations into specifications and the general system engineering process.

Engineers have often concentrated more on the technological aspects of system development while assuming that humans in the system will either adapt to whatever is given to them or will be trained to do the "right thing." When an accident occurs, it is blamed on the operator. This approach to safety, as argued above, is one of the reasons safety engineering is not as effective as it could be. The system design process needs to start by considering the human controller and continuing that perspective throughout development. The best way to reach that goal is to involve operators in the design decisions and safety analyses. Operators are sometimes left out of the conceptual design stages and only brought in later in development. To design safer systems, operators and maintainers must be included in the design process starting from the conceptual development stage and considerations of human error and preventing it should be at the forefront of the design effort.

Many companies, particularly in aerospace, use integrated product teams that include, among others, design engineers, safety engineers, human factors experts, potential users of the system (operators), and maintainers. But the development process used may not necessarily take maximum advantage of this potential for collaboration. The process outlined in part III tries to do that.

6.3.3   Operations

Once the system is built, it must be operated safely. System engineering creates the basic information needed to do this in the form of the safety constraints and operating assumptions upon which the safety of the design was based. These constraints and assumptions must be passed to operations in a form that they can understand and use.

Because changes in the physical components, human behavior, and the organizational safety control structure are almost guaranteed to occur over the life of the system, operations must manage change in order to ensure that the safety constraints are not violated. The requirements for safe operations are discussed in chapter 12.

It's now time to look at the changes in system engineering, operations, and management, based on STAMP, that can assist in engineering a safer world.

7 Fundamentals

All the parts of the process described in the following chapters start from the same fundamental system engineering activities. These include defining, for the system involved, accidents or losses, hazards, safety requirements and constraints, and the safety control structure.

7.1 Defining Accidents and Unacceptable Losses

The first step in any safety effort involves agreeing on the types of accidents or losses to be considered.

In general, the definition of an accident comes from the customer and occasionally from the government for systems that are regulated by government agencies. Other sources might be user groups, insurance companies, professional societies, industry standards, and other stakeholders. If the company or group developing the system is free to build whatever they want, then considerations of liability and the cost of accidents will come into play.

Definitions of basic terms differ greatly among industries and engineering disciplines. A set of basic definitions is used in this book (see appendix A) that reflect common usage in System Safety. An *accident* is defined as:

> **Accident:** An undesired or unplanned event that results in a loss, including loss of human life or human injury, property damage, environmental pollution, mission loss, etc.

An accident need not involve loss of life, but it does result in some loss that is unacceptable to the stakeholders. System Safety has always considered non-human losses, but for some reason, many other approaches to safety engineering have limited the definition of a loss to human death or injury. As an example of an inclusive definition, a spacecraft accident might include loss of the astronauts (if the spacecraft is manned), death or injury to support personnel or the public, non-accomplishment of the mission, major equipment damage (such as damage to launch facilities), environmental pollution of planets, and so on. An accident definition used in the design of an explorer spacecraft to characterize the icy moon of a planet in the Earth's solar system, for example, was [151]:

A1. Humans or human assets on earth are killed or damaged.

A2. Humans or human assets off of the earth are killed or damaged.

A3. Organisms on any of the moons of the outer planet (if they exist) are killed or mutated by biological agents of Earth origin.

    Rationale: Contamination of an icy outer planet moon with biological agents of Earth origin could have catastrophically adverse effects on any biological agents indigenous to the icy outer planet moon.

A4. The scientific data corresponding to the mission goals is not collected.

A5. The scientific data corresponding to the mission goals is rendered unusable (i.e., deleted or corrupted) before it can be fully investigated.

A6. Organisms of Earth origin are mistaken for organisms indigenous to any of the moons of the outer planet in future missions to study the outer planet's moon.

    Rationale: Contamination of a moon of an outer planet with biological agents of Earth origin could lead to a situation in which a future mission discovers the biological agents and falsely concludes that they are indigenous to the moon of the outer planet.

A7. An incident during this mission directly causes another mission to fail to collect, return, or use the scientific data corresponding to its mission goals.

    Rationale: It is possible for this mission to interfere with the completion of other missions through denying the other missions access to the space exploration infrastructure (for example, overuse of limited Deep Space Network[1] (DSN) resources, causing another mission to miss its launch window because of damage to the launch pad during this mission, etc.)

Prioritizing or assigning a level of severity to the identified losses may be useful when tradeoffs among goals are required in the design process. As an example, consider an industrial robot to service the thermal tiles on the Space Shuttle, which

---

1. The Deep Space Network is an international network of large antennas and communication facilities that supports interplanetary spacecraft missions and radio and radar astronomy observations for the exploration of the solar system and the universe. The network also supports some Earth-orbiting missions.

is used as an example in chapter 9. The goals for the robot are (1) to inspect the thermal tiles for damage caused during launch, reentry, and transport of a Space Shuttle and (2) to apply waterproofing chemicals to the thermal tiles.

Level 1:

Al-1: Loss of the orbiter and crew (e.g., inadequate thermal protection)

Al-2: Loss of life or serious injury in the processing facility

Level 2:

A2–1: Damage to the orbiter or to objects in the processing facility that results in the delay of a launch or in a loss of greater than x dollars

A2–2: Injury to humans requiring hospitalization or medical attention and leading to long-term or permanent physical effects

Level 3:

A3–1: Minor human injury (does not require medical attention or requires only minimal intervention and does not lead to long-term or permanent physical effects)

A3–2: Damage to orbiter that does not delay launch and results in a loss of less than x dollars

A3–3: Damage to objects in the processing facility (both on the floor or suspended) that does not result in delay of a launch or a loss of greater than x dollars

A3–4: Damage to the mobile robot

Assumption: It is assumed that there is a backup plan in place for servicing the orbiter thermal tiles in case the tile processing robot has a mechanical failure and that the same backup measures can be used in the event the robot is out of commission due to other reasons.

The customer may also have a safety policy that must be followed by the contractor or those designing the thermal tile servicing robot. As an example, the following is similar to a typical NASA safety policy:

General Safety Policy: All hazards related to human injury or damage to the orbiter must be eliminated or mitigated by the system design. A reasonable effort must be made to eliminate or mitigate hazards resulting at most in damage to the robot or objects in the work area. For any hazards that cannot be eliminated, the hazard analysis as well as the design features and development procedures, including any tradeoff studies, must be documented and presented to the customer for acceptance.

7.2 System Hazards

The term *hazard* has been used in different ways. For example, in aviation, a hazard is often used to denote something in the environment of the system, for example a mountain, that is in the path of the aircraft. In contrast, in System Safety, a hazard is defined as within the system being designed (or its relationship to an environmental object) and not just in its environment. For example, an aircraft flying too close to a mountain would be a hazard.

> **Hazard:** A system state or set of conditions that, together with a particular set of worst-case environmental conditions, will lead to an accident (loss).

This definition requires some explanation. First, hazards may be defined in terms of conditions, as here, or in terms of events as long as one of these choices is used consistently. While there have been arguments about whether hazards are events or conditions, the distinction is irrelevant and either can be used. Figure 2.6 depicts the relationship between events and conditions: conditions lead to events which lead to conditions which lead to events. . . . The hazard for a chemical plant could be stated as the release of chemicals (an event) or chemicals in the atmosphere (a condition). The only difference is that events are limited in time while the conditions caused by the event persist over time until another event occurs that changes the prevailing conditions. For different purposes, one choice might be advantageous over the other.

Second, note that the word *failure* does not appear anywhere. Hazards are not identical to failures—failures can occur without resulting in a hazard and a hazard may occur without any precipitating failures. C. O. Miller, one of the founders of System Safety, cautioned that "distinguishing hazards from failures is implicit in understanding the difference between safety and reliability" [138].

Sometimes, hazards are defined as something that "has the potential to do harm" or that "can lead to an accident." The problem with this definition is that most every system state has the potential to do harm or can lead to an accident. An airplane that is in the air is in a hazardous state according to this definition, but there is little that the designer of an air traffic control system or an air transportation system, for example, can do about designing a system where the planes never leave the ground. For practical reasons, the definition should preclude states that the system must normally be in to accomplish the mission. By limiting the definition of hazard to states that the system should never be in (that is, closer to the accident or loss event), the designer has greater freedom and ability to design hazards out of the system. For air traffic control, the hazard would not be two planes in the air but two planes that violate minimum separation standards.

An accident is defined with respect to the environment of the system or component:

Hazard + Environmental Conditions ⇒ Accident (Loss)

As an example, a release of toxic chemicals or explosive energy will cause a loss *{environ-* only if there are people or structures in the vicinity. Weather conditions may affect *ment?!* whether a loss occurs in the case of a toxic release. If the appropriate environmental conditions do not exist, then there is no loss and, by definition, no accident. This type of non-loss event is commonly called an *incident*. When a hazard is defined as an event, then hazards and incidents are identical.

7.2.1 Drawing the System Boundaries

What constitutes a hazard, using the preceding definition, depends on where the boundaries of the system are drawn. A system is an abstraction, and the boundaries of the system can be drawn anywhere the person defining the system wants. Where the boundaries are drawn will determine which conditions are considered part of the hazard and which are considered part of the environment. Because this choice is arbitrary, the most useful way to define the boundaries, and thus the hazard, is to draw them to include the conditions related to the accident over which the system designer has some control. That is, if we expect designers to create systems that eliminate or control hazards and thus prevent accidents, then those hazards must be in their design space. This control requirement is the reason for distinguishing between hazards and accidents—accidents may involve aspects of the environment over which the system designer or operator has no control.

In addition, because of the recursive nature of the definition of a system— that is, a system at one level may be viewed as a subsystem of a larger system— higher-level systems will have control over the larger hazards. But once boundaries are drawn, system designers can be held responsible only for controlling the accident factors that they have the ability to control, including those that have been passed to them from system designers above them as component safety requirements to ensure the encompassing system hazards are eliminated or controlled.

Consider the chemical plant example. While the hazard could be defined as death or injury of residents around the plant (the loss event), there may be many factors involved in such a loss that are beyond the control of the plant designers and operators. One example is the atmospheric conditions at the time of the release, such as velocity and direction of the wind. Other factors in a potential accident or loss are the location of humans around the plant and community emergency preparedness, both of which may be under the control of the local or state government. The designers of the chemical plant have a responsibility to provide the information necessary for the design and operation of appropriate emergency preparedness equipment and procedures, but their primary design responsibility is the part of a potential accident that is under their design control, namely the design of the plant to prevent release of toxic chemicals.

In fact, the environmental conditions contributing to a loss event may change over time: potentially dangerous plants may be originally located far from population centers, for example, but over time human populations tend to encroach on such plants in order to live close to their jobs or because land may be cheaper in remote areas or near smelly plants. The chemical plant designer usually has no design control over these conditions so it is most convenient to draw the system boundaries around the plant and define the hazard as uncontrolled release of chemicals from the plant. If the larger sociotechnical system is being designed or analyzed for safety, which it should be, the number of potential hazards and actions to prevent them increases. Examples include controlling the location of plants or land use near them through local zoning laws, and providing for emergency evacuation and medical treatment.

Each component of the sociotechnical system may have different aspects of an accident under its control and is responsible for different parts of the accident process, that is, different hazards and safety constraints. In addition, several components may have responsibilities related to the same hazards. The designers of the chemical plant and relevant government regulatory agencies, for example, may both be concerned with plant design features potentially leading to inadvertent toxic chemical release. The government role, however, may be restricted to design and construction approvals and inspection processes, while the plant designers have basic design creation responsibilities.

As another example of the relationship between hazards and system boundaries, consider the air traffic control system. If an accident is defined as a collision between aircraft, then the appropriate hazard is the violation of minimum separation between aircraft. The designer of an airborne collision avoidance system or a more general air traffic control system theoretically has control over the separation between aircraft, but may not have control over other factors that determine whether two aircraft that get close together actually collide, such as visibility and weather conditions or the state of mind or attentiveness of the pilots. These are under the control of other system components such as air traffic control in directing aircraft away from poor weather conditions or the control of other air transportation system components in the selection and training of pilots, design of aircraft, and so on.

Although individual designers and system components are responsible for controlling only the hazards in their design space, a larger system safety engineering effort preceding component design will increase overall system safety while decreasing the effort, cost, and tradeoffs involved in component safety engineering. By considering the larger sociotechnical system and not just the individual technical components, the most cost-effective way to eliminate or control hazards can be identified. If only part of the larger system is considered, the compromises required to eliminate or control the system hazard in one piece of the overall system design may be much greater than would be necessary if other parts of the overall system were considered. For example, a particular hazard associated with launching a spacecraft might be controllable by the spacecraft design, by the physical launch infrastructure, by launch procedures, by the launch control system, or by a combination of these. If only the spacecraft design is considered in the drawing of system boundaries and the hazard identification process, hazard control may require more tradeoffs than if the hazard is partially or completely eliminated or controlled by design features in other parts of the system.

All that is being suggested here is that top-down system engineering is critical for engineering safety into complex systems. In addition, when a new component is introduced into an existing system, such as the introduction of a collision avoidance system in the aircraft, the impact of the addition on the safety of the aircraft itself as well as the safety of air traffic control and the larger air transportation system safety needs to be considered.

Another case is when a set of systems that already exist are combined to create a new system.[2] While the individual systems may have been designed to be safe within the system for which they were originally created, the safety constraints enforced in the components may not adequately control hazards in the combined system or may not control hazards that involve interactions among new and old system components.

The reason for this discussion is to explain why the definition of the hazards associated with a system is an arbitrary but important step in assuring system safety and why a system engineering effort that considers the larger sociotechnical system is necessary. One of the first steps in designing a system, after the definition of an accident or loss and the drawing of boundaries around the subsystems, is to identify the hazards that need to be eliminated or controlled by the designers of that system or subsystem.

7.2.2   Identifying the High-Level System Hazards

For practical reasons, a small set of high-level system hazards should be identified first. Starting with too large a list at the beginning, usually caused by including refinements and causes of the high-level hazards in the list, often leads to a disorganized and incomplete hazard identification and analysis process. Even the most complex system seldom has more than a dozen high-level hazards, and usually less than this.

---

2. Sometimes called a *system* of *systems*, although all systems are subsystems of larger systems.

Hazards are identified using the definition of an accident or loss along with additional safety criteria that may be imposed by regulatory or industry associations and practices. For example, the hazards associated with the outer planets explorer accident definition in section 7.1 might be defined as [151]:

H1. Inability of the mission to collect data [A4]

H2. Inability of the mission to return collected data [A5]

H3. Inability of the mission scientific investigators to use the returned data [A5]

H4. Contamination of the outer planet moon with biological agents of Earth origin on mission hardware [A6]

H5. Exposure of Earth life or human assets on Earth to toxic, radioactive, or energetic elements of the mission hardware [A1]

H6. Exposure of Earth life or human assets off Earth to toxic, radioactive, or energetic elements of the mission hardware [A2]

H7. Inability of other space exploration missions to use the shared space exploration infrastructure to collect, return, or use data [A7]

The numbers in the square brackets identify the accidents related to each of these hazards.

The high-level system hazards that might be derived from the accidents defined for the NASA thermal tile processing robot in section 7.1 might be:

H1. Violation of minimum separation between mobile base and objects (including orbiter and humans)

H2. Unstable robot base

H3. Movement of the robot base or manipulator arm causing injury to humans or damage to the orbiter

H4. Damage to the robot

H5. Fire or explosion

H6. Contact of human with DMES waterproofing chemical

H7. Inadequate thermal protection

During the design process, these high-level hazards will be refined as the design alternatives are considered. Chapter 9 provides more information about the refinement process and an example.

Aircraft collision control provides a more complex example. As noted earlier, the relevant accident is a collision between two airborne aircraft and the overall system hazard to be avoided is violation of minimum physical separation (distance) between aircraft.

One (but only one) of the controls used to avoid this type of accident is an airborne collision avoidance system like TCAS (Traffic alert and Collision Avoidance System), which is now required on most commercial aircraft. While the goal of TCAS is increased safety, TCAS itself introduces new hazards associated with its use. Some hazards that were considered during the design of TCAS are:

H1. TCAS causes or contributes to a near midair collision (NMAC), defined as a pair of controlled aircraft violating minimum separation standards.

H2. TCAS causes or contributes to a controlled maneuver into the ground.

H3. TCAS causes or contributes to the pilot losing control over the aircraft.

H4. TCAS interferes with other safety-related aircraft systems.

H5. TCAS interferes with the ground-based Air Traffic Control system (e.g., transponder transmissions to the ground or radar or radio services).

H6. TCAS interferes with an ATC advisory that is safety-related (e.g., avoiding a restricted area or adverse weather conditions).

Ground-based air traffic control also plays an important role in collision avoidance, although it has responsibility for a larger and different set of hazards:

H1. Controlled aircraft violate minimum separation standards (NMAC).

H2. An airborne controlled aircraft enters an unsafe atmospheric region.

H3. A controlled airborne aircraft enters restricted airspace without authorization.

H4. A controlled airborne aircraft gets too close to a fixed obstacle other than a safe point of touchdown on assigned runway (known as controlled flight into terrain or CFIT).

H5. A controlled airborne aircraft and an intruder in controlled airspace violate minimum separation.

H6. Loss of controlled flight or loss of airframe integrity.

H7. An aircraft on the ground comes too close to moving objects or collides with stationary objects or leaves the paved area.

H8. An aircraft enters a runway for which it does not have a clearance (called runway incursion).

Unsafe behavior (hazards) at the system level can be mapped into hazardous behaviors at the component or subsystem level. Note, however, that the reverse (bottom-up) process is not possible, that is, it is not possible to identify the system-level hazards by looking only at individual component behavior. Safety is a system property, not a component property. Consider an automated door system. One reasonable hazard when considering the door alone is the door closing on someone. The associated safety constraint is that the door must not close on anyone in the doorway. This hazard is relevant if the door system is used in any environment. If the door is in a building, another important hazard is not being able to get out of a dangerous environment, for example, if the building is on fire. Therefore, a reasonable design constraint would be that the door opens whenever a door open request is received. But if the door is used on a moving train, an additional hazard must be considered, namely, the door opening while the train is moving and between stations. In a moving train, different safety design constraints would apply compared to an automated door system in a building. Hazard identification is a top-down process that must consider the encompassing system and its hazards and potential accidents.

Let's assume that the automated door system is part of a train control system. The system-level train hazards related to train doors include a person being hit by closing doors, someone falling from a moving train or from a stationary train that is not properly aligned with a station platform, and passengers and staff being unable to escape from a dangerous environment in the train compartment. Tracing these system hazards into the related hazardous behavior of the automated door component of the train results in the following hazards:

1. Door is open when the train starts.
2. Door opens while train is in motion.
3. Door opens while not properly aligned with station platform.
4. Door closes while someone is in the doorway.
5. Door that closes on an obstruction does not reopen or reopened door does not reclose.
6. Doors cannot be opened for emergency evacuation between stations.

The designers of the train door controller would design to control these hazards. Note that constraints 3 and 6 are conflicting, and the designers will have to reconcile such conflicts. In general, attempts should first be made to eliminate hazards at the system level. If they cannot be eliminated or adequately controlled at the system level, then they must be refined into hazards to be handled by the system components.

Unfortunately, no tools exist for identifying hazards. It takes domain expertise and depends on subjective evaluation by those constructing the system. Chapter 13 in *Safeware* provides some common heuristics that may be helpful in the process. The good news is that identifying hazards is usually not a difficult process. The later steps in the hazard analysis process are where most of the mistakes and effort occurs.

There is also no right or wrong set of hazards, only a set that the system stakeholders agree is important to avoid. Some government agencies have mandated the hazards they want considered for the systems they regulate or certify. For example, the U.S. Department of Defense requires that producers of nuclear weapons consider four hazards:

1. Weapons involved in accident or incidents, or jettisoned weapons, produce a nuclear yield.

2. Nuclear weapons are deliberately prearmed, armed, launched, fired, or released without execution of emergency war orders or without being directed to do so by a competent authority.

3. Nuclear weapons are inadvertently prearmed, armed, launched, fired, or released.

4. Inadequate security is applied to nuclear weapons.

Sometimes user or professional associations define the hazards for the systems they use and that they want developers to eliminate or control. In most systems, however, the hazards to be considered are up to the developer and their customer(s).

7.3   System Safety Requirements and Constraints

After the system and component hazards have been identified, the next major goal is to specify the system-level safety requirements and design constraints necessary to prevent the hazards from occurring. These constraints will be used to guide the system design and tradeoff analyses.

The system-level constraints are refined and allocated to each component during the system engineering decomposition process. The process then iterates over the individual components as they are refined (and perhaps further decomposed) and as design decisions are made.

Figure 7.1 shows an example of the design constraints that might be generated from the automated train door hazards. Again, note that the third constraint potentially conflicts with the last one and the resolution of this conflict will be an important part of the system design process. Identifying these types of conflicts early in the design process will lead to better solutions. Choices may be more limited later on when it may not be possible or practical to change the early decisions.

As the design process progresses and design decisions are made, the safety requirements and constraints are further refined and expanded. For example, a safety constraint on TCAS is that it must not interfere with the ground-based air traffic control system. Later in the process, this constraint will be refined into more detailed constraints on the ways this interference might occur. Examples include

	HAZARD	SAFETY DESIGN CONSTRAINT
1	Train starts with door open	Train must not be capable of moving with any door open
2	Door opens while train is in motion	Doors must remain closed while train is in motion
3	Door opens while improperly aligned with station platform	Door must be capable of opening only after train is stopped and properly aligned with platform unless emergency exists (see hazard 6 below)
4	Door closes while someone is in the doorway	Door areas must be clear before door closing begins
5	Door that closes on an obstruction does not reopen or reopened door does not reclose	An obstructed door must reopen to permit removal of obstruction and then automatically reclose
6	Doors cannot be opened for emergency evacuation	Means must be provided to open doors anywhere when the train is stopped for emergency evacuation

Figure 7.1
Design constraints for train door hazards.

constraints on TCAS design to limit interference with ground-based surveillance radar, with distance-measuring equipment channels, and with radio services. Additional constraints include how TCAS can process and transmit information (see chapter 10).

Figure 7.2 shows the high-level requirements and constraints for some of the air traffic control hazards identified above. Comparing the ATC high-level constraints with the TCAS high-level constraints (figure 7.3) is instructive. Ground-based air traffic control has additional requirements and constraints related to aspects of the collision problem that TCAS cannot handle alone, as well as other hazards and potential aircraft accidents that it must control.

Some constraints on the two system components (ATC and TCAS) are closely related, such as the requirement to provide advisories that maintain safe separation between aircraft. This example of overlapping control raises important concerns about potential conflicts and coordination problems that need to be resolved. As noted in section 4.5, accidents often occur in the boundary areas between controllers and when multiple controllers control the same process. The inadequate resolution of the conflict between multiple controller responsibilities for aircraft separation contributed to the collision of two aircraft over the town of Überlingen (Germany)

HAZARD	SAFETY DESIGN CONSTRAINTS
1 A pair of controlled aircraft violate minimum separation standards	a. ATC must provide advisories that maintain safe separation between aircraft b. ATC must provide conflict alerts
2 A controlled aircraft enters an unsafe atmospheric region (icing conditions, windsher areas, thunderstorm cells)	a. ATC must not issue advisories that direct aircraft into areas with unsafe atmospheric conditions b. ATC must provide weather advisories and alerts to flight crews c. ATC must warn aircraft that enter an unsafe atmospheric region
3 A controlled aircraft enters restricted airspace without authorization	a. ATC must not issue advisories that direct an aircraft into restricted airspace unless avoiding a greater hazard b. ATC must provide timely warnings to aircraft to prevent their incursion into restricted airspace
4 A controlled aircraft gets too close to a fixed obstacle or terrain other than a safe point of touchdown on its assigned runway	ATC must provide advisories that maintain safe separation between aircraft and terrain or physical obstacles
5 A controlled aircraft and an intruder in controlled airspace violate minimum separation standards	ATC must provide alerts and advisories to avoid intruders if at all possible
6 Loss of controlled flight or loss of airframe integrity	a. ATC must not issue advisories outside the the safe performance envelope of the aircraft b. ATC advisories must not distract or disrupt the crew from maintaining safety of flight c. ATC must not issue advisories that the pilot or aircraft cannot fly or that degrade the continued safe flight of the aircraft d. ATC must not provide advisories that cause an aircraft to fall below the standard glidepath or intersect it at the wrong place

Figure 7.2
High-level requirements and design constraints for air traffic control.

	HAZARD	SAFETY DESIGN CONSTRAINT
1	TCAS causes or contributes to an NMAC (near midair collision)	a. TCAS must provide effective warnings and appropriate collision avoidace guidance on potentially dangerous threats and must provide them within an appropriate time limit b. TCAS must not cause or contribute to an NMAC that would not have occurred had the aircraft not carried TCAS)
2	TCAS causes or contributes to a controlled maneuver into the ground	TCAS must not cause or contribute to controlled flight into terrain
3	TCAS causes or contributes to a pilot losing control over the aircraft	a. TCAS must not disrupt the pilot and ATC operations during critical phases of flight nor disrupt aircraft operation b. TCAS must operate with an acceptably low level of unwanted or nuisance alarms. The unwanted alarm rate must be sufficiently low to pose no safety of flight hazard nor adversely affect the workload in the cockpit c. TCAS must not issue advisories outside the safe performance envelope of the aircraft and degrade the continued safe flight of the aircraft (e.g., reduce stall margins or result in stall warnings)
4	TCAS interferes with other safety-related aircraft systems	TCAS must not interfere with other safety-related aircraft systems or contribute to non-separation-related hazards
5	TCAS interferes with the ground ATC system (e.g., transponder, radar. or radar transmissions)	TCAS must not interfere with the ground ATC system or other aircraft transmissions to the ground ATC system
6	TCAS interferes with a safety-related ATC advisory (e.g., avoiding a restricted area or adverse weather conditions)	TCAS must generate advisories that require as little deviation as possible from ATC clearances

Figure 7.3
High-level design constraints for TCAS.

in July 2002 when TCAS and the ground air traffic controller provided conflicting advisories to the pilots. Potentially conflicting responsibilities must be carefully handled in system design and operations and identifying such conflicts are part of the new hazard analysis technique described in chapter 8.

Hazards related to the interaction among components, for example the interaction between attempts by air traffic control and by TCAS to prevent collisions, need to be handled in the safety control structure design, perhaps by mandating how the pilot is to select between conflicting advisories. There may be considerations in handling these hazards in the subsystem design that will impact the behavior of multiple subsystems and therefore must be resolved at a higher level and passed to them as constraints on their behavior.

7.4   The Safety Control Structure

The safety requirements and constraints on the physical system design shown in section 7.3 act as input to the standard system engineering process and must be incorporated into the physical system design and safety control structure. An example of how they are used is provided in chapter 10.

Additional system safety requirements and constraints, including those on operations and maintenance or upgrades will be used in the design of the safety control structure at the organizational and social system levels above the physical system. There is no one correct safety control structure: what is practical and effective will depend greatly on cultural and other factors. Some general principles that apply to all safety control structures are described in chapter 13. These principles need to be combined with specific system safety requirements and constraints for the particular system involved to design the control structure.

The process for engineering social systems is very similar to the regular system engineering process and starts, like any system engineering project, with identifying system requirements and constraints. The responsibility for implementing each requirement needs to be assigned to the components of the control structure, along with requisite authority and accountability, as in any management system; controls must be designed to ensure that the responsibilities can be carried out; and feedback loops created to assist the controller in maintaining accurate process models.

7.4.1   The Safety Control Structure for a Technical System

An example from the world of space exploration is used in this section, but many of the same requirements and constraints could easily be adapted for other types of technical system development and operations.

The requirements in this example were generated to perform a programmatic risk assessment of a new NASA management structure called Independent

Technical Authority (ITA) recommended in the report of the Columbia Accident Investigation Board. The risk analysis itself is described in the chapter on the new hazard analysis technique called STPA (chapter 8). But the first step in the safety or risk analysis is the same as for technical systems: to identify the system hazards to be avoided, to generate a set of requirements for the new management structure, and to design the control structure.

The new safety control structure for the NASA manned space program was introduced to improve the flawed engineering and management decision making leading to the Columbia loss. The hazard to be eliminated or mitigated was:

System Hazard: Poor engineering and management decision making leading to a loss.

Four high-level system safety requirements and constraints for preventing the hazard were identified and then refined into more specific requirements and constraints.

1. Safety considerations must be first and foremost in technical decision making.

   a. State-of-the art safety standards and requirements for NASA missions must be established, implemented, enforced, and maintained that protect the astronauts, the workforce, and the public.

   b. Safety-related technical decision making must be independent from programmatic considerations, including cost and schedule.

   c. Safety-related decision making must be based on correct, complete, and up-to-date information.

   d. Overall (final) decision making must include transparent and explicit consideration of both safety and programmatic concerns.

   e. The Agency must provide for effective assessment and improvement in safety-related decision making.

2. Safety-related technical decision making must be done by eminently qualified experts, with broad participation of the full workforce.

   a. Technical decision making must be credible (executed using credible personnel, technical requirements, and decision-making tools) .

   b. Technical decision making must be clear and unambiguous with respect to authority, responsibility, and accountability.

   c. All safety-related technical decisions, before being implemented by the Program, must have the approval of the technical decision maker assigned responsibility for that class of decisions.

    d. Mechanisms and processes must be created that allow and encourage all employees and contractors to contribute to safety-related decision making.

3. Safety analyses must be available and used starting in the early acquisition, requirements development, and design processes and continuing through the system life cycle.

    a. High-quality system hazard analyses must be created.

    b. Personnel must have the capability to produce high-quality safety analyses.

    c. Engineers and managers must be trained to use the results of hazard analyses in their decision making.

    d. Adequate resources must be applied to the hazard analysis process.

    e. Hazard analysis results must be communicated in a timely manner to those who need them. A communication structure must be established that includes contractors and allows communication downward, upward, and sideways (e.g., among those building subsystems).

    f. Hazard analyses must be elaborated (refined and extended) and updated as the design evolves and test experience is acquired.

    g. During operations, hazard logs must be maintained and used as experience is acquired. All in-flight anomalies must be evaluated for their potential to contribute to hazards.

4. The Agency must provide avenues for the full expression of technical conscience (for safety-related technical concerns) and provide a process for full and adequate resolution of technical conflicts as well as conflicts between programmatic and technical concerns.

    a. Communication channels, resolution processes, adjudication procedures must be created to handle expressions of technical conscience.

    b. Appeals channels must be established to surface complaints and concerns about aspects of the safety-related decision making and technical conscience structures that are not functioning appropriately.

Where do these requirements and constraints come from? Many of them are based on fundamental safety-related development, operations and management principles identified in various chapters of this book, particularly chapters 12 and 13. Others are based on experience, such as the causal factors identified in the Columbia and Challenger accident reports or other critiques of the NASA safety culture and of NASA safety management. The requirements listed obviously reflect the advanced technology and engineering domain of NASA and the space program that was the focus of the ITA program along with some of the unique aspects of the NASA culture. Other industries will have their own requirements. An example for the pharmaceutical industry is shown in the next section of this chapter.

There is unlikely to be a universal set of requirements that holds for every safety control structure beyond a small set of requirements too general to be very useful in a risk analysis. Each organization needs to determine what its particular safety goals are and the system requirements and constraints that are likely to ensure that it reaches them.

Clearly buy-in and approval of the safety goals and requirements by the stakeholders, such as management and the broader workforce as well as anyone overseeing the group being analyzed, such as a regulatory agency, is important when designing and analyzing a safety control structure.

Independent Technical Authority is a safety control structure used in the nuclear Navy SUBSAFE program described in chapter 14. In this structure, safety-related decision making is taken out of the hands of the program manager and assigned to a Technical Authority. In the original NASA implementation, the technical authority rested in the NASA Chief Engineer, but changes have since been made. The overall safety control structure for the original NASA ITA is shown in figure 7.4.[3]

For each component of the structure, information must be determined about its overall role, responsibilities, controls, process model requirements, coordination and communication requirements, contextual (environmental and behavior-shaping) factors that might bear on the component's ability to fulfill its responsibilities, and inputs and outputs to other components in the control structure. The responsibilities are shown in figure 7.5. A risk analysis on ITA and the safety control structure is described in chapter 8.

7.4.2   Safety Control Structures in Social Systems

Social system safety control structures often are not designed but evolve over time. They can, however, be analyzed for inherent risk and redesigned or "reengineered" to prevent accidents or to eliminate or control past causes of losses as determined in an accident analysis.

The reengineering process starts with the definition of the hazards to be eliminated or mitigated, system requirements and constraints necessary to increase safety, and the design of the current safety-control structure. Analysis can then be used to drive the redesign of the safety controls. But once again, just like every system that has been described so far in this chapter, the process starts by identifying the hazards

---

3. The control structure was later changed to have ITA under the control of the NASA center directors rather than the NASA chief engineer; therefore, this control structure does not reflect the actual implementation of ITA at NASA, but it was the design at the time of the hazard analysis described in chapter 8.

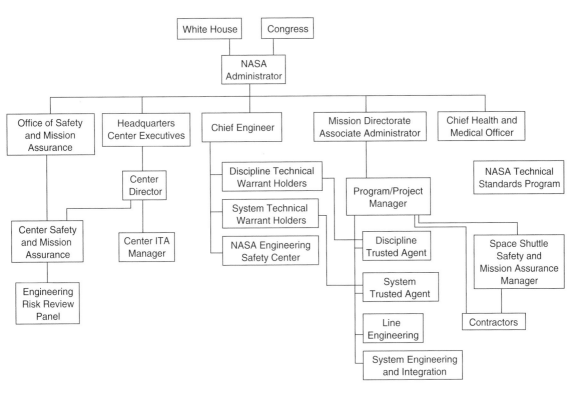

Figure 7.4
The NASA safety control structure under the original ITA design.

and safety requirements and constraints derived from them. The process is illustrated using drug safety.

Dozens of books have been written about the problems in the pharmaceutical industry. Everyone appears to have good intentions and are simply striving to optimize their performance within the existing incentive structure. The result is that the system has evolved to the point where each group's individual best interests do not necessarily add up to or are not aligned with the best interests of society as a whole. A safety control structure exists, but does not necessarily provide adequate satisfaction of the system-level goals, as opposed to the individual component goals.

This problem can be viewed as a classic system engineering problem: optimizing each component does not necessarily add up to a system optimum. Consider the air transportation system, as noted earlier. When each aircraft tries to optimize its path from its departure point to its destination, the overall system throughput may not be optimized when they all arrive in a popular hub at the same time. One goal of the air traffic control system is to control individual aircraft movement in order to

Executive Branch
   Appointment of NASA Administrator
   Setting of high-level goals and vision for NASA
   Creation of a draft budget appropriation for NASA

Congress
   Approval of NASA Administrator appointment
   NASA budget allocation
   Legislation affecting NASA operations

NASA Administrator
   Appointment of Chief Engineer (ITA) and head of Office of Safety and Mission Assurance
   Providing funding and authority to Chief Engineer to execute the Independent Technical Authority
   Demonstraion of commitment to safety over programmatic concerns through concrete actions
   Providing the directives and procedural requirements that define the ITA program
   Adjudication of differences between the Mission Directorate Associate Administrators and the
      Chief Engineer (ITA)

Chief Engineer
   Implementing ITA
   Effectiveness of the ITA program
   Communication channels with and among Warrant Holders
   Communication of decisions and lessons learned
   Establishment, monitoring, and approval of technical requirements, products, and policy and all
      changes, variances and waivers to the requirements
   Safety, risk, and trend analysis
   Independent assessment of flight (launch) readiness
   Conflict Resolution
   Developing a Technical Conscience throughout the engineering community

System Technical Warrant Holder
   Establishment and maintenance of technical policy, technical standards, requirements, and processes
      for a particular system or systems
   Technical product compliance with requirements, specifications, standards
   Primary interace between system and ITA (Chief Engineer)
   Assist Discipline Technical Warrant Holder in access to data, rationale, and other experts
   Production, quality, and use of FMEA/SIL, trending analysis, hazard and risk analyses
   Timely, day-to-day technical positions on issues pertaining to safe and reliable operations
   Establishing appropriate communication channels and networks
   Succession Planning
   Documentation of all methodologies, actions or closures, and decisions
   Sustaining the Agency knowledge base through communication of decisions and lessons learned
   Assessment of launch readiness from the standpoint of safe and reliable flight and operations
   Budget and resource requirements definition
   Maintaining competence
   Leading the technical conscience for the warranted system(s)

Figure 7.5
The responsibilities of the components in the NASA ITA safety control structure.

Discipline Technical Warrant Holder
  Interface to specialized knowledge within the Agency
  Assistance to System Technical Warrant Holders in carrying out their responsibilities
  Ownership of technical specifications and standards for warranted discipline (including system safety standards)
  Sustaining the Agency knowledge base in the warranted discipline
  Sustaining the general health of the warranted discipline throughout the Agency
  Succession Planning
  Leading the technical conscience for the warranted discipline
  Budget and resource requirements definition

Trusted Agents
  Screening: evaluate all changes and variances and perform all functions requested by Technical Warrant Holders
  Conducting daily business for System Technical Warrant Holder (represent on boards, meetings, committees)
  Providing information to Technical Warrant Holders about specific projects (e.g., safety analyses)

In-Line Engineers
  Provide unbiased technical positions to warrant holders, safety and mission assurance, trusted agents, and
    programs and projects
  Conduct system safety engineering (analyses and incorporation of results into design, development, and operations
  Evaluate contractor-produced analyses and incorporation of results into contractor products
  Act as the technical conscience of the Agency

  Chief Safety and Mission Assurance Officer (OSMA)
    Leadership, policy direction, functional oversight, and coordination of assurance activities across the Agency
    Assurance of safety and reliability on programs and projects
    Incident and accident investigation

  Center Safety and Mission Assurance (S&MA)
    Assure compliance with all requirements, standards, directives, policies, and procedures
    Perform quality (reliability and safety) assessments
    Participate in reviews
    Intervention in any activity to avoid an unnecessary safety risk
    Recommend a Safety, Reliability, and Quality Assurance plan for each project
    Chair Engineering Risk Review Panels at each Space Operations Center

  Lead Engineering Risk Review Panel Manager and Panels
    Conduct formal safety reviews of accepted and controlled hazards
    Oversee and resolve integrated hazards
    Assure compliance with requirements, accuracy of all data and hazard analyses, and proper classification
      of hazards

  Space Shuttle Program Safety and Mission Assurnace Manager
    Assure compliance with requirements in activities of prime contractors and technical support personnel from
      the NASA Centers
    Integrate and provide guidance for safety, reliability, and quality engineering activities performed by
      Space Operations Centers

Figure 7.5
(Continued)

Program/Project Managers
Communication of ITA understanding through program or project team
Prioritization safety over programmatic concerns among those reporting to him or her
Support of Trusted Agents
Provision of complete and timely data to Technical Warrant Holder
Compliance with System Technical Warrant Holders decisions

System Engineering and Integration Office
Integrated hazard analyses and anomaly investigation at system level
Communication of system-level, safety-related requirements and constraints to and from contractors
Update hazard analyses and maintain hazard logs during test and operations

Contractors
Production and use of hazard analyses in their designs
Communication of hazard information to NASA System Engineering and Integration

Center Director
Practice of technical conscience at the Center
Preservation of ITA financial and managerial independence at the Center
Support of ITA activities at the Center
Support of safety activities at the Center

Center ITA Manager
Administrative support for Technical Warrant Holders

NASA Engineering and Safety Center (NESC)
In-depth technical reviews, assessments, and analyses of high-risk projects
Selected mishap investigations
In-depth system engineering analyses

Headquarters Center Executives
Alignment of Center's organization and processes to support and maintain independence of ITA
Monitoring of ITA and expression and resolution of technical conscience at their Center
Oversight of safety and mission assurance at their Center

Mission Directorate Associate Administrators
Leadership and accountability for all engineering and technical work for their mission
Alignment of financial, personnel, and engineering infrastructure with ITA
Resolution of differences between warrant holders and program or project managers

NASA Technical Standards Program
Coordination of standards activities with ITA

Figure 7.5
(Continued)

optimize overall system throughput while trying to allow as much flexibility as possible for the individual aircraft and airlines to achieve their goals. The air traffic control system and the rules of operation of the air transportation system resolve conflicting goals when public safety is at stake. Each airline might want its own aircraft to land as quickly as possible, but the air traffic controllers ensure adequate spacing between aircraft to preserve safety margins. These same principles can be applied to non-engineered systems.

The ultimate goal is to determine how to reengineer or redesign the overall pharmaceutical safety control structure in a way that aligns incentives for the greater good of society. A well-designed system would make it easier for all stakeholders to do the right thing, both scientifically and ethically, while achieving their own goals as much as possible. By providing the decision makers with information about ways to achieve the overall system objectives and the tradeoffs involved, better decision making can result.

While system engineering is applicable to pharmaceutical (and more generally medical) safety and risk management, there are important differences from the classic engineering problem that require changes to the traditional system safety approaches. In most technical systems, managing risk is simpler because not doing something (e.g., not inadvertently launching the missile) is usually safe and the problem revolves around preventing the hazardous event (inadvertent launch): a risk/no risk situation. The traditional engineering approach identifies and evaluates the costs and potential effectiveness of different ways to eliminate or control the hazards involved in the operational system. Tradeoffs require comparing the costs of various solutions, including costs that involve reduction in desirable system functions or system reliability.

The problem in pharmaceutical safety is different: there is risk in prescribing a potentially unsafe drug, but there is also risk in not prescribing the drug (the patient dies from their medical condition): a risk/risk situation. The risks and benefits conflict in ways that greatly increase the complexity of decision making and the information needed to make decisions. New, more powerful system engineering techniques are required to deal with risk/risk decisions.

Once again, the basic goals, hazards, and safety requirements must first be identified [43].

System Goal: *To provide safe and effective pharmaceuticals to enhance the long-term health of the population.*

Important loss events (accidents) we are trying to avoid are:

1. Patients get a drug treatment that negatively impacts their health.
2. Patients do not get the treatment they need.

Three system hazards can be identified that are related to these loss events:

H1: The public is exposed to an unsafe drug.

    1. The drug is released with a label that does not correctly specify the conditions for its safe use.

    2. An approved drug is found to be unsafe and appropriate responses are not taken (warnings, withdrawals from the market, etc.)

    3. Patients are subjected to unacceptable risk during clinical trials.

H2: Drugs are taken unsafely.

    1. The wrong drug is prescribed for the indication.

    2. The pharmacist provides a different medication than was prescribed.

    3. Drugs are taken in an unsafe combination.

    4. Drugs are not taken according to directions (dosage, timing).

H3: Patients do not get an effective treatment they require.

    1. Safe and effective drugs are not developed, are not approved for use, or are withdrawn from the market.

    2. Safe and effective drugs are not affordable by those who need them.

    3. Unnecessary delays are introduced into development and marketing.

    4. Physicians do not prescribe needed drugs or patients have no access to those who could provide the drugs to them.

    5. Patients stop taking a prescribed drug due to perceived ineffectiveness or intolerable side effects.

From these hazards, a set of system requirements can be derived to prevent them:

1. Pharmaceutical products are developed to enhance long-term health.

    a. Continuous appropriate incentives exist to develop and market needed drugs.

    b. The scientific knowledge and technology needed to develop new drugs and optimize their use is available.

2. Drugs on the market are adequately safe and effective.

    a. Drugs are subjected to effective and timely safety testing.

    b. New drugs are approved by the FDA based upon a validated and reproducible decision-making process.

    c. The labels attached to drugs provide correct information about safety and efficacy.

    d. Drugs are manufactured according to good manufacturing practices.

    e. Marketed drugs are monitored for adverse events, side effects, and potential negative interactions. Long-term studies after approval are conducted to detect long-term effects and effects on subpopulations not in the original study.

    f. New information about potential safety risk is reviewed by an independent advisory board. Marketed drugs found to be unsafe after they are approved are removed, recalled, restricted, or appropriate risk/benefit information is provided.

3. Patients get and use the drugs they need for good health.

    a. Drug approval is not unnecessarily delayed.

    b. Drugs are obtainable by patients.

    c. Accurate information is available to support decision making about risks and benefits.

    d. Patients get the best intervention possible, practical, and reasonable for their health needs.

    e. Patients get drugs with the required dosage and purity.

4. Patients take the drugs in a safe and effective manner.

    a. Patients get correct instructions about dosage and follow them.

    b. Patients do not take unsafe combinations of drugs.

    c. Patients are properly monitored by a physician while they are being treated.

    d. Patients are not subjected to unacceptable risk during clinical trials.

In system engineering, the requirements may not be totally achievable in any practical design. For one thing, they may be conflicting among themselves (as was demonstrated in the train door example) or with other system (non-safety) requirements or constraints. The goal is to design a system or to evaluate and improve an existing system that satisfies the requirements as much as possible today and to continually improve the design over time using feedback and new scientific and engineering advances. Tradeoffs that must be made in the design process are carefully evaluated and considered and revisited when necessary.

Figure 7.6 shows the general pharmaceutical safety control structure in the United States. Each component's assigned responsibilities are those assumed in the design of the structure. In fact, at any time, they may not be living up to these responsibilities.

Congress provides guidance to the FDA by passing laws and providing directives, provides any necessary legislation to ensure drug safety, ensures that the FDA has

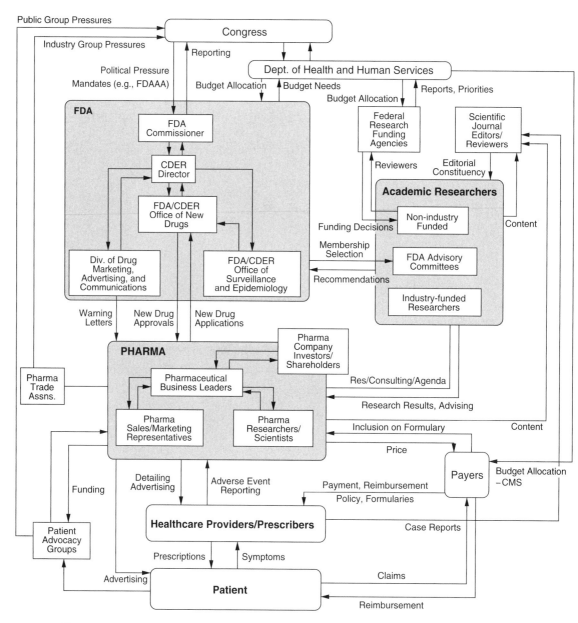

Figure 7.6
The U.S. pharmaceutical safety control structure.

enough funding to operate independently, provides legislative oversight on the effectiveness of FDA activities, and holds committee hearings and investigations of industry practices.

The FDA CDER (Center for Drug Evaluation and Research) ensures that the prescription, generic, and over-the-counter drug products are adequately available to the public and are safe and effective; monitors marketed drug products for unexpected health risks; and monitors and enforces the quality of marketed drug products. CDER staff members are responsible for selecting competent FDA advisory committee members, establishing and enforcing conflict of interest rules, and providing researchers with access to accurate and useful adverse event reports.

There are three major components within CDER. The Office of New Drugs (OND) is in charge of approving new drugs, setting drug labels and, when required, recalling drugs. More specifically, OND is responsible to:

- Oversee all U.S. human trials and development programs for investigational medical products to ensure safety of participants in clinical trials and provide oversight of the Institutional Review Boards (IRBs) that actually perform these functions for the FDA.
- Set the requirements and process for the approval of new drugs.
- Critically examine a sponsor's claim that a drug is safe for intended use (New Drug Application Safety Review). Impartially evaluate new drugs for safety and efficacy and approve them for sale if deemed appropriate.
- Upon approval, set the label for the drug.
- Not unnecessarily delay drugs that may have a beneficial effect.
- Require Phase IV (after-market) safety testing if there is a potential for long-term safety risk.
- Remove a drug from the market if new evidence shows that the risks outweigh the benefits.
- Update the label information when new information about drug safety is discovered.

The second office within the FDA CDER is the Division of Drug Marketing, Advertising, and Communications (DDMAC). This group provides oversight of the marketing and promotion of drugs. It reviews advertisements for accuracy and balance.

The third component of the FDA CDER is the Office of Surveillance and Epidemiology. This group is responsible for ongoing reviews of product safety, efficacy, and quality. It accomplishes this goal by performing statistical analysis of adverse event data it receives to determine whether there is a safety problem. This office reassesses risks based on new data learned after a drug is marketed and recommends ways to manage risk. Its staff members may also serve as consultants to OND with regard to drug safety issues. While they can recommend that a drug be removed from the market if new evidence shows significant risks, only OND can actually require that it be removed.

The FDA performs its duties with input from FDA Advisory Boards. These boards are made up of academic researchers whose responsibility is to provide independent advice and recommendations that are in the best interest of the general public. They must disclose any conflicts of interest related to subjects on which advice is being given.

Research scientists and centers are responsible for providing independent and objective research on a drug's safety, efficacy, and new uses and give their unbiased expert opinion when it is requested by the FDA. They should disclose all their conflicts of interest when publishing and take credit only for papers on which they have significantly contributed.

Scientific journals are responsible for publishing articles of high scientific quality and provide accurate and balanced information to doctors.

Payers and insurers pay the medical costs for the people insured as needed and only reimburse for drugs that are safe and effective. They control the use of drugs by providing formularies or lists of approved drugs for which they will reimburse claims.

Pharmaceutical developers and manufacturers also have responsibilities within the drug safety control structure. They must ensure that patients are protected from avoidable risks by providing safe and effective drugs, testing drugs for effectiveness, properly labeling their drugs, protecting patients during clinical trials by properly monitoring the trial, not promoting unsafe use of their drugs, removing a drug from the market if it is no longer considered safe, and manufacturing their drugs according to good manufacturing practice. They are also responsible for monitoring drugs for safety by running long-term, post-approval studies as required by the FDA; running new trials to test for potential hazards; and providing, maintaining, and incentivizing adverse-event reporting channels.

Pharmaceutical companies must also give accurate and up-to-date information to doctors and the FDA about drug safety by educating doctors, providing all available information about the safety of the drug to the FDA, and informing the FDA of potential new safety issues in a timely manner. Pharmaceutical companies also sponsor research for the development of new drugs and treatments.

Last, but not least, are the physicians and patients. Physicians have the responsibility to:

• Make treatment decisions based on the best interests of their clients
• Weigh the risks of treatment and non-treatment

- Prescribe drugs according to the limitations on the label
- Maintain up-to-date knowledge of the risk/benefit profile of the drugs they are prescribing
- Monitor the symptoms of their patients under treatment for adverse events and negative interactions
- Report adverse events potentially linked to the use of the drugs they prescribe

Patients are taking increasing responsibility for their own health in today's world, limited by what is practical. Traditionally they have been responsible to follow their physician's instructions and take drugs as prescribed, accede to the doctor's superior knowledge when appropriate, and go through physicians or appropriate channels to get prescription drugs.

As designed, this safety control structure looks strong and potentially effective. Unfortunately, it has not always worked the way it was supposed to work and the individual components have not always satisfied their responsibilities. Chapter 8 describes the use of the new hazard analysis technique, STPA, as well as other basic STAMP concepts in analyzing the potential risks in this structure.

8   STPA: A New Hazard Analysis Technique

Hazard analysis can be described as "investigating an accident before it occurs." The goal is to identify potential causes of accidents, that is, scenarios that can lead to losses, so they can be eliminated or controlled in design or operations *before* damage occurs.

The most widely used existing hazard analysis techniques were developed fifty years ago and have serious limitations in their applicability to today's more complex, software-intensive, sociotechnical systems. This chapter describes a new approach to hazard analysis, based on the STAMP causality model, called STPA (System-Theoretic Process Analysis).

8.1   Goals for a New Hazard Analysis Technique

Three hazard analysis techniques are currently used widely: Fault Tree Analysis, Event Tree Analysis, and HAZOP. Variants that combine aspects of these three techniques, such as Cause-Consequence Analysis (combining top-down fault trees and forward analysis Event Trees) and Bowtie Analysis (combining forward and backward chaining techniques) are also sometimes used. *Safeware* and other basic textbooks contain more information about these techniques for those unfamiliar with them. FMEA (Failure Modes and Effects Analysis) is sometimes used as a hazard analysis technique, but it is a bottom-up reliability analysis technique and has very limited applicability for safety analysis.

The primary reason for developing STPA was to include the new causal factors identified in STAMP that are not handled by the older techniques. More specifically, the hazard analysis technique should include design errors, including software flaws; component interaction accidents; cognitively complex human decision-making errors; and social, organizational, and management factors contributing to accidents. In short, the goal is to identify accident scenarios that encompass the entire accident process, not just the electromechanical components. While attempts have been made to add new features to traditional hazard analysis techniques to handle new technology, these attempts have had limited success because the underlying assumptions of the old techniques and the causality models on which they are based do not fit the characteristics of these new causal factors. STPA is based on the new causality assumptions identified in chapter 2.

An additional goal in the design of STPA was to provide guidance to the users in getting good results. Fault tree and event tree analysis provide little guidance to the analyst—the tree itself is simply the result of the analysis. Both the model of the system being used by the analyst and the analysis itself are only in the analyst's head. Analyst expertise in using these techniques is crucial, and the quality of the fault or event trees that result varies greatly.

HAZOP, widely used in the process industries, provides much more guidance to the analysts. HAZOP is based on a slightly different accident model than fault and event trees, namely that accidents result from deviations in system parameters, such as too much flow through a pipe or backflow when forward flow is required. HAZOP uses a set of guidewords to examine each part of a plant piping and wiring diagram, such as *more than*, *less than*, and *opposite*. Both guidance in performing the process and a concrete model of the physical structure of the plant are therefore available.

Like HAZOP, STPA works on a model of the system and has "guidewords" to assist in the analysis, but because in STAMP accidents are seen as resulting from inadequate control, the model used is a functional control diagram rather than a physical component diagram. In addition, the set of guidewords is based on lack of control rather than physical parameter deviations. While engineering expertise is still required, guidance is provided for the STPA process to provide some assurance of completeness in the analysis.

The third and final goal for STPA is that it can be used before a design has been created, that is, it provides the information necessary to guide the design process, rather than requiring a design to exist before the analysis can start. Designing safety into a system, starting in the earliest conceptual design phases, is the most cost-effective way to engineer safer systems. The analysis technique must also, of course, be applicable to existing designs or systems when safety-guided design is not possible.

8.2   The STPA Process

STPA (System-Theoretic Process Analysis) can be used at any stage of the system life cycle. It has the same general goals as any hazard analysis technique: accumulating information about how the behavioral safety constraints, which are derived from the system hazards, can be violated. Depending on when it is used, it provides the information and documentation necessary to ensure the safety constraints are enforced in system design, development, manufacturing, and operations, including the natural changes in these processes that will occur over time.

STPA uses a functional control diagram and the requirements, system hazards, and the safety constraints and safety requirements for the component as defined in chapter 7. When STPA is applied to an existing design, this information is available when the analysis process begins. When STPA is used for safety-guided design, only the system-level requirements and constraints may be available at the beginning of the process. In the latter case, these requirements and constraints are refined and traced to individual system components as the iterative design and analysis process proceeds.

STPA has two main steps:

1. Identify the potential for inadequate control of the system that could lead to a hazardous state. Hazardous states result from inadequate control or enforcement of the safety constraints, which can occur because:

   a. A control action required for safety is *not* provided or not followed.

   b. An unsafe control action *is* provided.

   c. A potentially safe control action is provided too early or too late, that is, at the wrong time or in the wrong sequence.

   d. A control action required for safety is stopped too soon or applied too long.

2. Determine how each potentially hazardous control action identified in step 1 could occur.

   a. For each unsafe control action, examine the parts of the control loop to see if they could cause it. Design controls and mitigation measures if they do not already exist or evaluate existing measures if the analysis is being performed on an existing design. For multiple controllers of the same component or safety constraint, identify conflicts and potential coordination problems.

   b. Consider how the designed controls could degrade over time and build in protection, including

      i. Management of change procedures to ensure safety constraints are enforced in planned changes.

      ii. Performance audits where the assumptions underlying the hazard analysis are the preconditions for the operational audits and controls so that unplanned changes that violate the safety constraints can be detected.

      iii. Accident and incident analysis to trace anomalies to the hazards and to the system design.

While the analysis can be performed in one step, dividing the process into discrete steps reduces the analytical burden on the safety engineers and provides a structured process for hazard analysis. The information from the first step (identifying the unsafe control actions) is required to perform the second step (identifying the causes of the unsafe control actions).

The assumption in this chapter is that the system design exists when STPA is performed. The next chapter describes safety-guided design using STPA and principles for safe design of control systems.

STPA is defined in this chapter using two examples. The first is a simple, generic interlock. The hazard involved is exposure of a human to a potentially dangerous energy source, such as high power. The power controller, which is responsible for turning the energy on or off, implements an interlock to prevent the hazard. In the physical controlled system, a door or barrier over the power source prevents exposure while it is active. To simplify the example, we will assume that humans cannot physically be inside the area when the barrier is in place—that is, the barrier is simply a cover over the energy source. The door or cover will be manually operated so the only function of the automated controller is to turn the power off when the door is opened and to turn it back on when the door is closed.

Given this design, the process starts from:

Hazard: Exposure to a high-energy source.

Constraint: The energy source must be off when the door is not closed.[1]

Figure 8.1 shows the control structure for this simple system. In this figure, the components of the system are shown along with the control instructions each component can provide and some potential feedback and other information or control sources for each component. Control operations by the automated controller include turning the power off and turning it on. The human operator can open and close the door. Feedback to the automated controller includes an indication of whether the door is open or not. Other feedback may be required or useful as determined during the STPA (hazard analysis) process.

The control structure for a second more complex example to be used later in the chapter, a fictional but realistic ballistic missile intercept system (FMIS), is shown in figure 8.2. Pereira, Lee, and Howard [154] created this example to describe their use of STPA to assess the risk of inadvertent launch in the U.S. Ballistic Missile Defense System (BMDS) before its first deployment and field test.

The BMDS is a layered defense to defeat all ranges of threats in all phases of flight (boost, midcourse, and terminal). The example used in this chapter is, for

---

1. The phrase "when the door is open" would be incorrect because a case is missing (a common problem): in the power controller's model of the controlled process, which enforces the constraint, the door may be open, closed, or the door position may be unknown to the controller. The phrase "is open or the door position is unknown" could be used instead. See section 9.3.2 for a discussion of why the difference is important.

HAZARD: Human exposed to high energy source

**SYSTEM SAFETY CONSTRAINT: The energy source must be off whenever
                                          the door is not completely closed.**

FUNCTIONAL REQUIREMENTS of the Power Controller:
   (1) Detect when the door is opened and turn off the power
   (2) When the door is closed, turn on the power

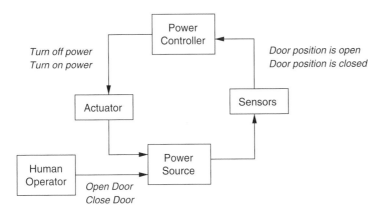

Figure 8.1
The control structure for a simple interlock system.

security reasons, changed from the real system, but it is realistic, and the problems identified by STPA in this chapter are similar to some that were found using STPA on the real system.

The U.S. BDMS system has a variety of components, including sea-based sensors in the Aegis shipborne platform; upgraded early warning systems; new and upgraded radars, ground-based midcourse defense, fire control, and communications; a Command and Control Battle Management and Communications component; and ground-based interceptors. Future upgrades will add features. Some parts of the system have been omitted in the example, such as the Aegis (ship-based) platform.

Figure 8.2 shows the control structure for the FMIS components included in the example. The command authority controls the operators by providing such things as doctrine, engagement criteria, and training. As feedback, the command authority gets the exercise results, readiness information, wargame results, and other information. The operators are responsible for controlling the launch of interceptors by sending instructions to the fire control subsystem and receiving status information as feedback.

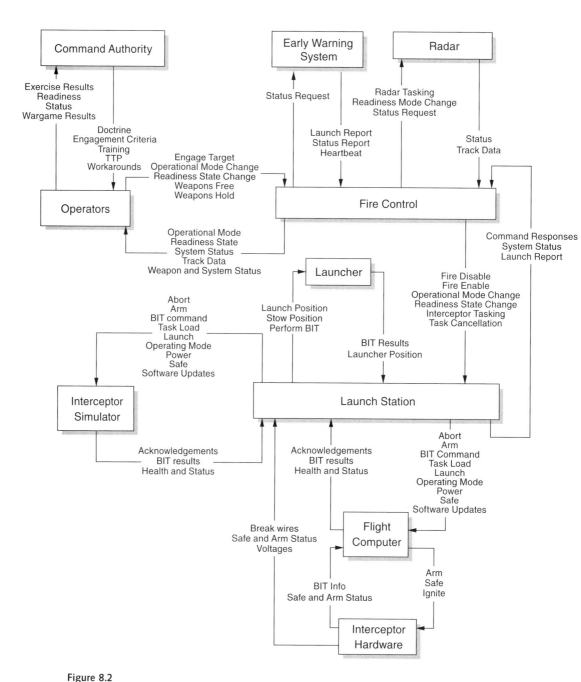

Figure 8.2
The control structure for a fictional ballistic missile defense system (FMIS) (adapted from Pereira, Lee, and Howard [154]).

Fire control receives instructions from the operators and information from the radars about any current threats. Using these inputs, fire control provides instructions to the launch station, which actually controls the launch of any interceptors. Fire control can enable firing, disable firing, and so forth, and, of course, it receives feedback from the launch station about the status of any previously provided control actions and the state of the system itself. The launch station controls the actual launcher and the flight computer, which in turn controls the interceptor hardware.

There is one other component of the system. To ensure operational readiness, the FMIS contains an interceptor simulator that periodically is used to mimic the flight computer in order to detect a failure in the system.

8.3   Identifying Potentially Hazardous Control Actions (Step 1)

Starting from the fundamentals defined in chapter 7, the first step in STPA is to assess the safety controls provided in the system design to determine the potential for inadequate control, leading to a hazard. The assessment of the hazard controls uses the fact that control actions can be hazardous in four ways (as noted earlier):

1. A control action required for safety is not provided or is not followed.
2. An unsafe control action is provided that leads to a hazard.
3. A potentially safe control action is provided too late, too early, or out of sequence.
4. A safe control action is stopped too soon or applied too long (for a continuous or nondiscrete control action).

For convenience, a table can be used to record the results of this part of the analysis. Other ways to record the information are also possible. In a classic System Safety program, the information would be included in the hazard log. Figure 8.3 shows the results of step 1 for the simple interlock example. The table contains four hazardous types of behavior:

1. A POWER OFF command is not given when the door is opened,
2. The door is opened and the controller waits too long to turn the power off;
3. A POWER ON command is given while the door is open, and
4. A POWER ON command is provided too early (when the door has not yet fully closed).

Incorrect but non-hazardous behavior is not included in the table. For example, not providing a POWER ON command when the power is off and the door is opened

Control Action	Not Providing Causes Hazard	Providing Causes Hazard	Wrong Timing or Order Causes Hazard	Stopped Too Soon or Applied Too Long
Power off	*Power not turned off when door opened*	Not Hazardous	*Door opened, controller waits too long to turn off power*	Not Applicable
Power on	Not Hazardous	*Power turned on while door opened*	*Power turned on too early; door not fully closed*	Not Applicable

Figure 8.3
Identifying hazardous system behavior.

or closed is not hazardous, although it may represent a quality-assurance problem. Another example of a mission assurance problem but not a hazard occurs when the power is turned off while the door is closed. Thomas has created a procedure to assist the analyst in considering the effect of all possible combinations of environmental and process variables for each control action in order to avoid missing any cases that should be included in the table [199a].

The final column of the table, *Stopped Too Soon or Applied Too Long*, is not applicable to the discrete interlock commands. An example where it does apply is in an aircraft collision avoidance system where the pilot may be told to climb or descend to avoid another aircraft. If the climb or descend control action is stopped too soon, the collision may not be avoided.

The identified hazardous behaviors can now be translated into safety constraints (requirements) on the system component behavior. For this example, four constraints must be enforced by the power controller (interlock):

1. The power must always be off when the door is open;
2. A POWER OFF command must be provided within x milliseconds after the door is opened;
3. A POWER ON command must never be issued when the door is open;
4. The POWER ON command must never be given until the door is fully closed.

For more complex examples, the mode in which the system is operating may determine the safety of the action or event. In that case, the operating mode may need to be included in the table, perhaps as an additional column. For example, some spacecraft mission control actions may only be hazardous during the launch or reentry phase of the mission.

In chapter 2, it was stated that many accidents, particularly component interaction accidents, stem from incomplete requirements specifications. Examples were

Command	Not Providing Causes Hazard	Providing Causes Hazard	Wrong Timing/Order Causes Hazard	Stopped Too Soon or Applied Too Long
Fire Enable	Not Hazardous	*Will accept interceptor tasking and can progress to a launch sequence*	*EARLY: Can inadvertently progress to an inadvertent launch*	Not Applicable
			OUT OF SEQUENCE: Disable comes before the enable	
...				

Figure 8.4
One row of the table identifying FMIS hazardous control actions.

provided such as missing constraints on the order of valve position changes in a batch chemical reactor and the conditions under which the descent engines should be shut down on the Mars Polar Lander spacecraft. The information provided in this first step of STPA can be used to identify the necessary constraints on component behavior to prevent the identified system hazards, that is, the safety requirements. In the second step of STPA, the information required by the component to properly implement the constraint is identified as well as additional safety constraints and information necessary to eliminate or control the hazards in the design or to design the system properly in the first place.

The FMIS system provides a less trivial example of step 1. Remember, the hazard is inadvertent launch. Consider the FIRE ENABLE command, which can be sent by the fire control module to the launch station to allow launch commands subsequently received by the launch station to be executed. As described in Pereira, Lee, and Howard [154], the FIRE ENABLE control command directs the launch station to enable the live fire of interceptors. Prior to receiving this command, the launch station will return an error message when it receives commands to fire an interceptor and will discard the fire commands.[2]

Figure 8.4 shows the results of performing STPA Step 1 on the FIRE ENABLE command. If this command is missing (column 2), a launch will not take place. While this omission might potentially be a mission assurance concern, it does not contribute to the hazard being analyzed (inadvertent launch).

---

2. Section 9.4.4 explains the safety-related reasons for breaking up potentially hazardous actions into multiple steps.

If the FIRE ENABLE command is provided to a launch station incorrectly, the launch station will transition to a state where it accepts interceptor tasking and can progress through a launch sequence. In combination with other incorrect or mistimed commands, this control action could contribute to an inadvertent launch.

A late FIRE ENABLE command will only delay the launch station's ability to process a launch sequence, which will not contribute to an inadvertent launch. A FIRE ENABLE command sent too early could open a window of opportunity for inadvertently progressing toward an inadvertent launch, similar to the incorrect FIRE ENABLE considered above. In the third case, a FIRE ENABLE command might be out of sequence with a FIRE DISABLE command. If this incorrect sequencing is possible in the system as designed and constructed, the system could be left capable of processing interceptor tasking and launching an interceptor when not intended.

Finally, the FIRE ENABLE command is a discrete command sent to the launch station to signal that it should allow processing of interceptor tasking. Because FIRE ENABLE is not a continuous command, the "stopped too soon" category does not apply.

8.4 Determining How Unsafe Control Actions Could Occur (Step 2)

Performing the first step of STPA provides the component safety requirements, which may be sufficient for some systems. A second step can be performed, however, to identify the scenarios leading to the hazardous control actions that violate the component safety constraints. Once the potential causes have been identified, the design can be checked to ensure that the identified scenarios have been eliminated or controlled in some way. If not, then the design needs to be changed. If the design does not already exist, then the designers at this point can try to eliminate or control the behaviors as the design is created, that is, use safety-guided design as described in the next chapter.

Why is the second step needed? While providing the engineers with the safety constraints to be enforced is necessary, it is not sufficient. Consider the chemical batch reactor described in section 2.1. The hazard is overheating of the reactor contents. At the system level, the engineers may decide (as in this design) to use water and a reflux condenser to control the temperature. After this decision is made, controls need to be enforced on the valves controlling the flow of catalyst and water. Applying step 1 of STPA determines that opening the valves out of sequence is dangerous, and the software requirements would accordingly be augmented with constraints on the order of the valve opening and closing instructions, namely that the water valve must be opened before the catalyst valve and the catalyst valve must be closed before the water valve is closed or, more generally, that the water valve must always be open when the catalyst valve is opened. If the software already exists, the hazard analysis would ensure that this ordering of commands has been enforced in the software. Clearly, building the software to enforce this ordering is a great deal easier than proving the ordering is true after the software already exists.

But enforcing these safety constraints is not enough to ensure safe software behavior. Suppose the software has commanded the water valve to open but something goes wrong and the valve does not actually open or it opens but water flow is restricted in some way (the *no flow* guideword in HAZOP). Feedback is needed for the software to determine if water is flowing through the pipes and the software needs to check this feedback before opening the catalyst valve. The second step of STPA is used to identify the ways that the software safety constraint, even if provided to the software engineers, might still not be enforced by the software logic and system design. In essence, step 2 identifies the scenarios or paths to a hazard found in a classic hazard analysis. This step is the usual "magic" one that creates the contents of a fault tree, for example. The difference is that guidance is provided to help create the scenarios and more than just failures are considered.

To create causal scenarios, the control structure diagram must include the process models for each component. If the system exists, then the content of these models should be easily determined by looking at the system functional design and its documentation. If the system does not yet exist, the analysis can start with a best guess and then be refined and changed as the analysis process proceeds.

For the high power interlock example, the process model is simple and shown in figure 8.5. The general causal factors, shown in figure 4.8 and repeated here in figure 8.6 for convenience, are used to identify the scenarios.

8.4.1   Identifying Causal Scenarios

Starting with each hazardous control action identified in step 1, the analysis in step 2 involves identifying how it could happen. To gather information about how the hazard could occur, the parts of the control loop for each of the hazardous control actions identified in step 1 are examined to determine if they could cause or contribute to it. Once the potential causes are identified, the engineers can design controls and mitigation measures if they do not already exist or evaluate existing measures if the analysis is being performed on an existing design.

Each potentially hazardous control action must be considered. As an example, consider the unsafe control action of not turning off the power when the door is opened. Figure 8.7 shows the results of the causal analysis in a graphical form. Other ways of documenting the results are, of course, possible.

The hazard in figure 8.7 is that the door is open but the power is not turned off. Looking first at the controller itself, the hazard could result if the requirement is not passed to the developers of the controller, the requirement is not implemented

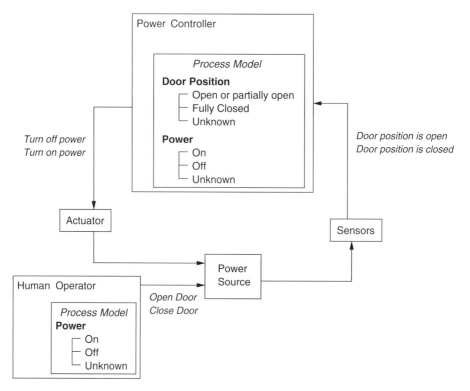

Figure 8.5
The process model for the high-energy controller.

correctly, or the process model incorrectly shows the door closed and/or the power off when that is not true. Working around the loop, the causal factors for each of the loop components are similarly identified using the general causal factors shown in figure 8.6. These causes include that the POWER OFF command is sent but not received by the actuator, the actuator received the command but does not implement it (actuator failure), the actuator delays in implementing the command, the POWER ON and POWER OFF commands are received or executed in the wrong order, the door open event is not detected by the door sensor or there is an unacceptable delay in detecting it, the sensor fails or provides spurious feedback, and the feedback about the state of the door or the power is not received by the controller or is not incorporated correctly into the process model.

More detailed causal analysis can be performed if a specific design is being considered. For example, the features of the communication channels used will determine the potential way that commands or feedback could be lost or delayed.

Once the causal analysis is completed, each of the causes that cannot be shown to be physically impossible must be checked to determine whether they are

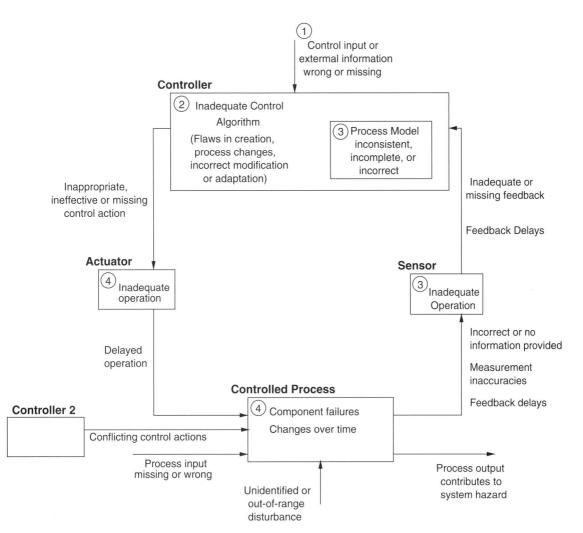

Figure 8.6
The causal factors to be considered to create scenarios in step 3.

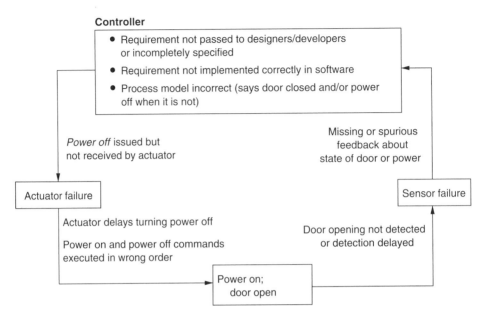

HAZARD: Door opened, power not turned off.

Figure 8.7
Example of step 2b STPA analysis for the high power interlock.

adequately handled in the design (if the design exists) or design features added to control them if the design is being developed with support from the analysis.

The first step in designing for safety is to try to eliminate the hazard completely. In this example, the hazard can be eliminated by redesigning the system to have the circuit run through the door in such a way that the circuit is broken as soon as the door opens. Let's assume, however, that for some reason this design alternative is rejected, perhaps as impractical. Design precedence then suggests that the next best alternatives in order are to reduce the likelihood of the hazard occurring, to prevent the hazard from leading to a loss, and finally to minimize damage. More about safe design can be found in chapters 16 and 17 of *Safeware* and chapter 9 of this book.

Because design almost always involves tradeoffs with respect to achieving multiple objectives, the designers may have good reasons not to select the most effective way to control the hazard but one of the other alternatives instead. It is important that the rationale behind the choice is documented for future analysis, certification, reuse, maintenance, upgrades, and other activities.

For this simple example, one way to mitigate many of the causes is to add a light that identifies whether the power supply is on or off. How do human operators know that the power has been turned off before inserting their hands into the high-energy power source? In the original design, they will most likely assume it is off because they have opened the door, which may be an incorrect assumption. Additional feedback and assurance can be attained from the light. In fact, protection systems in automated factories commonly are designed to provide humans in the vicinity with aural or visual information that they have been detected by the protection system. Of course, once a change has been made, such as adding a light, that change must then be analyzed for new hazards or causal scenarios. For example, a light bulb can burn out. The design might ensure that the safe state (the power is off) is represented by the light being on rather than the light being off, or two colors might be used. Every solution for a safety problem usually has its own drawbacks and limitations and therefore they will need to be compared and decisions made about the best design given the particular situation involved.

In addition to the factors shown in figure 8.6, the analysis must consider the impact of having two controllers of the same component whenever this occurs in the system safety control structure. In the friendly fire example in chapter 5, for example, confusion existed between the two AWACS operators responsible for tracking aircraft inside and outside of the no-fly-zone about who was responsible for aircraft in the boundary area between the two. The FMIS example below contains such a scenario. An analysis must be made to determine that no path to a hazard exists because of coordination problems.

The FMIS system provides a more complex example of STPA step 2. Consider the FIRE ENABLE command provided by fire control to the launch station. In step 1, it was determined that if this command is provided incorrectly, the launch station will transition to a state where it accepts interceptor tasking and can progress through a launch sequence. In combination with other incorrect or mistimed control actions, this incorrect command could contribute to an inadvertent launch.

The following are two examples of causal factors identified using STPA step 2 as potentially leading to the hazardous state (violation of the safety constraint). Neither of these examples involves component failures, but both instead result from unsafe component interactions and other more complex causes that are for the most part not identifiable by current hazard analysis methods.

In the first example, the FIRE ENABLE command can be sent inadvertently due to a missing case in the requirements—a common occurrence in accidents where software is involved.

The FIRE ENABLE command is sent when the fire control receives a WEAPONS FREE command from the operators and the fire control system has at least one active track. An active track indicates that the radars have detected something that might be an incoming missile. Three criteria are specified for declaring a track inactive: (1) a given period passes with no radar input, (2) the total predicted impact time elapses for the track, and (3) an intercept is confirmed. Operators are allowed to deselect any of these options. One case was not considered by the designers: if an operator deselects all of the options, no tracks will be marked as inactive. Under these conditions, the inadvertent entry of a WEAPONS FREE command would send the FIRE ENABLE command to the launch station immediately, even if there were no threats currently being tracked by the system.

Once this potential cause is identified, the solution is obvious—fix the software requirements and the software design to include the missing case. While the operator might instead be warned not to deselect all the options, this kind of human error is possible and the software should be able to handle the error safely. Depending on humans not to make mistakes is an almost certain way to guarantee that accidents will happen.

The second example involves confusion between the regular and the test software. The FMIS undergoes periodic system operability testing using an interceptor simulator that mimics the interceptor flight computer. The original hazard analysis had identified the possibility that commands intended for test activities could be sent to the operational system. As a result, the system status information provided by the launch station includes whether the launch station is connected only to missile simulators or to any live interceptors. If the fire control computer detects a change in this state, it will warn the operator and offer to reset into a matching state. There is, however, a small window of time before the launch station notifies the fire control component of the change. During this time interval, the fire control software could send a FIRE ENABLE command intended for test to the live launch station. This latter example is a coordination problem arising because there are multiple controllers of the launch station and two operating modes (e.g., testing and live fire). A potential mode confusion problem exists where the launch station can think it is in one mode but really be in the other one. Several different design changes could be used to prevent this hazardous state.

In the use of STPA on the real missile defense system, the risks involved in integrating separately developed components into a larger system were assessed, and several previously unknown scenarios for inadvertent launch were identified. Those conducting the assessment concluded that the STPA analysis and supporting data provided management with a sound basis on which to make risk acceptance decisions [154]. The assessment results were used to plan mitigations for open safety risks deemed necessary to change before deployment and field-testing of the system. As system changes are proposed, they are assessed by updating the control structure diagrams and assessment analysis results.

8.4.2   Considering the Degradation of Controls over Time

A final step in STPA is to consider how the designed controls could degrade over time and to build in protection against it. The mechanisms for the degradation could be identified and mitigated in the design: for example, if corrosion is identified as a potential cause, a stronger or less corrosive material might be used. Protection might also include planned performance audits where the assumptions underlying the hazard analysis are the preconditions for the operational audits and controls. For example, an assumption for the interlock system with a light added to warn the operators is that the light is operational and operators will use it to determine whether it is safe to open the door. *Performance audits* might check to validate that the operators know the purpose of the light and the importance of not opening the door while the warning light is on. Over time, operators might create workarounds to bypass this feature if it slows them up too much in their work or if they do not understand the purpose, the light might be partially blocked from view because of workplace changes, and so on. The assumptions and required audits should be identified during the system design process and then passed to the operations team.

Along with performance audits, *management of change procedures* need to be developed and the STPA analysis revisited whenever a planned change is made in the system design. Many accidents occur after changes have been made in the system. If appropriate documentation is maintained along with the rationale for the control strategy selected, this reanalysis should not be overly burdensome. How to accomplish this goal is discussed in chapter 10.

Finally, after accidents and incidents, the design and the hazard analysis should be revisited to determine why the controls were not effective. The hazard of foam damaging the thermal surfaces of the Space Shuttle had been identified during design, for example, but over the years before the *Columbia* loss the process for updating the hazard analysis after anomalies occurred in flight was eliminated. The Space Shuttle standard for hazard analyses (NSTS 22254, Methodology for Conduct of Space Shuttle Program Hazard Analyses) specified that hazards be revisited only when there was a new design or the design was changed: There was no process for updating the hazard analyses when anomalies occurred or even for determining whether an anomaly was related to a known hazard [117].

Chapter 12 provides more information about the use of the STPA results during operations.

8.5 Human Controllers

Humans in the system can be treated in the same way as automated components in step 1 of STPA, as was seen in the interlock system above where a person controlled the position of the door. The causal analysis and detailed scenario generation for human controllers, however, is much more complex than that of electromechanical devices and even software, where at least the algorithm is known and can be evaluated. Even if operators are given a procedure to follow, for reasons discussed in

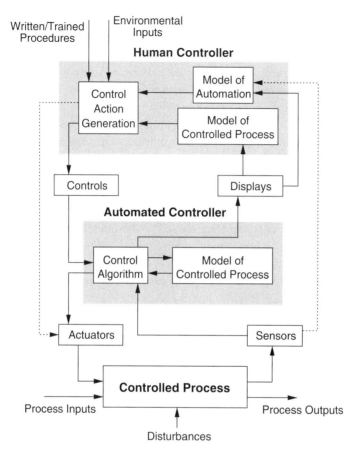

Figure 8.8
A human controller controlling an automated controller controlling a physical process.

chapter 2, it is very likely that the operator may feel the need to change the procedure over time.

The first major difference between human and automated controllers is that humans need an additional process model. All controllers need a model of the process they are controlling directly, but human controllers also need a model of any process, such as an oil refinery or an aircraft, they are indirectly controlling through an automated controller. If the human is being asked to supervise the automated controller or to monitor it for wrong or dangerous behavior then he or she needs to have information about the state of both the automated controller and the controlled process. Figure 8.8 illustrates this requirement. The need for an additional process model explains why supervising an automated system requires extra training and skill. A wrong assumption is sometimes made that if the human is supervising a computer, training requirements are reduced but this belief is untrue. Human skill levels and required knowledge almost always go up in this situation.

Figure 8.8 includes dotted lines to indicate that the human controller may need direct access to the process actuators if the human is to act as a backup to the automated controller. In addition, if the human is to monitor the automation, he or she will need direct input from the sensors to detect when the automation is confused and is providing incorrect information as feedback about the state of the controlled process.

The system design, training, and operational procedures must support accurate creation and updating of the extra process model required by the human supervisor. More generally, when a human is supervising an automated controller, there are extra analysis and design requirements. For example, the control algorithm used by the automation must be learnable and understandable. Inconsistent behavior or unnecessary complexity in the automation function can lead to increased human error. Additional design requirements are discussed in the next chapter.

With respect to STPA, the extra process model and complexity in the system design requires additional causal analysis when performing step 2 to determine the ways that both process models can become inaccurate.

The second important difference between human and automated controllers is that, as noted by Thomas [199], while automated systems have basically static control algorithms (although they may be updated periodically), humans employ dynamic control algorithms that they change as a result of feedback and changes in goals. Human error is best modeled and understood using feedback loops, not as a chain of directly related events or errors as found in traditional accident causality models. Less successful actions are a natural part of the search by operators for optimal performance [164].

Consider again figure 2.9. Operators are often provided with procedures to follow by designers. But designers are dealing with their own models of the controlled process, which may not reflect the actual process as constructed and changed over time. Human controllers must deal with the system as it exists. They update their process models using feedback, just as in any control loop. Sometimes humans use experimentation to understand the behavior of the controlled system and its current state and use that information to change their control algorithm. For example, after picking up a rental car, drivers may try the brakes and the steering system to get a feel for how they work before driving on a highway.

If human controllers suspect a failure has occurred in a controlled process, they may experiment to try to diagnose it and determine a proper response. Humans also use experimentation to determine how to optimize system performance. The driver's control algorithm may change over time as the driver learns more about the automated system and learns how to optimize the car's behavior. Driver goals and motivation may also change over time. In contrast, automated controllers by necessity must be designed with a single set of requirements based on the designer's model of the controlled process and its environment.

Thomas provides an example [199] using cruise control. Designers of an automated cruise control system may choose a control algorithm based on their model of the vehicle (such as weight, engine power, response time), the general design of roadways and vehicle traffic, and basic engineering design principles for propulsion and braking systems. A simple control algorithm might control the throttle in proportion to the difference between current speed (monitored through feedback) and desired speed (the goal).

Like the automotive cruise control designer, the human driver also has a process model of the car's propulsion system, although perhaps simpler than that of the automotive control expert, including the approximate rate of car acceleration for each accelerator position. This model allows the driver to construct an appropriate control algorithm for the current road conditions (slippery with ice or clear and dry) and for a given goal (obeying the speed limit or arriving at the destination at a required time). Unlike the static control algorithm designed into the automated cruise control, the human driver may dynamically change his or her control algorithm over time based on changes in the car's performance, in goals and motivation, or driving experience.

The differences between automated and human controllers lead to different requirements for hazard analysis and system design. Simply identifying human "failures" or errors is not enough to design safer systems. Hazard analysis must identify the specific human behaviors that can lead to the hazard. In some cases, it may be possible to identify why the behaviors occur. In either case, we are not able to "redesign" humans. Training can be helpful, but not nearly enough—training can do only so much in avoiding human error even when operators are highly trained and skilled. In many cases, training is impractical or minimal, such as automobile drivers. The only real solution lies in taking the information obtained in the hazard analysis about worst-case human behavior and using it in the design of the other system components and the system as a whole to eliminate, reduce, or compensate for that behavior. Chapter 9 discusses why we need human operators in systems and how to design to eliminate or reduce human errors.

STPA as currently defined provides much more useful information about the cause of human errors than traditional hazard analysis methods, but augmenting STPA could provide more information for designers. Stringfellow has suggested some additions to STPA for human controllers [195]. In general, engineers need better tools for including humans in hazard analyses in order to cope with the unique aspects of human control.

8.6   Using STPA on Organizational Components of the Safety Control Structure

The examples above focus on the lower levels of safety control structures, but STPA can also be used on the organizational and management components. Less experimentation has been done on applying it at these levels, and, once again, more needs to be done.

Two examples are used in this section: one was a demonstration for NASA of risk analysis using STPA on a new management structure proposed after the *Columbia* accident. The second is pharmaceutical safety. The fundamental activities of identifying system hazards, safety requirements and constraints, and of documenting the safety control structure were described for these two examples in chapter 7. This section starts from that point and illustrates the actual risk analysis process.

8.6.1   Programmatic and Organizational Risk Analysis

The Columbia Accident Investigation Board (CAIB) found that one of the causes of the *Columbia* loss was the lack of independence of the safety program from the Space Shuttle program manager. The CAIB report recommended that NASA institute an Independent Technical Authority (ITA) function similar to that used in SUBSAFE (see chapter 14), and individuals with SUBSAFE experience were recruited to help design and implement the new NASA Space Shuttle program organizational structure. After the program was designed and implementation started, a risk analysis of the program was performed to assist in a planned review of the program's effectiveness. A classic programmatic risk analysis, which used experts to identify the risks in the program, was performed. In parallel, a group at MIT developed a process to use STAMP as a foundation for the same type of programmatic risk analysis to understand the risks and vulnerabilities of this new organizational structure and recommend improvements [125].[3] This section describes the STAMP-based process and results as an example of what can be done for other systems and other emergent properties. Laracy [108] used a similar process to examine transportation system security, for example.

The STAMP-based analysis rested on the basic STAMP concept that most major accidents do not result simply from a unique set of proximal, physical events but from the migration of the organization to a state of heightened risk over time as safeguards and controls are relaxed due to conflicting goals and tradeoffs. In such a high-risk state, events are bound to occur that will trigger an accident. In both the *Challenger* and *Columbia* losses, organizational risk had been increasing to unacceptable levels for quite some time as behavior and decision-making evolved in

---

3. Many people contributed to the analysis described in this section, including Nicolas Dulac, Betty Barrett, Joel Cutcher-Gershenfeld, John Carroll, and Stephen Friedenthal.

response to a variety of internal and external performance pressures. Because risk increased slowly, nobody noticed, that is, the *boiled frog* phenomenon. In fact, confidence and complacency were increasing at the same time as risk due to the lack of accidents.

The goal of the STAMP-based analysis was to apply a classic system safety engineering process to the analysis and redesign of this organizational structure. Figure 8.9 shows the basic process used, which started with a preliminary hazard analysis to identify the system hazards and the safety requirements and constraints. In the second step, a STAMP model of the ITA safety control structure was created (as designed by NASA; see figure 7.4) and a gap analysis was performed to map the identified safety requirements and constraints to the assigned responsibilities in the safety control structure and identify any gaps. A detailed hazard analysis using STPA was then performed to identify the system risks and to generate recommendations for improving the designed new safety control structure and for monitoring the implementation and long-term health of the new program. Only enough of the modeling and analysis is included here to allow the reader to understand the process. The complete modeling and analysis effort is documented elsewhere [125].

The hazard identification, system safety requirements, and safety control structure for this example are described in section 7.4.1, so the example starts from this basic information.

8.6.2   Gap Analysis

In analyzing an existing organizational or social safety control structure, one of the first steps is to determine where the responsibility for implementing each requirement rests and to perform a *gap analysis* to identify holes in the current design, that is, requirements that are not being implemented (enforced) anywhere. Then the safety control structure needs to be evaluated to determine whether it is potentially effective in enforcing the system safety requirements and constraints.

A mapping was made between the system-level safety requirements and constraints and the individual responsibilities of each component in the NASA safety control structure to see where and how requirements are enforced. The ITA program was at the time being carefully defined and documented. In other situations, where such documentation may be lacking, interview or other techniques may need to be used to elicit how the organizational control structure actually works. In the end, complete documentation should exist in order to maintain and operate the system safely. While most organizations have job descriptions for each employee, the safety-related responsibilities are not necessarily separated out or identified, which can lead to unidentified gaps or overlaps.

As an example, in the ITA structure the responsibility for the system-level safety requirement:

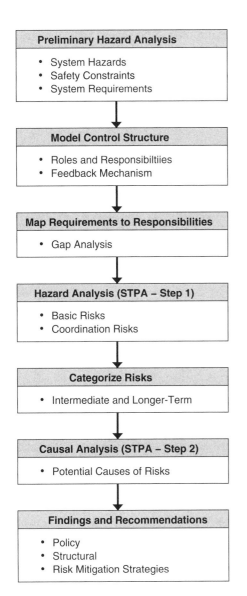

Figure 8.9
The basic process used in the NASA ITA risk analysis.

1a. State-of-the art safety standards and requirements for NASA missions must be established, implemented, enforced, and maintained that protect the astronauts, the workforce, and the public

was assigned to the NASA Chief Engineer but the Discipline Technical Warrant Holders, the Discipline Trusted Agents, the NASA Technical Standards Program, and the headquarters Office of Safety and Mission Assurance also play a role in implementing this Chief Engineer responsibility. More specifically, system requirement *1a* was implemented in the control structure by the following responsibility assignments:

- **Chief Engineer:** Develop, monitor, and maintain technical standards and policy.
- **Discipline Technical Warrant Holders:**
  - Recommend priorities for development and updating of technical standards.
  - Approve all new or updated NASA Preferred Standards within their assigned discipline (the NASA Chief Engineer retains Agency approval)
  - Participate in (lead) development, adoption, and maintenance of NASA Preferred Technical Standards in the warranted discipline.
  - Participate as members of technical standards working groups.
- **Discipline Trusted Agents:** Represent the Discipline Technical Warrant Holders on technical standards committees
- **NASA Technical Standards Program:** Coordinate with Technical Warrant Holders when creating or updating standards
- **NASA Headquarters Office Safety and Mission Assurance:**
  - Develop and improve generic safety, reliability, and quality process standards and requirements, including FMEA, risk, and the hazard analysis process.
  - Ensure that safety and mission assurance policies and procedures are adequate and properly documented.

Once the mapping is complete, a gap analysis can be performed to ensure that each system safety requirement and constraint is embedded in the organizational design and to find holes or weaknesses in the design. In this analysis, concerns surfaced, particularly about requirements not reflected in the defined ITA organizational structure.

As an example, one omission detected was appeals channels for complaints and concerns about the components of the ITA structure itself that may not function appropriately. All channels for expressing what NASA calls "technical conscience" go through the warrant holders, but there was no defined way to express concerns about the warrant holders themselves or about aspects of ITA that are not working well.

A second example was the omission in the documentation of the ITA implementation plans of the person(s) who was to be responsible to see that engineers and managers are trained to use the results of hazard analyses in their decision making. More generally, a distributed and ill-defined responsibility for the hazard analysis process made it difficult to determine responsibility for ensuring that adequate resources are applied; that hazard analyses are elaborated (refined and extended) and updated as the design evolves and test experience is acquired; that hazard logs are maintained and used as experience is acquired; and that all anomalies are evaluated for their hazard potential. Before ITA, many of these responsibilities were assigned to each Center's Safety and Mission Assurance Office, but with much of this process moving to engineering (which is where it should be) under the new ITA structure, clear responsibilities for these functions need to be specified. One of the basic causes of accidents in STAMP is multiple controllers with poorly defined or overlapping responsibilities.

A final example involved the ITA program assessment process. An assessment of how well ITA is working is part of the plan and is an assigned responsibility of the chief engineer. The official risk assessment of the ITA program performed in parallel with the STAMP-based one was an implementation of that chief engineer's responsibility and was planned to be performed periodically. We recommended the addition of specific organizational structures and processes for implementing a continual learning and improvement process and making adjustments to the design of ITA itself when necessary outside of the periodic review.

8.6.3 Hazard Analysis to Identify Organizational and Programmatic Risks

A risk analysis to identify ITA programmatic risks and to evaluate these risks periodically had been specified as one of the chief engineer's responsibilities. To accomplish this goal, NASA identified the programmatic risks using a classic process using experts in risk analysis interviewing stakeholders and holding meetings where risks were identified and discussed. The STAMP-based analysis used a more formal, structured approach.

Risks in STAMP terms can be divided into two types: (1) basic inadequacies in the way individual components in the control structure fulfill their responsibilities and (2) risks involved in the coordination of activities and decision making that can lead to unintended interactions and consequences.

Basic Risks

Applying the four types of inadequate control identified in STPA and interpreted for the hazard, which in this case is unsafe decision-making leading to an accident, ITA has four general types of risks:

1. Unsafe decisions are made or approved by the chief engineer or warrant holders.

2. Safe decisions are disallowed (e.g., overly conservative decision making that undermines the goals of NASA and long-term support for ITA).

3. Decision making takes too long, minimizing impact and also reducing support for the ITA.

4. Good decisions are made by the ITA, but do not have adequate impact on system design, construction, and operation.

The specific potentially unsafe control actions by those in the ITA safety control structure that could lead to these general risks are the ITA programmatic risks. Once identified, they must be eliminated or controlled just like any unsafe control actions.

Using the responsibilities and control actions defined for the components of the safety control structure, the STAMP-based risk analysis applied the four general types of inadequate control actions, omitting those that did not make sense for the particular responsibility or did not impact risk. To accomplish this, the general responsibilities must be refined into more specific control actions.

As an example, the chief engineer is responsible as the ITA for the technical standards and system requirements and all changes, variances, and waivers to the requirements, as noted earlier. The control actions the chief engineer has available to implement this responsibility are:

- To develop, monitor, and maintain technical standards and policy.
- In coordination with programs and projects, to establish or approve the technical requirements and ensure they are enforced and implemented in the programs and projects (ensure the design is compliant with the requirements).
- To approve all changes to the initial technical requirements.
- To approve all variances (waivers, deviations, exceptions to the requirements.
- Etc.

Taking just one of these, the control responsibility to develop, monitor, and maintain technical standards and policy, the risks (potentially inadequate or unsafe control actions) identified using STPA step 1 include:

1. General technical and safety standards are not created.

2. Inadequate standards and requirements are created.

3. Standards degrade over time due to external pressures to weaken them. The process for approving changes is flawed.

4. Standards are not changed over time as the environment changes.

As another example, the chief engineer cannot perform all these duties himself, so he has a network of people below him in the hierarchy to whom he delegates or "warrants" some of the responsibilities. The chief engineer retains responsibility for ensuring that the warrant holders perform their duties adequately as in any hierarchical management structure.

The chief engineer responsibility to approve all variances and waivers to technical requirements is assigned to the System Technical Warrant Holder (STWH). The risks or potentially unsafe control actions of the STWH with respect to this responsibility are:

- An unsafe engineering variance or waiver is approved.

- Designs are approved without determining conformance with safety requirements. Waivers become routine.

- Reviews and approvals take so long that ITA becomes a bottleneck. Mission achievement is threatened. Engineers start to ignore the need for approvals and work around the STWH in other ways.

Although a long list of risks was identified in this experimental application of STPA to a management structure, many of the risks for different participants in the ITA process were closely related. The risks listed for each participant are related to his or her particular role and responsibilities and therefore those with related roles or responsibilities will generate related risks. The relationships were made clear in the earlier step tracing from system requirements to the roles and responsibilities for each of the components of the ITA.

Coordination Risks

Coordination risks arise when multiple people or groups control the same process. The types of unsafe interactions that may result include: (1) both controllers assume that the other is performing the control responsibilities, and as a result nobody does, or (2) controllers provide conflicting control actions that have unintended side effects.

Potential coordination risks are identified by the mapping from the system requirements to the component requirements used in the gap analysis described earlier. When similar responsibilities related to the same system requirement are identified, the potential for new coordination risks needs to be considered.

As an example, the original ITA design documentation was ambiguous about who had the responsibility for performing many of the safety engineering functions. Safety engineering had previously been the responsibility of the Center Safety and Mission Assurance Offices but the plan envisioned that these functions would shift to the ITA in the new organization leading to several obvious risks.

Another example involves the transition of responsibility for the production of standards to the ITA from the NASA Headquarters Office of Safety and Mission Assurance (OSMA). In the plan, some of the technical standards responsibilities were retained by OSMA, such as the technical design standards for human rating spacecraft and for conducting hazard analyses, while others were shifted to the ITA without a clear demarcation of who was responsible for what. At the same time, responsibilities for the assurance that the plans are followed, which seems to logically belong to the mission assurance group, were not cleanly divided. Both overlaps raised the potential for some functions not being accomplished or conflicting standards being produced.

8.6.4 Use of the Analysis and Potential Extensions

While risk mitigation and control measures could be generated from the list of risks themselves, the application of step 2 of STPA to identify causes of the risks will help to provide better control measures in the same way STPA step 2 plays a similar role in physical systems. Taking the responsibility of the System Technical Warrant Holder to approve all variances and waivers to technical requirements in the example above, potential causes for approving an unsafe engineering variance or waiver include: inadequate or incorrect information about the safety of the action, inadequate training, bowing to pressure about programmatic concerns, lack of support from management, inadequate time or resources to evaluate the requested variance properly, and so on. These causal factors were generated using the generic factors in figure 8.6 but defined in a more appropriate way. Stringfellow has examined in more depth how STPA can be applied to organizational factors [195].

The analysis can be used to identify potential changes to the safety control structure (the ITA program) that could eliminate or mitigate identified risks. General design principles for safety are described in the next chapter.

A goal of the NASA risk analysis was to determine what to include in a planned special assessment of the ITA early in its existence. To accomplish the same goal, the MIT group categorized their identified risks as (1) immediate, (2) long-term, or (3) controllable by standard ongoing processes. These categories were defined in the following way:

Immediate concern: An immediate and substantial concern that should be part of a near-term assessment.

Longer-term concern: A substantial longer-term concern that should potentially be part of future assessments; as the risk will increase over time or cannot be evaluated without future knowledge of the system or environment behavior.

Standard process: An important concern that should be addressed through standard processes, such as inspections, rather than an extensive special assessment procedure.

This categorization allowed identifying a manageable subset of risks to be part of the planned near-term risk assessment and those that could wait for future assessments or could be controlled by on-going procedures. For example, it is important to assess immediately the degree of "buy-in" to the ITA program. Without such support, ITA cannot be sustained and the risk of dangerous decision making is very high. On the other hand, the ability to find appropriate successors to the current warrant holders is a longer-term concern identified in the STAMP-based risk analysis that would be difficult to assess early in the existence of the new ITA control structure. The performance of the current technical warrant holders, for example, is one factor that will have an impact on whether the most qualified people will want the job in the future.

8.6.5 Comparisons with Traditional Programmatic Risk Analysis Techniques

The traditional risk analysis performed by NASA on ITA identified about one hundred risks. The more rigorous, structured STAMP-based analysis—done independently and without any knowledge of the results of the NASA process—identified about 250 risks, all the risks identified by NASA plus additional ones. A small part of the difference was related to the consideration by the STAMP group of more components in the safety control structure, such as the NASA administrator, Congress, and the Executive Branch (White House). There is no way to determine whether the other additional risks identified by the STAMP-based process were simply missed in the NASA analysis or were discarded for some reason.

The NASA analysis did not include a causal analysis of the risks and thus no comparison is possible. Their goal was to determine what should be included in the upcoming ITA risk assessment process and thus was narrower than the STAMP demonstration risk analysis effort.

8.7 Reengineering a Sociotechnical System: Pharmaceutical Safety and the Vioxx Tragedy

The previous section describes the use of STPA on the management structure of an organization that develops and operates high-tech systems. STPA and other types of analysis are potentially also applicable to social systems. This section provides an example using pharmaceutical safety.

Couturier has performed a STAMP-based causal analysis of the incidents associated with the introduction and withdrawal of Vioxx [43]. Once the causes of such losses are determined, changes need to be made to prevent a recurrence. Many suggestions for changes as a result of the Vioxx losses (for example, [6, 66, 160, 190]) have been proposed. After the Vioxx recall, three main reports were written by the Government Accountability Office (GAO) [73], the Institute of Medicine (IOM) [16], and one commissioned by Merck. The publication of these reports led to two waves of changes, the first initiated within the FDA and the second by Congress in the form of a new set of rules called FDAAA (FDA Amendments Act). Couturier [43, 44], with inputs from others,[4] used the Vioxx events to demonstrate how these proposed and implemented policy and structural changes could be analyzed to predict their potential effectiveness using STAMP.

8.7.1 The Events Surrounding the Approval and Withdrawal of Vioxx

Vioxx (Rofecoxib) is a prescription COX-2 inhibitor manufactured by Merck. It was approved by the Food and Drug Administration (FDA) in May 1999 and was widely used for pain management, primarily from osteoarthritis. Vioxx was one of the major sources of revenue for Merck while on the market: It was marketed in more than eighty countries with worldwide sales totaling $2.5 billion in 2003.

In September 2004, Merck voluntarily withdrew the drug from the market because of safety concerns: The drug was suspected to increase the risk of cardiovascular events (heart attacks and stroke) for the patients taking it long term at high dosages. Vioxx was one of the most widely used drugs ever to be withdrawn from the market. According to an epidemiological study done by Graham, an FDA scientist, Vioxx has been associated with more than 27,000 heart attacks or deaths and may be the "single greatest drug safety catastrophe in the history of this country or the history of the world" [76].

The important question to be considered is how did such a dangerous drug get on the market and stay there so long despite warnings of problems and how can this type of loss be avoided in the future.

The major events that occurred in this saga start with the discovery of the Vioxx molecule in 1994. Merck sought FDA approval in November 1998.

In May 1999 the FDA approved Vioxx for the relief of osteoarthritis symptoms and management of acute pain. Nobody had suggested that the COX-2 inhibitors are more effective than the classic NSAIDS in relieving pain, but their selling point had been that they were less likely to cause bleeding and other digestive tract complications. The FDA was not convinced and required that the drug carry a warning on its label about possible digestive problems. By December, Vioxx had more than 40 percent of the new prescriptions in its class.

In order to validate their claims about Rofecoxib having fewer digestive system complications, Merck launched studies to prove their drugs should not be lumped with other NSAIDS. The studies backfired.

In January 1999, before Vioxx was approved, Merck started a trial called VIGOR (Vioxx Gastrointestinal Outcomes Research) to compare the efficacy and adverse

---

4. Many people provided input to the analysis described in this section, including Stan Finkelstein, John Thomas, John Carroll, Margaret Stringfellow, Meghan Dierks, Bruce Psaty, David Wierz, and various other reviewers.

effects of Rofecoxib and Naproxen, an older nonsteroidal anti-inflammatory drug or NSAID. In March 2000, Merck announced that the VIGOR trial had shown that Vioxx was safer on the digestive tract than Naproxen, but it doubled the risk of cardiovascular problems. Merck argued that the increased risk resulted not because Vioxx caused the cardiovascular problems but that Celebrex (the Naproxen used in the trial) protected against them. Merck continued to minimize unfavorable findings for Vioxx up to a month before withdrawing it from the market in 2004.

Another study, ADVANTAGE, was started soon after the VIGOR trial. ADVANTAGE had the same goal as VIGOR, but it targeted osteoarthritis, whereas VIGOR was for rheumatoid arthritis. Although the ADVANTAGE trial did demonstrate that Vioxx was safer on the digestive track than Naproxen, it failed to show that Rofecoxib had any advantage over Naproxen in terms of pain relief. Long after the report on ADVANTAGE was published, it turned out that its first author had no involvement in the study until Merck presented him with a copy of the manuscript written by Merck authors. This turned out to be one of the more prominent recent examples of ghostwriting of journal articles where company researchers wrote the articles and included the names of prominent researchers as authors [178].

In addition, Merck documents later came to light that appear to show the ADVANTAGE trial emerged from the Merck marketing division and was actually a "seeding" trial, designed to market the drug by putting "its product in the hands of practicing physicians, hoping that the experience of treating patients with the study drug and a pleasant, even profitable interaction with the company will result in more loyal physicians who prescribe the drug" [83].

Although the studies did demonstrate that Vioxx was safer on the digestive track than Naproxen, they also again unexpectedly found that the COX-2 inhibitor doubled the risk of cardiovascular problems. In April 2002, the FDA required that Merck note a possible link to heart attacks and strokes on Vioxx's label. But it never ordered Merck to conduct a trial comparing Vioxx with a placebo to determine whether a link existed. In April 2000 the FDA recommended that Merck conduct an animal study with Vioxx to evaluate cardiovascular safety, but no such study was ever conducted.

For both the VIGOR and ADVANTAGE studies, claims have been made that cardiovascular events were omitted from published reports [160]. In May 2000 Merck published the results from the VIGOR trial. The data included only seventeen of the twenty heart attacks the Vioxx patients had. When the omission was later detected, Merck argued that the events occurred after the trial was over and therefore did not have to be reported. The data showed a four times higher risk of heart attacks compared with Naproxen. In October 2000, Merck officially told the FDA about the other three heart attacks in the VIGOR study.

Merck marketed Vioxx heavily to doctors and spent more than $100 million a year on direct-to-the-consumer advertising using popular athletes including Dorothy Hamill and Bruce Jenner. In September 2001, the FDA sent Merck a letter warning the company to stop misleading doctors about Vioxx's effect on the cardiovascular system.

In 2001, Merck started a new study called APPROVe (Adenomatous Polyp PRevention On Vioxx) in order to expand its market by showing the efficacy of Vioxx on colorectal polyps. APPROVe was halted early when the preliminary data showed an increased relative risk of heart attacks and strokes after eighteen months of Vioxx use. The long-term use of Rofecoxib resulted in nearly twice the risk of suffering a heart attack or stroke compared to patients receiving a placebo.

David Graham, an FDA researcher, did an analysis of a database of 1.4 million Kaiser Permanente members and found that those who took Vioxx were more likely to suffer a heart attack or sudden cardiac death than those who took Celebrex, Vioxx's main rival. Graham testified to a congressional committee that the FDA tried to block publication of his findings. He described an environment "where he was 'ostracized'; 'subjected to veiled threats' and 'intimidation.'" Graham gave the committee copies of email that support his claims that his superiors at the FDA suggested watering down his conclusions [178].

Despite all their efforts to deny the risks associated with Vioxx, Merck withdrew the drug from the market in September 2004. In October 2004, the FDA approved a replacement drug for Vioxx by Merck, called Arcoxia.

Because of the extensive litigation associated with Vioxx, many questionable practices in the pharmaceutical industry have come to light [6]. Merck has been accused of several unsafe "control actions" in this sequence of events, including not accurately reporting trial results to the FDA, not having a proper control board (DSMB) overseeing the safety of the patients in at least one of the trials, misleading marketing efforts, ghostwriting journal articles about Rofecoxib studies, and paying publishers to create fake medical journals to publish favorable articles [45]. Post-market safety studies recommended by the FDA were never done, only studies directed at increasing the market.

8.7.2 Analysis of the Vioxx Case

The hazards, system safety requirements and constraints, and documentation of the safety control structure for pharmaceutical safety were shown in chapter 7. Using these, Couturier performed several types of analysis.

He first traced the system requirements to the responsibilities assigned to each of the components in the safety control structure, that is, he performed a gap analysis as described above for the NASA ITA risk analysis. The goal was to check that at least one controller was responsible for enforcing each of the safety requirements, to identify when multiple controllers had the same responsibility, and to study each of the controllers independently to determine if they are capable of carrying out their assigned responsibilities.

In the gap analysis, no obvious gaps or missing responsibilities were found, but multiple controllers are in charge of enforcing some of the same safety requirements. For example, the FDA, the pharmaceutical companies, and physicians are all responsible for monitoring drugs for adverse events. This redundancy is helpful if the controllers work together and share the information they have. Problems can occur, however, if efforts are not coordinated and gaps occur.

The assignment of responsibilities does not necessarily mean they are carried out effectively. As in the NASA ITA analysis, potentially inadequate control actions can be identified using STPA step 1, potential causes identified using step 2, and controls to protect against these causes designed and implemented. Contextual factors must be considered such as external or internal pressures militating against effective implementation or application of the controls. For example, given the financial incentives involved in marketing a blockbuster drug—Vioxx in 2003 provided $2.5 billion, or 11 percent of Merck's revenue [66]—it may be unreasonable to expect pharmaceutical companies to be responsible for drug safety without strong external oversight and controls or even to be responsible at all: Suggestions have been made that responsibility for drug development and testing be taken away from the pharmaceutical manufacturers [67].

Controllers must also have the resources and information necessary to enforce the safety constraints they have been assigned. Physicians need information about drug safety and efficacy that is independent from the pharmaceutical company representatives in order to adequately protect their patients. One of the first steps in performing an analysis of the drug safety control structure is to identify the contextual factors that can influence whether each component's responsibilities are carried out and the information required to create an accurate process model to support informed decision making in exercising the controls they have available to carry out their responsibilities.

Couturier also used the drug safety control structure, system safety requirements and constraints, the events in the Vioxx losses, and STPA and system dynamics models (see appendix D) to investigate the potential effectiveness of the changes implemented after the Vioxx events to control the marketing of unsafe drugs and the impact of the changes on the system as a whole. For example, the Food and Drug Amendments Act of 2007 (FDAAA) increased the responsibilities of the FDA and provided it with new authority. Couturier examined the recommendations from the FDAAA, the IOM report, and those generated from his STAMP causal analysis of the Vioxx events.

System dynamics modeling was used to show the relationship among the contextual factors and unsafe control actions and the reasons why the safety control structure migrated toward ineffectiveness over time. Most modeling techniques provide

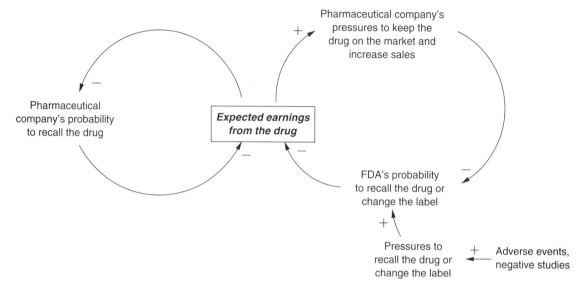

Figure 8.10
Overview of reinforcing pressures preventing the recall of drugs.

only direct relationships (arrows), which are inadequate to understand the indirect relationships between causal factors. System dynamics provides a way to show such indirect and nonlinear relationships. Appendix D explains this modeling technique.

First, system dynamics models were created to model the contextual influences on the behavior of each component (patients, pharmaceutical companies, the FDA, and so on) in the pharmaceutical safety control structure. Then the models were combined to assist in understanding the behavior of the system as a whole and the interactions among the components. The complete analysis can be found in [43] and a shorter paper on some of the results [44]. An overview and some examples are provided here.

Figure 8.10 shows a simple model of two types of pressures in this system that militate against drugs being recalled. The loop on the left describes pressures within the pharmaceutical company related to drug recalls while the loop on the right describes pressures on the FDA related to drug recalls.

Once a drug has been approved, the pharmaceutical company, which invested large resources in developing, testing, and marketing the drug, has incentives to maximize profits from the drug and keep it on the market. Those pressures are accentuated in the case of expected blockbuster drugs where the company's financial well-being potentially depends on the success of the product. This goal creates a reinforcing loop within the company to try to keep the drug on the market. The company also has incentives to pressure the FDA to increase the number of approved

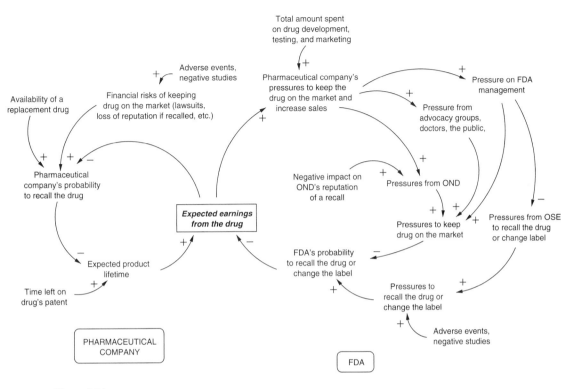

Figure 8.11
A more detailed model of reinforcing pressures preventing the recall of drugs.

indications, and thus purchasers, resist label changes, and prevent drug recalls. If the company is successful at preventing recalls, the expectations for the drug increase, creating another reinforcing loop. External pressures to recall the drug limit the reinforcing dynamics, but they have a lot of inertia to overcome.

Figure 8.11 includes more details, more complex feedback loops, and more outside pressures, such as the availability of a replacement drug, the time left on the drug's patent, and the amount of time spent on drug development. Pressures on the FDA from the pharmaceutical companies are elaborated including the pressures on the Office of New Drugs (OND) through PDUFA fees,[5] pressures from advisory boards

5. The Prescription Drug Use Fee Act (PDUFA) was first passed by Congress in 1992. It allows the FDA to collect fees from the pharmaceutical companies to pay the expenses for the approval of new drugs. In return, the FDA agrees to meet drug review performance goals. The main goal of PDUFA is to accelerate the drug review process. Between 1993 and 2002, user fees allowed the FDA to increase by 77 percent the number of personnel assigned to review applications. In 2004, more than half the funding for the CDEH was coming from user fees [148]. A growing group of scientists and regulators have expressed fears that in allowing the FDA to be sponsored by the pharmaceutical companies, the FDA has shifted its priorities to satisfying the companies, its "client," instead of protecting the public.

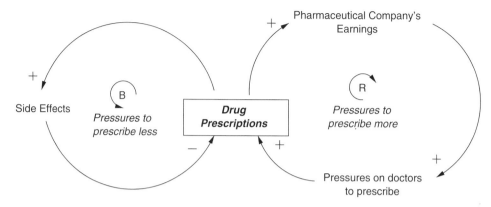

Figure 8.12
Overview of influences on physician prescriptions.

to keep the drug (which are, in turn, subject to pressures from patient advocacy groups and lucrative consulting contracts with the pharmaceutical companies), and pressures from the FDA Office of Surveillance and Epidemiology (OSE) to recall the drug.

Figures 8.12 and 8.13 show the pressures leading to overprescribing drugs. The overview in figure 8.12 has two primary feedback loops. The loop on the left describes pressures to lower the number of prescriptions based on the number of adverse events and negative studies. The loop on the right shows the pressures within the pharmaceutical company to increase the number of prescriptions based on company earnings and marketing efforts.

For a typical pharmaceutical product, more drug prescriptions lead to higher earnings for the drug manufacturer, part of which can be used to pay for more advertising to get doctors to continue to prescribe the drug. This reinforcing loop is usually balanced by the adverse effects of the drug. The more the drug is prescribed, the more likely is observation of negative side effects, which will serve to balance the pressures from the pharmaceutical companies. The two loops then theoretically reach a dynamic equilibrium where drugs are prescribed only when their benefits outweigh the risks.

As demonstrated in the Vioxx case, delays within a loop can significantly alter the behavior of the system. By the time the first severe side effects were discovered, millions of prescriptions had been given out. The balancing influences of the side-effects loop were delayed so long that they could not effectively control the reinforcing pressures coming from the pharmaceutical companies. Figure 8.13 shows how additional factors can be incorporated including the quality of collected data, the market size, and patient drug requests.

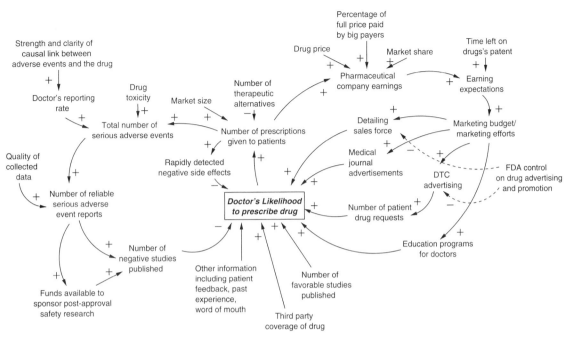

Figure 8.13
A more detailed model of physician prescription behavior.

Couturier incorporated into the system dynamics models the changes that were proposed by the IOM after the Vioxx events, the changes actually implemented in FDAAA, and the recommendations coming out of the STAMP-based causal analysis. One major difference was that the STAMP-based recommendations had a broader scope. While the IOM and FDAAA changes focused on the FDA, the STAMP analysis considered the contributions of all the components of the pharmaceutical safety control structure to the Vioxx events and the STAMP causal analysis led to recommendations for changes in nearly all of them.

Couturier concluded, not surprisingly, that most of the FDAAA changes are useful and will have the intended effects. He also determined that a few may be counterproductive and others need to be added. The added ones come from the fact that the IOM recommendations and the FDAAA focus on a single component of the system (the FDA). The FDA does not operate in a vacuum, and the proposed changes do not take into account the safety role played by other components in the system, particularly physicians. As a result, the pressures that led to the erosion of the overall system safety controls were left unaddressed and are likely to lead to changes in the system static and dynamic safety controls that will undermine the improvements implemented by FDAAA. See Couturier [43] for the complete results.

A potential contribution of such an analysis is the ability to consider the impact of multiple changes within the entire safety control structure. Less than effective controls may be implemented when they are created piecemeal to fix a current set of adverse events. Existing pressures and influences, not changed by the new procedures, can defeat the intent of the changes by leading to unintended and counterbalancing actions in the components of the safety control structure. STAMP-based analysis suggest how to reengineer the safety control structure as a whole to achieve the system goals, including both enhancing the safety of current drugs while at the same time encouraging the development of new drugs.

8.8   Comparison of STPA with Traditional Hazard Analysis Techniques

Few formal comparisons have been made yet between STPA and traditional techniques such as fault tree analysis and HAZOP. Theoretically, because STAMP extends the causality model underlying the hazard analysis, non-failures and additional causes should be identifiable, as well as the failure-related causes found by the traditional techniques. The few comparisons that have been made, both informal and formal, have confirmed this hypothesis.

In the use of STPA on the U.S. missile defense system, potential paths to inadvertent launch were identified that had not been identified by previous analyses or in extensive hazard analyses on the individual components of the system [BMDS]. Each element of the system had an active safety program, but the complexity and coupling introduced by their integration into a single system created new subtle and complex hazard scenarios. While the scenarios identified using STPA included those caused by potential component failures, as expected, scenarios were also identified that involved unsafe interactions among the components without any components actually failing—each operated according to its specified requirements, but the interactions could lead to hazardous system states. In the evaluation of this effort, two other advantages were noted:

1. The effort was bounded and predictable and assisted the engineers in scoping their efforts. Once all the control actions have been examined, the assessment is complete.

2. As the control structure is developed and the potential inadequate control actions are identified, they were able to prioritize required changes according to which control actions have the greatest role in keeping the system from transitioning to a hazardous state.

A paper published on this effort concluded:

The STPA safety assessment methodology . . . provided an orderly, organized fashion in which to conduct the analysis. The effort successfully assessed safety risks arising from the integration of the Elements. The assessment provided the information necessary to characterize the residual safety risk of hazards associated with the system. The analysis and supporting data provided management a sound basis on which to make risk acceptance decisions. Lastly, the assessment results were also used to plan mitigations for open safety risks. As changes are made to the system, the differences are assessed by updating the control structure diagrams and assessment analysis templates.

Another informal comparison was made in the ITA (Independent Technical Authority) analysis described in section 8.6. An informal review of the risks identified by using STPA showed that they included all the risks identified by the informal NASA risk analysis process using the traditional method common to such analyses. The additional risks identified by STPA appeared on the surface to be as important as those identified by the NASA analysis. As noted, there is no way to determine whether the less formal NASA process identified additional risks and discarded them for some reason or simply missed them.

A more careful comparison has also been made. JAXA (the Japanese Space Agency) and MIT engineers compared the use of STPA on a JAXA unmanned spacecraft (HTV) to transfer cargo to the International Space Station (ISS). Because human life is potentially involved (one hazard is collision with the International Space Station), rigorous NASA hazard analysis standards using fault trees and other analyses had been employed and reviewed by NASA. In an STPA analysis of the HTV used in an evaluation of the new technique for potential use at JAXA, all of the hazard causal factors identified by the fault tree analysis were identified also by STPA [88]. As with the BMDS comparison, additional causal factors were identified by STPA alone. These additional causal factors again involved those related to more sophisticated types of errors beyond simple component failures and those related to software and human errors.

Additional independent comparisons (not done by the author or her students) have been made between accident causal analysis methods comparing STAMP and more traditional methods. The results are described in chapter 11 on accident analysis based on STAMP.

8.9 Summary

Some new approaches to hazard and risk analysis based on STAMP and systems theory have been suggested in this chapter. We are only beginning to develop such techniques and hopefully others will work on alternatives and improvements. The only thing for sure is that applying the techniques developed for simple electromechanical systems to complex, human and software-intensive systems without fundamentally changing the foundations of the techniques is futile. New ideas are desperately needed if we are going to solve the problems and respond to the changes in the world of engineering described in chapter 1.

9 Safety-Guided Design

In the examples of STPA in the last chapter, the development of the design was assumed to occur independently. Most of the time, hazard analysis is done after the major design decisions have been made. But STPA can be used in a proactive way to help guide the design and system development, rather than as simply a hazard analysis technique on an existing design. This integrated design and analysis process is called *safety-guided design* (figure 9.1).

As the systems we build and operate increase in size and complexity, the use of sophisticated system engineering approaches becomes more critical. Important system-level (emergent) properties, such as safety, must be built into the design of these systems; they cannot be effectively added on or simply measured afterward. Adding barriers or protection devices after the fact is not only enormously more expensive, it is also much less effective than designing safety in from the beginning (see *Safeware*, chapter 16). This chapter describes the process of safety-guided design, which is enhanced by defining accident prevention as a control problem rather than a "prevent failures" problem. The next chapter shows how safety engineering and safety-guided design can be integrated into basic system engineering processes.

9.1 The Safety-Guided Design Process

One key to having a cost-effective safety effort is to embed it into a system engineering process from the very beginning and to design safety into the system as the design decisions are made. Once again, the process starts with the fundamental activities in chapter 7. After the hazards and system-level safety requirements and constraints have been identified; the design process starts:

1. Try to eliminate the hazards from the conceptual design.
2. If any of the hazards cannot be eliminated, then identify the potential for their control at the system level.

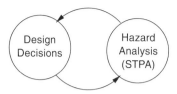

Figure 9.1
Safety-guided design entails tightly intertwining the design decisions and their analysis to support better decision making.

3. Create a system control structure and assign responsibilities for enforcing safety constraints. Some guidance for this process is provided in the operations and management chapters.

4. Refine the constraints and design in parallel.

   a. Identify potentially hazardous control actions by each of system components that would violate system design constraints using STPA step 1. Restate the identified hazard control actions as component design constraints.

   b. Using STPA Step 2, determine what factors could lead to a violation of the safety constraints.

   c. Augment the basic design to eliminate or control potentially unsafe control actions and behaviors.

   d. Iterate over the process, that is, perform STPA steps 1 and 2 on the new augmented design and continue to refine the design until all hazardous scenarios are eliminated, mitigated, or controlled.

The next section provides an example of the process. The rest of the chapter discusses safe design principles for physical processes, automated controllers, and human controllers.

9.2   An Example of Safety-Guided Design for an Industrial Robot

The process of safety-guided design and the use of STPA to support it is illustrated here with the design of an experimental Space Shuttle robotic Thermal Tile Processing System (TTPS) based on a design created for a research project at CMU [57].

The goal of the TTPS system is to inspect and waterproof the thermal protection tiles on the belly of the Space Shuttle, thus saving humans from a laborious task, typically lasting three to four months, that begins within minutes after the Shuttle lands and ends just prior to launch. Upon landing at either the Dryden facility in California or Kennedy Space Center in Florida, the orbiter is brought to either the Mate-Demate Device (MDD) or the Orbiter Processing Facility (OPF). These large structures provide access to all areas of the orbiters.

The Space Shuttle is covered with several types of heat-resistant tiles that protect the orbiter's aluminum skin during the heat of reentry. While the majority of the upper surfaces are covered with flexible insulation blankets, the lower surfaces are covered with silica tiles. These tiles have a glazed coating over soft and highly porous silica fibers. The tiles are 95 percent air by volume, which makes them extremely light but also makes them capable of absorbing a tremendous amount of water. Water in the tiles causes a substantial weight problem that can adversely affect launch and orbit capabilities for the shuttles. Because the orbiters may be exposed to rain during transport and on the launch pad, the tiles must be waterproofed. This task is accomplished through the use of a specialized hydrophobic chemical, DMES, which is injected into each tile. There are approximately 17,000 lower surface tiles covering an area that is roughly 25m × 40m.

In the standard process, DMES is injected into a small hole in each tile by a handheld tool that pumps a small quantity of chemical into the nozzle. The nozzle is held against the tile and the chemical is forced through the tile by a pressurized nitrogen purge for several seconds. It takes about 240 hours to waterproof the tiles on an orbiter. Because the chemical is toxic, human workers have to wear heavy suits and respirators while injecting the chemical and, at the same time, maneuvering in a crowded work area. One goal for using a robot to perform this task was to eliminate a very tedious, uncomfortable, and potentially hazardous human activity.

The tiles must also be inspected. A goal for the TTPS was to inspect the tiles more accurately than the human eye and therefore reduce the need for multiple inspections. During launch, reentry, and transport, a number of defects can occur on the tiles in the form of scratches, cracks, gouges, discoloring, and erosion of surfaces. The examination of the tiles determines if they need to be replaced or repaired. The typical procedures involve visual inspection of each tile to see if there is any damage and then assessment and categorization of the defects according to detailed check-lists. Later, work orders are issued for repair of individual tiles.

Like any design process, safety-guided design starts with identifying the goals for the system and the constraints under which the system must operate. The high-level goals for the TTPS are to:

1. Inspect the thermal tiles for damage caused during launch, reentry, and transport

2. Apply waterproofing chemicals to the thermal tiles

Environmental constraints delimit how these goals can be achieved and identifying those constraints, particularly the safety constraints, is an early goal in safety-guided design.

The environmental constraints on the system design stem from physical properties of the Orbital Processing Facility (OPF) at KSC, such as size constraints on the physical system components and the necessity of any mobile robotic components to deal with crowded work areas and for humans to be in the area. Example work area environmental constraints for the TTPS are:

EA1: The work areas of the Orbiter Processing Facility (OPF) can be very crowded. The facilities provide access to all areas of the orbiters through the use of intricate platforms that are laced with plumbing, wiring, corridors, lifting devices, and so on. After entering the facility, the orbiters are jacked up and leveled. Substantial structure then swings around and surrounds the orbiter on all sides and at all levels. With the exception of the jack stands that support the orbiters, the floor space directly beneath the orbiter is initially clear but the surrounding structure can be very crowded.

EA2: The mobile robot must enter the facility through personnel access doors 1.1 meters (42″) wide. The layout within the OPF allows a length of 2.5 meters (100″) for the robot. There are some structural beams whose heights are as low as 1.75 meters (70″), but once under the orbiter the tile heights range from about 2.9 meters to 4 meters. The compact roll-in form of the mobile system must maneuver these spaces and also raise its inspection and injection equipment up to heights of 4 meters to reach individual tiles while still meeting a 1 millimeter accuracy requirement.

EA3: Additional constraints involve moving around the crowded workspace. The robot must negotiate jack stands, columns, work stands, cables, and hoses. In addition, there are hanging cords, clamps, and hoses. Because the robot might cause damage to the ground obstacles, cable covers will be used for protection and the robot system must traverse these covers.

Other design constraints on the TTPS include:

- Use of the TTPS must not negatively impact the flight schedules of the orbiters more than that of the manual system being replaced.

- Maintenance costs of the TTPS must not exceed x dollars per year.

- Use of the TTPS must not cause or contribute to an unacceptable loss (accident) as defined by Shuttle management.

As with many systems, prioritizing the hazards by severity is enough in this case to assist the engineers in making decisions during design. Sometimes a preliminary hazard analysis is performed using a risk matrix to determine how much effort will be put into eliminating or controlling the hazards and in making tradeoffs in design. Likelihood, at this point, is unknowable but some type of surrogate, like mitigatibility, as demonstrated in section 10.3.4, could be used. In the TTPS example, severity plus the NASA policy described earlier is adequate. To decide not to consider some of the hazards at all would be pointless and dangerous at this stage of development as likelihood is not determinable. As the design proceeds and decisions must be made, specific additional information may be found to be useful and acquired at that time. After the system design is completed, if it is determined that some hazards cannot be adequately handled or the compromises required to handle them are too great; then the limitations would be documented (as described in chapter 10) and decisions would have to be made at that point about the risks of using the system. At that time, however, the information necessary to make those decisions will more likely be available than before the development process begins.

After the hazards are identified, system-level safety-related requirements and design constraints are derived from them. As an example, for hazard H7 (inadequate thermal protection), a system-level safety design constraint is that the mobile robot processing must not result in any tiles being missed in the inspection or waterproofing process. More detailed design constraints will be generated during the safety-guided design process.

To get started, a general system architecture must be selected (figure 9.2). Let's assume that the initial TTPS architecture consists of a mobile base on which tools will be mounted, including a manipulator arm that performs the processing and contains the vision and waterproofing tools. This very early decision may be changed after the safety-guided design process starts, but some very basic initial assumptions are necessary to get going. As the concept development and detailed design process proceeds, information generated about hazards and design tradeoffs may lead to changes in the initial configuration. Alternatively, multiple design configurations may be considered in parallel.

In the initial candidate architecture (control structure), a decision is made to introduce a human operator in order to supervise robot movement as so many of the hazards are related to movement. At the same time, it may be impractical for an operator to monitor all the activities so the first version of the system architecture is to have the TTPS control system in charge of the non-movement activities and to have both the TTPS and the control room operator share control of movement. The safety-guided design process, including STPA, will identify the implications of this decision and will assist in analyzing the allocation of tasks to the various components to determine the safety tradeoffs involved.

In the candidate starting architecture (control structure), there is an automated robot work planner to provide the overall processing goals and tasks for the

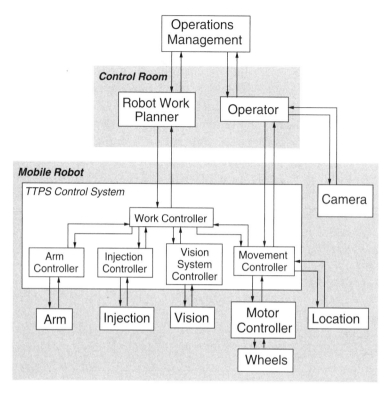

Figure 9.2
A candidate structure for the TTPS.

TTPS. A location system is needed to provide information to the movement controller about the current location of the robot. A camera is used to provide information to the human controller, as the control room will be located at a distance from the orbiter. The role of the other components should be obvious.

The proposed design has two potential movement controllers, so coordination problems will have to be eliminated. The operator could control all movement, but that may be considered impractical given the processing requirements. To assist with this decision process, engineers may create a *concept of operations* and perform a human task analysis [48, 122].

The safety-guided design process, including STPA, will identify the implications of the basic decisions in the candidate tasks and will assist in analyzing the allocation of tasks to the various components to determine the safety tradeoffs involved.

The design process is now ready to start. Using the information already specified, particularly the general functional responsibilities assigned to each component, designers will identify potentially hazardous control actions by each of the system components that could violate the safety constraints, determine the causal factors that could lead to these hazardous control actions, and prevent or control them in the system design. The process thus involves a top-down identification of scenarios in which the safety constraints could be violated. The scenarios can then be used to guide more detailed design decisions.

In general, safety-guided design involves first attempting to eliminate the hazard from the design and, if that is not possible or requires unacceptable tradeoffs, reducing the likelihood the hazard will occur, reducing the negative consequences of the hazard if it does occur, and implementing contingency plans for limiting damage. More about design procedures is presented in the next section.

As design decisions are made, an STPA-based hazard analysis is used to inform these decisions. Early in the system design process, little information is available, so the hazard analysis will be very general at first and will be refined and augmented as additional information emerges through the system design activities.

For the example, let's focus on the robot instability hazard. The first goal should be to eliminate the hazard in the system design. One way to eliminate potential instability is to make the robot base so heavy that it cannot become unstable, no matter how the manipulator arm is positioned. A heavy base, however, could increase the damage caused by the base coming into contact with a human or object or make it difficult for workers to manually move the robot out of the way in an emergency situation. An alternative solution is to make the base long and wide so the moment created by the operation of the manipulator arm is compensated by the moments created by base supports that are far from the robot's center of mass. A long and wide base could remove the hazard but may violate the environmental constraints in the facility layout, such as the need to maneuver through doors and in the crowded OPF.

The environmental constraint EA2 above implies a maximum length for the robot of 2.5 meters and a width no larger than 1.1 meter. Given the required maximum extension length of the manipulator arm and the estimated weight of the equipment that will need to be carried on the mobile base, a calculation might show that the length of the robot base is sufficient to prevent any longitudinal instability, but that the width of the base is not sufficient to prevent lateral instability.

If eliminating the hazard is determined to be impractical (as in this case) or not desirable for some reason, the alternative is to identify ways to control it. The decision to try to control it may turn out not to be practical or later may seem less satisfactory than increasing the weight (the solution earlier discarded). All decisions should remain open as more information is obtained about alternatives and back-tracking is an option.

At the initial stages in design, we identified only the general hazards—for example, instability of the robot base and the related system design constraint that the mobile base must not be capable of falling over under worst-case operational conditions. As design decisions are proposed and analyzed, they will lead to additional refinements in the hazards and the design constraints.

For example, a potential solution to the stability problem is to use lateral stabilizer legs that are deployed when the manipulator arm is extended but must be retracted when the robot base moves. Let's assume that a decision is made to at least consider this solution. That potential design decision generates a new refined hazard from the high-level stability hazard (H2):

> **H2.1:** The manipulator arm is extended while the stabilizer legs are not fully extended.

Damage to the mobile base or other equipment around the OPF is another potential hazard introduced by the addition of the legs if the mobile base moves while the stability legs are extended. Again, engineers would consider whether this hazard could be eliminated by appropriate design of the stability legs. If it cannot, then that is a second additional hazard that must be controlled in the design with a corresponding design constraint that the mobile base must not move with the stability legs extended.

There are now two new refined hazards that must be translated into design constraints:

1. The manipulator arm must never be extended if the stabilizer legs are not extended.
2. The mobile base must not move with the stability legs extended.

STPA can be used to further refine these constraints and to evaluate the resulting designs. In the process, the safety control structure will be refined and perhaps changed. In this case, a controller must be identified for the stabilizer legs, which were previously not in the design. Let's assume that the legs are controlled by the TTPS movement controller (figure 9.3).

Using the augmented control structure, the remaining activities in STPA are to identify potentially hazardous control actions by each of the system components that could violate the safety constraints, determine the causal factors that could lead to these hazardous control actions, and prevent or control them in the system design. The process thus involves a top-down identification of scenarios in which the safety

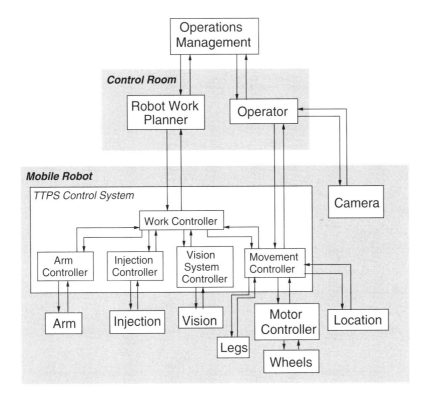

Figure 9.3
A refined control structure for the TTPS.

constraints could be violated so that they can be used to guide more detailed design decisions.

The unsafe control actions associated with the stability hazard are shown in figure 9.4. Movement and thermal tile processing hazards are also identified in the table. Combining similar entries for H1 in the table leads to the following unsafe control actions by the leg controller with respect to the instability hazard:

1. The leg controller does not command a deployment of the stabilizer legs before the arm is extended.

2. The leg controller commands a retraction of the stabilizer legs before the manipulator arm is fully stowed.

3. The leg controller commands a retraction of the stabilizer legs after the arm has been extended or commands a retraction of the stabilizer legs before the manipulator arm is stowed.

HAZARD1: Arm extended while legs retracted

HAZARD2: Legs extended during movement

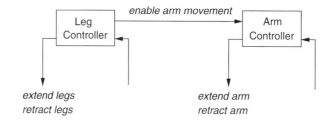

Command	Not Providing Causes Hazard	Providing Causes Hazard	Timing/Sequencing Causes Hazard	Stopped Too Soon or Applied Too Long
extend legs	Legs not extended before arm extended **H1**	Extend legs during movement **H2**	Extend arm before legs extended **H1**	Stop before fully extended **H1**
retract legs	Not retracted before movement **H2**	Retract while arm extended **H1**	Retract legs before arm fully stowed **H1**	Stop while still partially extended **H1**

Command	Not Providing Causes Hazard	Providing Causes Hazard	Timing/Sequencing Causes Hazard	Stopped Too Soon or Applied Too Long
extend arm	Not Hazardous	Extend arm when legs retracted **H1**	Extend arm before legs fully extended **H1**	(tile processing hazard)
retract arm	Not retracted before movement **H2**	(tile processing hazard)	(tile processing hazard)	Stop retraction before arm fully stowed and movement starts or legs retracted **H1 H2**

Figure 9.4
STPA step 1 for the stability and movement hazards related to the leg and arm control.

4. The leg controller stops extension of the stabilizer legs before they are fully extended.

and by the arm controller:

1. The arm controller extends the manipulator arm when the stabilizer legs are not extended or before they are fully extended.

The inadequate control actions can be restated as system safety constraints on the controller behavior (whether the controller is automated or human):

1. The leg controller must ensure the stabilizer legs are fully extended before arm movements are enabled.

2. The leg controller must not command a retraction of the stabilizer legs when the manipulator arm is not in a fully stowed position.

3. The leg controller must command a deployment of the stabilizer legs before arm movements are enabled; the leg controller must not command a retraction of the stabilizer legs before the manipulator arm is stowed.

4. The leg controller must not stop the leg extension until the legs are fully extended.

Similar constraints will be identified for all hazardous commands: for example, the arm controller must not extend the manipulator arm before the stabilizer legs are fully extended.

These system safety constraints might be enforced through physical interlocks, human procedures, and so on. Performing STPA step 2 will provide information during detailed design (1) to evaluate and compare the different design choices, (2) to design the controllers and design fault tolerance features for the system, and (3) to guide the test and verification procedures (or training for humans). As design decisions and safety constraints are identified, the functional specifications for the controllers can be created.

To produce detailed scenarios for the violation of safety constraints, the control structure is augmented with process models. The preliminary design of the process models comes from the information necessary to ensure the system safety constraints hold. For example, the constraint that the arm controller must not enable manipulator movement before the stabilizer legs are completely extended implies there must be some type of feedback to the arm controller to determine when the leg extension has been completed.

While a preliminary functional decomposition of the system components is created to start the process, as more information is obtained from the hazard analysis and the system design continues, this decomposition may be altered to optimize fault tolerance and communication requirements. For example, at this point the need

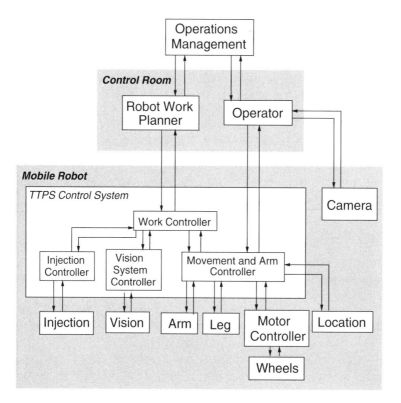

Figure 9.5
A further refined control structure for the TTPS.

for the process models of the leg and arm controllers to be consistent and the communication required to achieve this goal may lead the designers to decide to combine the leg and arm controllers (figure 9.5).

Causal factors for the stability hazard being violated can be determined using STPA step 2. Feedback about the position of the legs is clearly critical to ensure that the process model of the state of the stabilizer legs is consistent with the actual state. The movement and arm controller cannot assume the legs are extended simply because a command was issued to extend them. The command may not be executed or may only be executed partly. One possible scenario, for example, involves an external object preventing the complete extension of the stabilizer legs. In that case, the robot controller (either human or automated) may assume the stabilizer legs are extended because the extension motors have been powered up (a common type of design error). Subsequent movement of the manipulator arm would then violate the identified safety constraints. Just as the analysis assists in refining the component safety constraints (functional requirements), the causal analysis can be used to further refine those requirements and to design the control algorithm, the control loop components, and the feedback necessary to implement them.

Many of the causes of inadequate control actions are so common that they can be restated as general design principles for safety-critical control loops. The requirement for feedback about whether a command has been executed in the previous paragraph is one of these. The rest of this chapter presents those general design principles.

9.3   Designing for Safety

Hazard analysis using STPA will identify application-specific safety design constraints that must be enforced by the control algorithm. For the thermal-tile processing robot, a safety constraint identified above is that the manipulator arm must never be extended if the stabilizer legs are not fully extended. Causal analysis (step 2 of STPA) can identify specific causes for the constraint to be violated and design features can be created to eliminate or control them.

More general principles of safe control algorithm functional design can also be identified by using the general causes of accidents as defined in STAMP (and used in STPA step 2), general engineering principles, and common design flaws that have led to accidents in the past.

Accidents related to software or system logic design often result from incompleteness and unhandled cases in the functional design of the controller. This incompleteness can be considered a requirements or functional design problem. Some requirements completeness criteria were identified in *Safeware* and specified using a state machine model. Here those criteria plus additional design criteria are translated into functional design principles for the components of the control loop.

In STAMP, accidents are caused by inadequate control. The controllers can be human or physical. This section focuses on design principles for the components of the control loop that are important whether a human is in the loop or not. Section 9.4 describes extra safety-related design principles that apply for systems that include human controllers. We cannot "design" human controllers, but we can design the environment or context in which they operate, and we can design the procedures they use, the control loops in which they operate, the processes they control, and the training they receive.

9.3.1   Controlled Process and Physical Component Design

Protection against component failure accidents is well understood in engineering. Principles for safe design of common hardware systems (including sensors and actuators) with standard safety constraints are often systematized and encoded in checklists for an industry, such as mechanical design or electrical design. In addition, most engineers have learned about the use of redundancy and overdesign (safety margins) to protect against component failures.

These standard design techniques are still relevant today but provide little or no protection against component interaction accidents. The added complexity of redundant designs may even increase the occurrence of these accidents. Figure 9.6 shows the design precedence described in *Safeware*. The highest precedence is to eliminate the hazard. If the hazard cannot be eliminated, then its likelihood of occurrence should be reduced, the likelihood of it leading to an accident should be reduced and, at the lowest precedence, the design should reduce the potential damage incurred. Clearly, the higher the precedence level, the more effective and less costly will be the safety design effort. As there is little that is new here that derives from using the STAMP causality model, the reader is referred to *Safeware* and standard engineering references for more information.

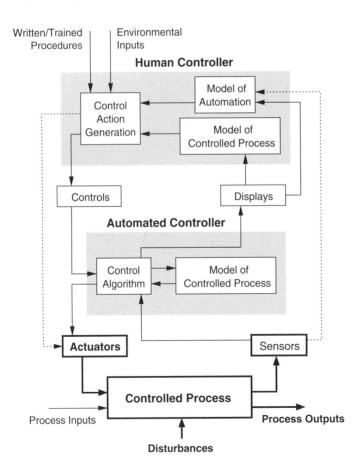

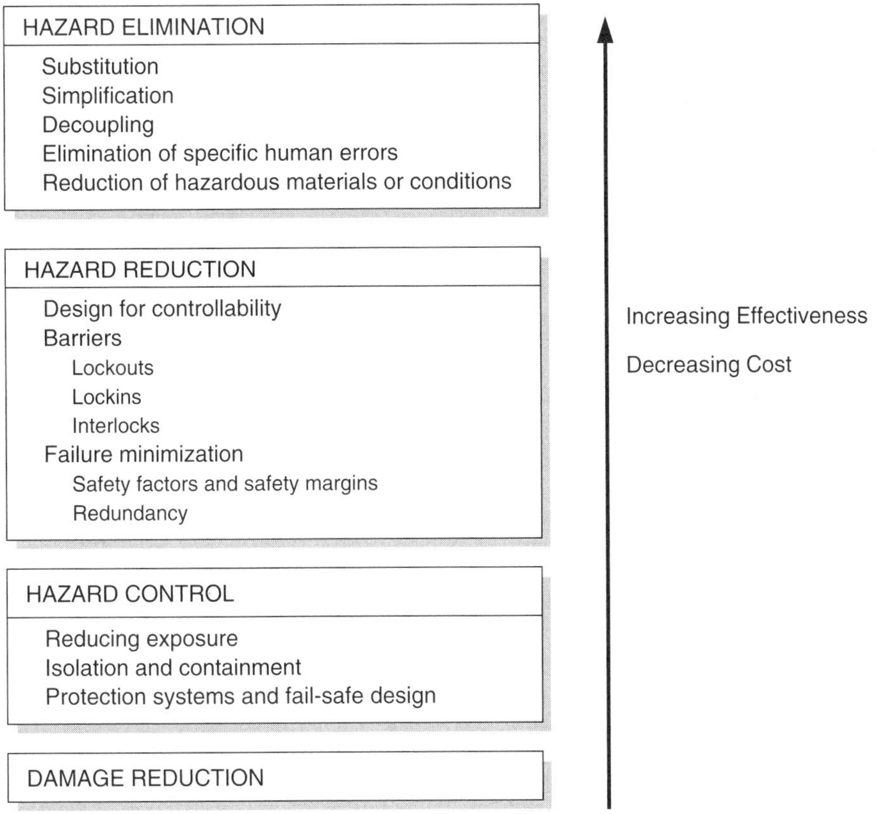

HAZARD ELIMINATION

   Substitution
   Simplification
   Decoupling
   Elimination of specific human errors
   Reduction of hazardous materials or conditions

HAZARD REDUCTION

   Design for controllability
   Barriers
      Lockouts
      Lockins
      Interlocks
   Failure minimization
      Safety factors and safety margins
      Redundancy

HAZARD CONTROL

   Reducing exposure
   Isolation and containment
   Protection systems and fail-safe design

DAMAGE REDUCTION

Increasing Effectiveness

Decreasing Cost

Figure 9.6
Basic system safety design precedence.

9.3.2   Functional Design of the Control Algorithm

Design for safety includes more than simply the physical components but also the control components. We start by considering the design of the control algorithm.

The controller algorithm is responsible for processing inputs and feedback, initializing and updating the process model, and using the process model plus other knowledge and inputs to produce control outputs. Each of these is considered in turn.

Designing and Processing Inputs and Feedback

The basic function of the algorithm is to implement a feedback control loop, as defined by the controller responsibilities, along with appropriate checks to detect internal or external failures or errors.

Feedback is critical for safe control. Without feedback, controllers do not know whether their control actions were received and performed properly or whether

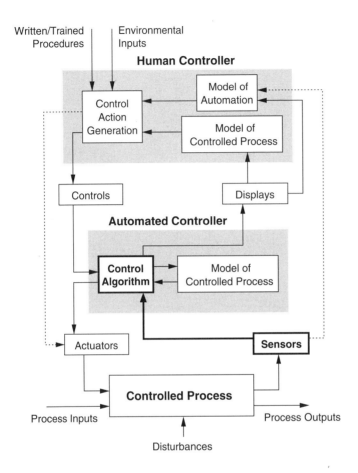

these commands were effective in achieving the controllers' goals. Feedback is also critical in detecting errors and failures, both errors in the controllers' own actions and failures or faults in the controlled system. Finally, feedback is important in updating process models and in learning about the system and how it will respond to a variety of situations.

Updating process models requires feedback about the current state of the system and any changes that occur. In a system where rapid response is necessary, timing requirements must be placed on the feedback information that the controller uses to make decisions. In addition, when task performance requires or implies a need for the controller to assess timeliness of information, the feedback should include time and date information.

Hazard analysis using STPA will provide information about the types of feedback needed and when. Some additional guidance can be provided to the designer, once again, using general safety design principles.

The controller must be designed to respond appropriately to the arrival of any possible (i.e., detectable by the sensors) input at any time as well as the lack of an expected input over a given time period. Humans are better (and more flexible) than automated controllers at this task. Often automation is not designed to handle input arriving unexpectedly, for example, a target detection report from a radar that was previously sent a message to shut down.

All inputs should be checked for out-of-range or unexpected values and a response designed into the control algorithm. A surprising number of losses still occur due to software not being programmed to handle unexpected inputs.

In addition, the time bounds (minimum and maximum) for every input should be checked and appropriate behavior provided in case the input does not arrive within these bounds. There should also be a response for the non-arrival of an input within a given amount of time (a timeout) for every variable in the process model. The controller must also be designed to respond to excessive inputs (overload conditions) in a safe way.

Because sensors and input channels can fail, there should be a minimum-arrival-rate check for each physically distinct communication path, and the controller should have the ability to query its environment with respect to inactivity over a given communication path. Traditionally these queries are called *sanity* or *health checks*. Care needs to be taken, however, to ensure that the design of the response to a health check is distinct from the normal inputs and that potential hardware failures cannot impact the sanity checks. As an example of the latter, in June 1980 warnings were received at the U.S. command and control headquarters that a major nuclear attack had been launched against the United States [180]. The military prepared for retaliation, but the officers at command headquarters were able to ascertain from direct contact with warning sensors that no incoming missile had been detected and the alert was canceled. Three days later, the same thing happened again. The false alerts were caused by the failure of a computer chip in a multiplexor system that formats messages sent out continuously to command posts indicating that communication circuits are operating properly. This *health check* message was designed to report that there were 000 ICBMs and 000 SLBMs detected. Instead, the integrated circuit failure caused some of the zeros to be replaced with twos. After the problem was diagnosed, the message formats were changed to report only the status of the communication system and nothing about detecting ballistic missiles. Most likely, the developers thought it would be easier to have one common message format but did not consider the impact of erroneous hardware behavior.

STAMP identifies inconsistency between the process model and the actual system state as a common cause of accidents. Besides incorrect feedback, as in the example early warning system, a common way for the process model to become inconsistent with the state of the actual process is for the controller to assume that an output command has been executed when it has not. The TTPS controller, for example, assumes that because it has sent a command to extend the stabilizer legs, the legs will, after a suitable amount of time, be extended. If commands cannot be executed for any reason, including time outs, controllers have to know about it. To detect errors and failures in the actuators or controlled process, there should be an input (feedback) that the controller can use to detect the effect of any output on the process.

a way to verify that the control law is working.

This feedback, however, should not simply be an indication that the command arrived at the controlled process—for example, the command to open a valve was received by the valve, but that the valve actually opened. An explosion occurred in a U.S. Air Force system due to overpressurization when a relief valve failed to open after the operator sent a command to open it. Both the position indicator light and open indicator light were illuminated on the control board. Believing the primary valve had opened, the operator did not open the secondary valve, which was to be used if the primary valve failed. A post-accident examination discovered that the indicator light circuit was wired to indicate presence of a signal at the valve, but it did not indicate valve position. The indicator therefore showed only that the activation button had been pushed, not that the valve had opened. An extensive quantitative safety analysis of this design had assumed a low probability of simultaneous failure for the two relief valves, but it ignored the possibility of a design error in the electrical wiring; the probability of the design error was not quantifiable. Many other accidents have involved a similar design flaw, including Three Mile Island.

When the feedback associated with an output is received, the controller must be able to handle the normal response as well as deal with feedback that is missing, too late, too early, or has an unexpected value.

Initializing and Updating the Process Model

Because the process model is used by the controller to determine what control commands to issue and when, the accuracy of the process model with respect to the controlled process is critical. As noted earlier, many software-related losses have resulted from such inconsistencies. STPA will identify which process model variables are critical to safety; the controller design must ensure that the controller receives and processes updates for these variables in a timely manner.

Sometimes normal updating of the process model is done correctly by the controller, but problems arise in initialization at startup and after a temporary shutdown. The process model must reflect the actual process state at initial startup and after a restart. It seems to be common, judging from the number of incidents and accidents that have resulted, for software designers to forget that the world

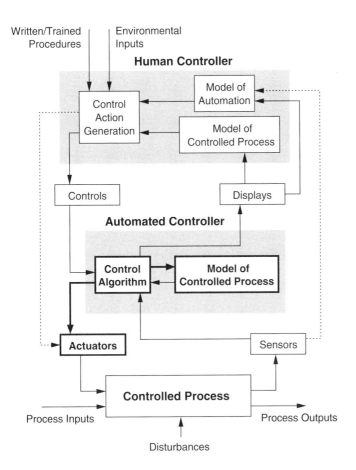

continues to change even though the software may not be operating. When the computer controlling a process is temporarily shut down, perhaps for maintenance or updating of the software, it may restart with the assumption that the controlled process is still in the state it was when the software was last operating. In addition, assumptions may be made about when the operation of the controller will be started, which may be violated. For example, an assumption may be made that a particular aircraft system will be powered up and initialized before takeoff and appropriate default values used in the process model for that case. In the event it was not started at that time or was shut down and then restarted after takeoff, the default startup values in the process model may not apply and may be hazardous.

Consider the mobile tile-processing robot at the beginning of this chapter. The mobile base may be designed to allow manually retracting the stabilizer legs if an emergency occurs while the robot is servicing the tiles and the robot must be physically moved out of the way. When the robot is restarted, the controller may assume that the stabilizer legs are still extended and arm movements may be commanded that would violate the safety constraints.

The use of an *unknown* value can assist in protecting against this type of design flaw. At startup and after temporary shutdown, process variables that reflect the state of the controlled process should be initialized with the value *unknown* and updated when new feedback arrives. This procedure will result in resynchronizing the process model and the controlled process state. The control algorithm must also account, of course, for the proper behavior in case it needs to use a process model variable that has the *unknown* value.

Just as timeouts must be specified and handled for basic input processing as described earlier, the maximum time the controller waits until the first input after startup needs to be determined and what to do if this time limit is violated. Once again, while human controllers will likely detect such a problem eventually, such as a failed input channel or one that was not restarted on system startup, computers will patiently wait forever if they are not given instructions to detect such a timeout and to respond to it.

In general, the system and control loop should start in a safe state. Interlocks may need to be initialized or checked to be operational at system startup, including startup after temporarily overriding the interlocks.

Finally the behavior of the controller with respect to input received before startup, after shutdown, or while the controller is temporarily disconnected from the process (offline) must be considered and it must be determined if this information can be safely ignored or how it will be stored and later processed if it cannot. One factor in the loss of an aircraft that took off from the wrong runway at Lexington Airport, for example, is that information about temporary changes in the airport taxiways was not reflected in the airport maps provided to the crew. The information about the changes, which was sent by the National Flight Data Center, was received by the map-provider computers at a time when they were not online, leading to airport charts that did not match the actual state of the airport. The document control system software used by the map provider was designed to only make reports of information received during business hours Monday through Friday [142].

Producing Outputs

The primary responsibility of the process controller is to produce commands to fulfill its control responsibilities. Again, the STPA hazard analysis and safety-guided design process will produce the application-specific behavioral safety requirements and constraints on controller behavior to ensure safety. But some general guidelines are also useful.

One general safety constraint is that the behavior of an automated controller should be deterministic: it should exhibit only one behavior for arrival of any input

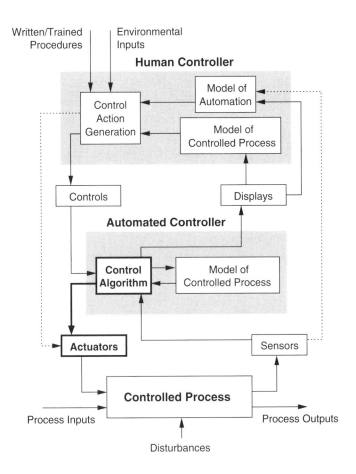

in a particular state. While it is easy to design software with nondeterministic behavior and, in some cases, actually has some advantages from a software point of view, nondeterministic behavior makes testing more difficult and, more important, much more difficult for humans to learn how an automated system works and to monitor it. If humans are expected to control or monitor an automated system or an automated controller, then the behavior of the automation should be deterministic.

Just as inputs can arrive faster than they can be processed by the controller, the absorption rate of the actuators and recipients of output from the controller must be considered. Again, the problem usually arises when a fast output device (such as a computer) is providing input to a slower device, such as a human. Contingency action must be designed when the output absorption rate limit is exceeded.

Three additional general considerations in the safe design of controllers are data age, latency, and fault handling.

Data age: No inputs or output commands are valid forever. The control loop design must account for inputs that are no longer valid and should not be used by the controller and for outputs that cannot be executed immediately. All inputs used in the generation of output commands must be properly limited in the time they are used and marked as obsolete once that time limit has been exceeded. At the same time, the design of the control loop must account for outputs that are not executed within a given amount of time. As an example of what can happen when data age is not properly handled in the design, an engineer working in the cockpit of a B-1A aircraft issued a CLOSE WEAPONS BAY DOOR command during a test. At the time, a mechanic working on the door had activated a mechanical inhibit on it. The CLOSE DOOR command was not executed, but it remained active. Several hours later, when the door maintenance was completed, the mechanical inhibit was removed. The door closed unexpectedly, killing the worker [64].

Latency: Latency is the time interval during which receipt of new information cannot change an output even though it arrives prior to the output. While latency time can be reduced by using various types of design techniques, it cannot be eliminated completely. Controllers need to be informed about the arrival of feedback affecting previously issued commands and, if possible, provided with the ability to undo or to mitigate the effects of the now unwanted command.

Fault-handling: Most accidents involve off-nominal processing modes, including startup and shutdown and fault handling. The design of the control loop should assist the controller in handling these modes and the designers need to focus particular attention on them.

The system design may allow for performance degradation and may be designed to fail into safe states or to allow partial shutdown and restart. Any fail-safe behavior that occurs in the process should be reported to the controller. In some cases, automated systems have been designed to fail so gracefully that human controllers are not aware of what is going on until they need to take control and may not be prepared to do so. Also, hysteresis needs to be provided in the control algorithm for transitions between off-nominal and nominal processing modes to avoid *ping-ponging* when the conditions that caused the controlled process to leave the normal state still exist or recur.

Hazardous functions have special requirements. Clearly, interlock failures should result in the halting of the functions they are protecting. In addition, the control algorithm design may differ after failures are detected, depending on whether the controller outputs are hazard-reducing or hazard-increasing. A hazard-increasing output is one that moves the controlled process to a more hazardous state, for example, arming a weapon. A hazard-reducing output is a command that leads to a reduced risk state, for example, safing a weapon or any other command whose purpose is to maintain safety.

If a failure in the control loop, such as a sensor or actuator, could inhibit the production of a hazard-reducing command, there should be multiple ways to trigger such commands. On the other hand, multiple inputs should be required to trigger commands that can lead to hazardous states so they are not inadvertently issued. Any failure should inhibit the production of a hazard-increasing command. As an example of the latter condition, loss of the ability of the controller to receive input, such as failure of a sensor, that might inhibit the production of a hazardous output should prevent such an output from being issued.

9.4   Special Considerations in Designing for Human Controllers

The design principles in section 9.3 apply when the controller is automated or human, particularly when designing procedures for human controllers to follow. But humans do not always follow procedures, nor should they. We use humans to control systems because of their flexibility and adaptability to changing conditions and to the incorrect assumptions made by the designers. Human error is an inevitable and unavoidable consequence. But appropriate design can assist in reducing human error and increasing safety in human-controlled systems.

Human error is not random. It results from basic human mental abilities and physical skills combined with the features of the tools being used, the tasks assigned, and the operating environment. We can use what is known about human mental abilities and design the other aspects of the system—the tools, the tasks, and the operating environment—to reduce and control human error to a significant degree. The previous section described general principles for safe design. This section focuses on additional design principles that apply when humans control, either directly or indirectly, safety-critical systems.

9.4.1   Easy but Ineffective Approaches

One simple solution for engineers is to simply use human factors checklists. While many such checklists exist, they often do not distinguish among the qualities they enhance, which may not be related to safety and may even conflict with safety. The only way such universal guidelines could be useful is if all design qualities were complementary and achieved in exactly the same way, which is not the case. Qualities are conflicting and require design tradeoffs and decisions about priorities.

Usability and safety, in particular, are often conflicting; an interface that is easy to use may not necessarily be safe. As an example, a common guideline is to ensure that a user must enter data only once and that the computer can access that data if needed later for the same task or for different tasks [192]. Duplicate entry, however, is required for the computer to detect entry errors unless the errors are so extreme that they violate reasonableness criteria. A small slip usually cannot be detected and such entry errors have led to many accidents. Multiple entry of critical data can prevent such losses.

As another example, a design that involves displaying data or instructions on a screen for an operator to check and verify by pressing the *enter* button minimizes the typing an operator must do. Over time, however, and after few errors are detected, operators will get in the habit of pressing the enter key multiple times in rapid succession. This design feature has been implicated in many losses. For example, the Therac-25 was a linear accelerator that overdosed multiple patients during radiation therapy. In the original Therac-25 design, operators were required to enter the treatment parameters at the treatment site as well as on the computer console. After the operators complained about the duplication, the parameters entered at the treatment site were instead displayed on the console and the operator needed only to press the *return* key if they were correct. Operators soon became accustomed to pushing the return key quickly the required number of times without checking the parameters carefully.

The second easy but not very effective solution is to write procedures for human operators to follow and then assume the engineering job is done. Enforcing the following of procedures is unlikely, however, to lead to a high level of safety.

Dekker notes what he called the "Following Procedures Dilemma" [50]. Operators must balance between adapting procedures in the face of unanticipated conditions versus sticking to procedures rigidly when cues suggest they should be adapted. If human controllers choose the former, that is, they adapt procedures when it appears the procedures are wrong, a loss may result when the human controller does not have complete knowledge of the circumstances or system state. In this case, the humans will be blamed for deviations and nonadherence to the procedures. On the other hand, if they stick to procedures (the control algorithm provided) rigidly when the procedures turn out to be wrong, they will be blamed for their inflexibility and the application of the rules in the wrong context. Hindsight bias is often involved in identifying what the operator should have known and done.

Insisting that operators always follow procedures does not guarantee safety although it does usually guarantee that there is someone to blame—either for following the procedures or for not following them—when things go wrong. Safety comes from controllers being skillful in judging when and how procedures apply. As discussed in chapter 12, organizations need to monitor adherence to procedures not simply to enforce compliance but to understand how and why the gap between procedures and practice grows and to use that information to redesign both the system and the procedures [50].

Section 8.5 of chapter 8 describes important differences between human and automated controllers. One of these differences is that the control algorithm used by humans is dynamic. This dynamic aspect of human control is why humans are kept in systems. They provide the flexibility to deviate from procedures when it turns out the assumptions underlying the engineering design are wrong. But with this flexibility comes the possibility of unsafe changes in the dynamic control algorithm and raises new design requirements for engineers and system designers to understand the reason for such unsafe changes and prevent them through appropriate system design.

Just as engineers have the responsibility to understand the hazards in the physical systems they are designing and to control and mitigate them, engineers also must understand how their system designs can lead to human error and how they can design to reduce errors.

Designing to prevent human error requires some basic understanding about the role humans play in systems and about human error.

9.4.2   The Role of Humans in Control Systems

Humans can play a variety of roles in a control system. In the simplest cases, they create the control commands and apply them directly to the controlled process. For a variety of reasons, particularly speed and efficiency, the system may be designed with a computer between the human controller and the system. The computer may exist only in the feedback loop to process and present data to the human operator. In other systems, the computer actually issues the control instructions with the human operator either providing high-level supervision of the computer or simply monitoring the computer to detect errors or problems.

An unanswered question is what is the best role for humans in safety-critical process control. There are three choices beyond direct control: the human can monitor an automated control system, the human can act as a backup to the automation, or the human and automation can both participate in the control through some type of partnership. These choices are discussed in depth in *Safeware* and are only summarized here.

Unfortunately for the first option, humans make very poor monitors. They cannot sit and watch something without active control duties for any length of time and maintain vigilance. Tasks that require little active operator behavior may result in lowered alertness and can lead to complacency and overreliance on the automation. Complacency and lowered vigilance are exacerbated by the high reliability and low failure rate of automated systems.

But even if humans could remain vigilant while simply sitting and monitoring a computer that is performing the control tasks (and usually doing the right thing), Bainbridge has noted the irony that automatic control systems are installed because they can do the job better than humans, but then humans are assigned the task of monitoring the automated system [14]. Two questions arise:

1. The human monitor needs to know what the correct behavior of the controlled or monitored process should be; however, in complex modes of operation—for example, where the variables in the process have to follow a particular trajectory over time—evaluating whether the automated control system is performing correctly requires special displays and information that may only be available from the automated system being monitored. How will human monitors know when the computer is wrong if the only information they have comes from that computer? In addition, the information provided by an automated controller is more indirect, which may make it harder for humans to get a clear picture of the system: Failures may be silent or masked by the automation.

2. If the decisions can be specified fully, then a computer can make them more quickly and accurately than a human. How can humans monitor such a system? Whitfield and Ord found that, for example, air traffic controllers' appreciation of the traffic situation was reduced at the high traffic levels made feasible by using computers [198]. In such circumstances, humans must monitor the automated controller at some metalevel, deciding whether the computer's decisions are acceptable rather than completely correct. In case of a disagreement, should the human or the computer be the final arbiter?

Employing humans as backups is equally ineffective. Controllers need to have accurate process models to control effectively, but not being in active control leads to a degradation of their process models. At the time they need to intervene, it may take a while to "get their bearings"—in other words, to update their process models so that effective and safe control commands can be given. In addition, controllers need both manual and cognitive skills, but both of these decline in the absence of practice. If human backups need to take over control from automated systems, they may be unable to do so effectively and safely. Computers are often introduced into safety-critical control loops because they increase system reliability, but at the same time, that high reliability can provide little opportunity for human controllers to practice and maintain the skills and knowledge required to intervene when problems *do* occur.

It appears, at least for now, that humans will have to provide direct control or will have to share control with automation unless adequate confidence can be established in the automation to justify eliminating monitors completely. Few systems exist today where such confidence can be achieved when safety is at stake. The problem then becomes one of finding the correct partnership and allocation of tasks between humans and computers. Unfortunately, this problem has not been solved, although some guidelines are presented later.

One of the things that make the problem difficult is that it is not just a matter of splitting responsibilities. Computer control is changing the cognitive demands on human controllers. Humans are increasingly supervising a computer rather than directly monitoring the process, leading to more cognitively complex decision making. Automation logic complexity and the proliferation of control modes are confusing humans. In addition, whenever there are multiple controllers, the requirements for cooperation and communication are increased, not only between the human and the computer but also between humans interacting with the same computer, for example, the need for coordination among multiple people making entries to the computer. The consequences can be increased memory demands, new skill and knowledge requirements, and new difficulties in the updating of the human's process models.

A basic question that must be answered and implemented in the design is who will have the final authority if the human and computers disagree about the proper control actions. In the loss of an Airbus 320 while landing at Warsaw in 1993, one of the factors was that the automated system prevented the pilots from activating the braking system until it was too late to prevent crashing into a bank built at the end of the runway. This automation feature was a protection device included to prevent the reverse thrusters accidentally being deployed in flight, a presumed cause of a previous accident. For a variety of reasons, including water on the runway causing the aircraft wheels to hydroplane, the criteria used by the software logic to determine that the aircraft had landed were not satisfied by the feedback received by the automation [133]. Other incidents have occurred where the pilots have been confused about who is in control, the pilot or the automation, and found themselves fighting the automation [181].

One common design mistake is to set a goal of automating everything and then leaving some miscellaneous tasks that are difficult to automate for the human controllers to perform. The result is that the operator is left with an arbitrary collection of tasks for which little thought was given to providing support, particularly support for maintaining accurate process models. The remaining tasks may, as a consequence, be significantly more complex and error-prone. New tasks may be added, such as maintenance and monitoring, that introduce new types of errors. Partial automation, in fact, may not reduce operator workload but merely change the type of demands on the operator, leading to potentially increased workload. For example, cockpit automation may increase the demands on the pilots by creating a lot of data entry tasks during approach when there is already a lot to do. These automation interaction tasks also create "heads down" work at a time when increased monitoring of nearby traffic is necessary.

By taking away the easy parts of the operator's job, automation may make the more difficult ones even harder [14]. One causal factor here is that taking away or changing some operator tasks may make it difficult or even impossible for the operators to receive the feedback necessary to maintain accurate process models.

When designing the automation, these factors need to be considered. A basic design principle is that automation should be designed to augment human abilities, not replace them, that is, to aid the operator, not to take over.

To design safe automated controllers with humans in the loop, designers need some basic knowledge about human error related to control tasks. In fact, Rasmussen has suggested that the term *human error* be replaced by considering such events as *human–task mismatches*.

9.4.3 Human Error Fundamentals

Human error can be divided into the general categories of slips and mistakes [143, 144]. Basic to the difference is the concept of *intention* or desired action. A *mistake* is an error in the intention, that is, an error that occurs during the planning of an action. A *slip*, on the other hand, is an error in carrying out the intention. As an example, suppose an operator decides to push button A. If the operator instead pushes button B, then it would be called a slip because the action did not match the intention. If the operator pushed A (carries out the intention correctly), but it turns out that the intention was wrong, that is, button A should *not* have been pushed, then this is called a mistake.

Designing to prevent slips involves applying different principles than designing to prevent mistakes. For example, making controls look very different or placing them far apart from each other may reduce slips, but not mistakes. In general, designing to reduce mistakes is more difficult than reducing slips, which is relatively straightforward.

One of the difficulties in eliminating planning errors or mistakes is that such errors are often only visible in hindsight. With the information available at the time, the decisions may seem reasonable. In addition, planning errors are a necessary side effect of human problem-solving ability. Completely eliminating mistakes or planning errors (if possible) would also eliminate the need for humans as controllers.

Planning errors arise from the basic human cognitive ability to solve problems. Human error in one situation is human ingenuity in another. Human problem solving rests on several unique human capabilities, one of which is the ability to create hypotheses and to test them and thus create new solutions to problems not previously considered. These hypotheses, however, may be wrong. Rasmussen has suggested that human error is often simply unsuccessful experiments in an unkind environment, where an unkind environment is defined as one in which it is not possible for the human to correct the effects of inappropriate variations in performance before they lead to unacceptable consequences [166]. He concludes that human performance is a balance between a desire to optimize skills and a willingness to accept the risk of exploratory acts.

A second basic human approach to problem solving is to try solutions that worked in other circumstances for similar problems. Once again, this approach is not always successful but the inapplicability of old solutions or plans (learned procedures) may not be determinable without the benefit of hindsight.

The ability to use these problem-solving methods provides the advantages of human controllers over automated controllers, but success is not assured. Designers, if they understand the limitations of human problem solving, can provide assistance in the design to avoid common pitfalls and enhance human problem solving. For example, they may provide ways for operators to obtain extra information or to test hypotheses safely. At the same time, there are some additional basic human cognitive characteristics that must be considered.

Hypothesis testing can be described in terms of basic feedback control concepts. Using the information in the process model, the controller generates a hypothesis about the controlled process. A test composed of control actions is created to generate feedback useful in evaluating the hypothesis, which in turn is used to update the process model and the hypothesis.

When controllers have no accurate diagnosis of a problem, they must make provisional assessments of what is going on based on uncertain, incomplete, and often contradictory information [50]. That provisional assessment will guide their information gathering, but it may also lead to over attention to confirmatory evidence when processing feedback and updating process models while, at the same time, discounting information that contradicts their current diagnosis. Psychologists call this phenomenon *cognitive fixation*. The alternative is called *thematic vagabonding*, where the controller jumps around from explanation to explanation, driven by the loudest or latest feedback or alarm and never develops a coherent assessment of what is going on. Only hindsight can determine whether the controller should have abandoned one explanation for another: Sticking to one assessment can lead to more progress in many situations than jumping around and not pursuing a consistent planning process.

Plan continuation is another characteristic of human problem solving related to cognitive fixation. Commitment to a preliminary diagnosis can lead to sticking with the original plan even though the situation has changed and calls for a different plan. Orisanu [149] notes that early cues that suggest an initial plan is correct are usually very strong and unambiguous, helping to convince people to continue the plan. Later feedback that suggests the plan should be abandoned is typically more ambiguous and weaker. Conditions may deteriorate gradually. Even when controllers receive and acknowledge this feedback, the new information may not change their plan, especially if abandoning the plan is costly in terms of organizational and economic consequences. In the latter case, it is not surprising that controllers will seek and focus on confirmatory evidence and will need a lot of contradictory evidence to justify changing their plan.

Cognitive fixation and plan continuation are compounded by stress and fatigue. These two factors make it more difficult for controllers to juggle multiple hypotheses about a problem or to project a situation into the future by mentally simulating the effects of alternative plans [50].

Automated tools can be designed to assist the controller in planning and decision making, but they must embody an understanding of these basic cognitive limitations and assist human controllers in overcoming them. At the same time, care must be taken that any simulation or other planning tools to assist human problem solving do not rest on the same incorrect assumptions about the system that led to the problems in the first place.

Another useful distinction is between errors of omission and errors of commission. Sarter and Woods [181] note that in older, less complex aircraft cockpits, most pilot errors were *errors of commission* that occurred as a result of a pilot control action. Because the controller, in this case the pilot, took a direct action, he or she is likely to check that the intended effect of the action has actually occurred. The short feedback loops allow the operators to repair most errors before serious consequences result. This type of error is still the prevalent one for relatively simple devices.

In contrast, studies of more advanced automation in aircraft find that *errors of omission* are the dominant form of error [181]. Here the controller does not implement a control action that is required. The operator may not notice that the automation has done something because that automation behavior was not explicitly invoked by an operator action. Because the behavioral changes are not expected, the human controller is less likely to pay attention to relevant indications and feedback, particularly during periods of high workload.

Errors of omission are related to the change of human roles in systems from direct controllers to monitors, exception handlers, and supervisors of automated controllers. As their roles change, the cognitive demands may not be reduced but instead may change in their basic nature. The changes tend to be more prevalent at high-tempo and high-criticality periods. So while some types of human errors have declined, new types of errors have been introduced.

The difficulty and perhaps impossibility of eliminating human error does not mean that greatly improved system design in this respect is not possible. System design can be used to take advantage of human cognitive capabilities and to minimize the errors that may result from them. The rest of the chapter provides some principles to create designs that better support humans in controlling safety-critical processes and reduce human errors.

9.4.4 Providing Control Options

If the system design goal is to make humans responsible for safety in control systems, then they must have adequate flexibility to cope with undesired and unsafe behavior and not be constrained by inadequate control options. Three general design principles apply: design for redundancy, design for incremental control, and design for error tolerance.

Design for redundant paths: One helpful design feature is to provide multiple physical devices and logical paths to ensure that a single hardware failure or software error cannot prevent the operator from taking action to maintain a safe system state and avoid hazards. There should also be multiple ways to change

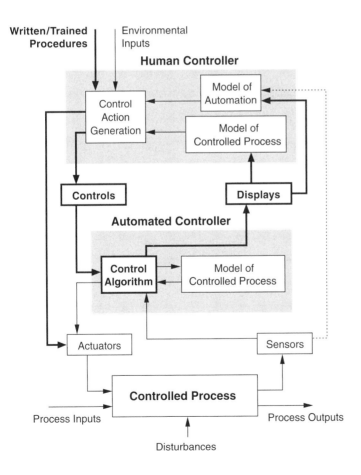

from an unsafe to a safe state, but only one way to change from an unsafe to a safe state.

Design for incremental control: Incremental control makes a system easier to control, both for humans and computers, by performing critical steps incrementally rather than in one control action. The common use of incremental *arm*, *aim*, *fire* sequences is an example. The controller should have the ability to observe the system and get feedback to test the validity of the assumptions and models upon which the decisions are made. The system design should also provide the controller with compensating control actions to allow modifying or aborting previous control actions before significant damage is done. An important consideration in designing for controllability in general is to lower the time pressures on the controllers, if possible.

The design of incremental control algorithms can become complex when a human controller is controlling a computer, which is controlling the actual physical process, in a stressful and busy environment, such as a military aircraft. If one of the commands in an incremental control sequence cannot be executed within a specified period of time, the human operator needs to be informed about any delay or postponement or the entire sequence should be canceled and the operator informed. At the same time, interrupting the pilot with a lot of messages that may not be critical at a busy time could also be dangerous. Careful analysis is required to determine when multistep controller inputs can be preempted or interrupted before they are complete and when feedback should occur that this happened [90].

Design for error tolerance: Rasmussen notes that people make errors all the time, but we are able to detect and correct them before adverse consequences occur [165]. System design can limit people's ability to detect and recover from their errors. He defined a system design goal of *error tolerant systems*. In these systems, errors are observable (within an appropriate time limit) and they are reversible before unacceptable consequences occur. The same applies to computer errors: they should be observable and reversible.

The general goal is to allow controllers to monitor their own performance. To achieve this goal, the system design needs to:

1. Help operators monitor their actions and recover from errors.

2. Provide feedback about actions operators took and their effects, in case the actions were inadvertent. Common examples are echoing back operator inputs or requiring confirmation of intent.

3. Allow for recovery from erroneous actions. The system should provide control options, such as compensating or reversing actions, and enough time for recovery actions to be taken before adverse consequences result.

Incremental control, as described earlier, is a type of error-tolerant design technique.

9.4.5 Matching Tasks to Human Characteristics

In general, the designer should tailor systems to human requirements instead of the opposite. Engineered systems are easier to change in their behavior than are humans.

Because humans without direct control tasks will lose vigilance, the design should combat lack of alertness by designing human tasks to be stimulating and varied, to provide good feedback, and to require active involvement of the human controllers in most operations. Maintaining manual involvement is important, not just for alertness but also in getting the information needed to update process models.

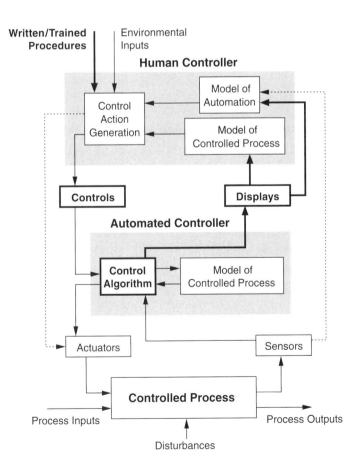

Maintaining active engagement in the tasks means that designers must distinguish between providing help to human controllers and taking over. The human tasks should not be oversimplified and tasks involving passive or repetitive actions should be minimized. Allowing latitude in how tasks are accomplished will not only reduce monotony and error proneness, but can introduce flexibility to assist operators in improvising when a problem cannot be solved by only a limited set of behaviors. Many accidents have been avoided when operators jury-rigged devices or improvised procedures to cope with unexpected events. Physical failures may cause some paths to become nonfunctional and flexibility in achieving goals can provide alternatives.

Designs should also be avoided that require or encourage *management by exception*, which occurs when controllers wait for alarm signals before taking action. Management by exception does not allow controllers to prevent disturbances by looking for early warnings and trends in the process state. For operators to anticipate undesired events, they need to continuously update their process models. Experiments by Swaanenburg and colleagues found that management by exception is not the strategy adopted by human controllers as their normal supervisory mode [196]. Avoiding management by exception requires active involvement in the control task and adequate feedback to update process models. A display that provides only an overview and no detailed information about the process state, for example, may not provide the information necessary for detecting imminent alarm conditions.

Finally, if designers expect operators to react correctly to emergencies, they need to design to support them in these tasks and to help fight some basic human tendencies described previously such as cognitive fixation and plan continuation. The system design should support human controllers in decision making and planning activities during emergencies.

9.4.6 Designing to Reduce Common Human Errors

Some human errors are so common and unnecessary that there is little excuse for not designing to prevent them. Care must be taken though that the attempt to reduce erroneous actions does not prevent the human controller from intervening in an emergency when the assumptions made during design about what should and should not be done turn out to be incorrect.

One fundamental design goal is to make safety-enhancing actions easy, natural, and difficult to omit or do wrong. In general, the design should make it more difficult for the human controller to operate unsafely than safely. If safety-enhancing actions are easy, they are less likely to be bypassed intentionally or accidentally. Stopping an unsafe action or leaving an unsafe state should be possible with a single keystroke that moves the system into a safe state. The design should make fail-safe actions easy and natural, and difficult to avoid, omit, or do wrong.

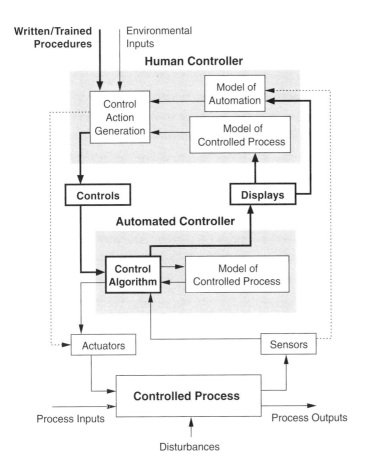

In contrast, two or more unique operator actions should be required to start any potentially hazardous function or sequence of functions. Hazardous actions should be designed to minimize the potential for inadvertent activation; they should not, for example, be initiated by pushing a single key or button (see the preceding discussion of incremental control).

The general design goal should be to enhance the ability of the human controller to act safely while making it more difficult to behave unsafely. Initiating a potentially unsafe process change, such as a spacecraft launch, should require multiple keystrokes or actions while stopping a launch should require only one.

Safety may be enhanced by using procedural safeguards, where the operator is instructed to take or avoid specific actions, or by designing safeguards into the system. The latter is much more effective. For example, if the potential error involves leaving out a critical action, either the operator can be instructed to always take that action or the action can be made an integral part of the process. A typical error during maintenance is not to return equipment (such as safety interlocks) to the operational mode. The accident sequence at Three Mile Island was initiated by such an error. An action that is isolated and has no immediate relation to the "gestalt" of the repair or testing task is easily forgotten. Instead of stressing the need to be careful (the usual approach), change the system by integrating the act physically into the task, make detection a physical consequence of the tool design, or change operations planning or review. That is, change design or management rather than trying to change the human [162].

To enhance decision making, references should be provided for making judgments, such as marking meters with safe and unsafe limits. Because humans often revert to stereotype and cultural norms, such norms should be followed in design. Keeping things simple, natural, and similar to what has been done before (not making gratuitous design changes) is a good way to avoid errors when humans are working under stress, are distracted, or are performing tasks while thinking about something else.

To assist in preventing sequencing errors, controls should be placed in the sequence in which they are to be used. At the same time, similarity, proximity, interference, or awkward location of critical controls should be avoided. Where operators have to perform different classes or types of control actions, sequences should be made as dissimilar as possible.

Finally, one of the most effective design techniques for reducing human error is to design so that the error is not physically possible or so that errors are obvious. For example, valves can be designed so they cannot be interchanged by making the connections different sizes or preventing assembly errors by using asymmetric or male and female connections. Connection errors can also be made obvious by color coding. Amazingly, in spite of hundreds of deaths due to misconnected tubes in hospitals that have occurred over decades, such as a feeding tube inadvertently connected to a tube that is inserted in a patient's vein, regulators, hospitals, and tube manufacturers have taken no action to implement this standard safety design technique [80].

9.4.7 Support in Creating and Maintaining Accurate Process Models

Human controllers who are supervising automation have two process models to maintain: one for the process being controlled by the automation and one for the automated controller itself. The design should support human controllers in maintaining both of these models. An appropriate goal here is to provide humans with the facilities to experiment and learn about the systems they are controlling, either directly or indirectly. Operators should also be allowed to maintain manual involvement to update process models, to maintain skills, and to preserve self-confidence. Simply observing will degrade human supervisory skills and confidence.

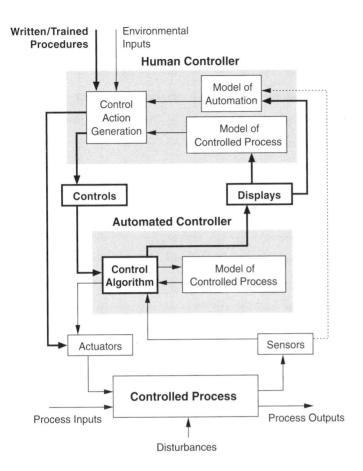

When human controllers are supervising automated controllers, the automation has extra design requirements. The control algorithm used by the automation must be learnable and understandable. Two common design flaws in automated controllers are inconsistent behavior by the automation and unintended side effects.

Inconsistent Behavior
Carroll and Olson define a consistent design as one where a similar task or goal is associated with similar or identical actions [35]. Consistent behavior on the part of the automated controller makes it easier for the human providing supervisory control to learn how the automation works, to build an appropriate process model for it, and to anticipate its behavior.

An example of inconsistency, detected in an A320 simulator study, involved an aircraft go-around below 100 feet above ground level. Sarter and Woods found that pilots failed to anticipate and realize that the autothrust system did not arm when they selected takeoff/go-around (TOGA) power under these conditions because it did so under all other circumstances where TOGA power is applied [181].

Another example of inconsistent automation behavior, which was implicated in an A320 accident, is a protection function that is provided in all automation configurations except the specific mode (in this case altitude acquisition) in which the autopilot was operating [181].

Human factors for critical systems have most extensively been studied in aircraft cockpit design. Studies have found that consistency is most important in high-tempo, highly dynamic phases of flight where pilots have to rely on their automatic systems to work as expected without constant monitoring. Even in more low-pressure situations, consistency (or predictability) is important in light of the evidence from pilot surveys that their normal monitoring behavior may change on high-tech flight decks [181].

Pilots on conventional aircraft use a highly trained instrument-scanning pattern of recurrently sampling a given set of basic flight parameters. In contrast, some A320 pilots report that they no longer scan anymore but allocate their attention within and across cockpit displays on the basis of expected automation states and behaviors. Parameters that are not expected to change may be neglected for a long time [181]. If the automation behavior is not consistent, errors of omission may occur where the pilot does not intervene when necessary.

In section 9.3.2, determinism was identified as a safety design feature for automated controllers. Consistency, however, requires more than deterministic behavior. If the operator provides the same inputs but different outputs (behaviors) result for some reason other than what the operator has done (or may even know about), then the behavior is inconsistent from the operator viewpoint even though it is deterministic. While the designers may have good reasons for including inconsistent behavior in the automated controller, there should be a careful tradeoff made with the potential hazards that could result.

Unintended Side Effects

Incorrect process models can result when an action intended to have one effect has an additional side effect not easily anticipated by the human controller. An example occurred in the Sarter and Woods A320 aircraft simulator study cited earlier. Because the approach to the destination airport is such a busy time for the pilots and the automation requires so much heads down work, pilots often program the automation as soon as the air traffic controllers assign them a runway. Sarter and Woods found that the experienced pilots in their study were not aware that entering a runway change *after* entering data for the assigned approach results in the deletion by the automation of all the previously entered altitude and speed constraints, even though they may still apply.

Once again, there may be good reason for the automation designers to include such side effects, but they need to consider the potential for human error that can result.

Mode Confusion

Modes define mutually exclusive sets of automation behaviors. Modes can be used to determine how to interpret inputs or to define required controller behavior. Four general types of modes are common: controller operating modes, supervisory modes, display modes, and controlled process modes.

Controller operating modes define sets of related behavior in the controller, such as shutdown, nominal behavior, and fault-handling.

Supervisory modes determine who or what is controlling the component at any time when multiple supervisors can assume control responsibilities. For example, a flight guidance system in an aircraft may be issued direct commands by the pilot(s) or by another computer that is itself being supervised by the pilot(s). The movement controller in the thermal tile processing system might be designed to be in either manual supervisory mode (by a human controller) or automated mode (by the TTPS task controller). Coordination of control actions among multiple supervisors can be defined in terms of these supervisory modes. Confusion about the current supervisory mode can lead to hazardous system behavior.

A third type of common mode is a *display mode*. The display mode will affect the information provided on the display and how the user interprets that information.

A final type of mode is the operating mode of the controlled process. For example, the mobile thermal tile processing robot may be in a moving mode (between work areas) or in a work mode (in a work area and servicing tiles, during which time it may be controlled by a different controller). The value of this mode may determine whether various operations—for example, extending the stabilizer legs or the manipulator arm—are safe.

Early automated systems had a fairly small number of independent modes. They provided a passive background on which the operator would act by entering target data and requesting system operations. They also had only one overall mode setting for each function performed. Indications of currently active mode and of transitions between modes could be dedicated to one location on the display.

The consequences of breakdown in mode awareness were fairly small in these system designs. Operators seemed able to detect and recover from erroneous actions relatively quickly before serious problems resulted. Sarter and Woods conclude that, in most cases, mode confusion in these simpler systems are associated with errors of commission, that is, with errors that require a controller action in order for the problem to occur [181]. Because the human controller has taken an explicit action, he or she is likely to check that the intended effect of the action has actually occurred. The short feedback loops allow the controller to repair most errors quickly, as noted earlier.

The flexibility of advanced automation allows designers to develop more complicated, mode-rich systems. The result is numerous mode indications often spread over multiple displays, each containing just that portion of mode status data corresponding to a particular system or subsystem. The designs also allow for interactions across modes. The increased capabilities of automation can, in addition, lead to increased delays between user input and feedback about system behavior.

These new mode-rich systems increase the need for and difficulty of maintaining mode awareness, which can be defined in STAMP terms as keeping the controlled-system operating mode in the controller's process model consistent with the actual controlled system mode. A large number of modes challenges human ability to maintain awareness of active modes, armed modes, interactions between environmental status and mode behavior, and interactions across modes. It also increases the difficulty of error or failure detection and recovery.

Calling for systems with fewer or less complex modes is probably unrealistic. Simplifying modes and automation behavior often requires tradeoffs with precision or efficiency and with marketing demands from a diverse set of customers [181]. Systems with accidental (unnecessary) complexity, however, can be redesigned to reduce the potential for human error without sacrificing system capabilities. Where tradeoffs with desired goals are required to eliminate potential mode confusion errors, system and interface design, informed by hazard analysis, can help find solutions that require the fewest tradeoffs. For example, accidents most often occur during transitions between modes, particularly normal and nonnormal modes, so they should have more stringent design constraints applied to them.

Understanding more about particular types of mode confusion errors can assist with design. Two common types leading to problems are interface interpretation modes and indirect mode changes.

Interface Interpretation Mode Confusion: Interface mode errors are the classic form of mode confusion error:

1. *Input-related errors:* The software interprets user-entered values differently than intended.

2. *Output-related errors:* The software maps multiple conditions onto the same output, depending on the active controller mode, and the operator interprets the interface incorrectly.

A common example of an input interface interpretation error occurs with many word processors where the user may think they are in INSERT mode but instead they are in INSERT AND DELETE mode or in COMMAND mode and their input is interpreted in a different way and results in different behavior than they intended.

A more complex example occurred in what is believed to be a cause of an A320 aircraft accident. The crew directed the automated system to fly in the TRACK/FLIGHT PATH ANGLE mode, which is a combined mode related to both lateral (TRACK) and vertical (FLIGHT PATH ANGLE) navigation:

> When they were given radar vectors by the air traffic controller, they may have switched from the TRACK to the HDG SEL mode to be able to enter the heading requested by the controller. However, pushing the button to change the lateral mode also automatically changes the vertical mode from FLIGHT PATH ANGLE to VERTICAL SPEED—the mode switch button affects both lateral and vertical navigation. When the pilots subsequently entered "33" to select the desired flight path angle of 3.3 degrees, the automation interpreted their input as a desired vertical speed of 3300 ft. This was not intended by the pilots who were not aware of the active "interface mode" and failed to detect the problem. As a consequence of the too steep descent, the airplane crashed into a mountain [181].

An example of an output interface mode problem was identified by Cook et al. [41] in a medical operating room device with two operating modes: warmup and normal. The device starts in warmup mode when turned on and changes from normal mode to warmup mode whenever either of two particular settings is adjusted by the operator. The meaning of alarm messages and the effect of controls are different in these two modes, but neither the current device operating mode nor a change in mode is indicated to the operator. In addition, four distinct alarm-triggering conditions are mapped onto two alarm messages so that the same message has different meanings depending on the operating mode. In order to understand what internal condition triggered the message, the operator must infer which malfunction is being indicated by the alarm.

Several design constraints can assist in reducing interface interpretation errors. At a minimum, any mode used to control interpretation of the supervisory interface should be annunciated to the supervisor. More generally, the current operating mode of the automation should be displayed at all times. In addition, any change of operating mode should trigger a change in the current operating mode reflected in the interface and thus displayed to the operator, that is, the annunciated mode must be consistent with the internal mode.

A stronger design choice, but perhaps less desirable for various reasons, might be not to condition the interpretation of the supervisory interface on modes at all. Another possibility is to simplify the relationships between modes, for example in the A320, the lateral and vertical modes might be separated with respect to the heading select mode. Other alternatives are to make the required inputs different to lessen confusion (such as 3.3 and 3,300 rather than 33), or the mode indicator on the control panel could be made clearer as to the current mode. While simply annunciating the mode may be adequate in some cases, annunciations can easily to missed for a variety of reasons and additional design features should be considered.

Mode Confusion Arising from Indirect Mode Changes: Indirect mode changes occur when the automation changes mode without an explicit instruction or direct command by the operator. Such transitions may be triggered on conditions in the automation, such as preprogrammed envelope protection. They may also result from sensor input to the computer about the state of the computer-controlled process, such as achievement of a preprogrammed target or an armed mode with a preselected mode transition. An example of the latter is a mode in which the autopilot might command leveling off of the plane once a particular altitude is reached: the operating mode of the aircraft (leveling off) is changed when the altitude is reached without a direct command to do so by the pilot. In general, the problem occurs when activating one mode can result in the activation of different modes depending on the system status at the time.

There are four ways to trigger a mode change:

1. The automation supervisor explicitly selects a new mode.

2. The automation supervisor enters data (such as a target altitude) or a command that leads to a mode change:

    a. Under all conditions.

    b. When the automation is in a particular state

    c. When the automation's controlled system model or environment is in a particular state.

3. The automation supervisor does not do anything, but the automation logic changes mode as a result of a change in the system it is controlling.

4. The automation supervisor selects a mode change but the automation does something else, either because of the state of the automation at the time or the state of the controlled system.

Again, errors related to mode confusion are related to problems that human supervisors of automated controllers have in maintaining accurate process models. Changes in human controller behavior in highly automated systems, such as the changes in pilot scanning behavior described earlier, are also related to these types of mode confusion error.

Behavioral expectations about the automated controller behavior are formed based on the human supervisors' knowledge of the input to the automation and on their process models of the automation. Gaps or misconceptions in this model may interfere with predicting and tracking indirect mode transitions or with understanding the interactions among modes.

An example of an accident that has been attributed to an indirect mode change occurred while an A320 was landing in Bangalore, India [182]. The pilot's selection of a lower altitude while the automation was in the ALTITUDE ACQUISITION mode resulted in the activation of the OPEN DESCENT mode, where speed is controlled only by the pitch of the aircraft and the throttles go to idle. In that mode, the automation ignores any preprogrammed altitude constraints. To maintain pilot-selected speed without power, the automation had to use an excessive rate of descent, which led to the aircraft crashing short of the runway.

Understanding how this could happen is instructive in understanding just how complex mode logic can get. There are three different ways to activate OPEN DESCENT mode on the A320:

1. Pull the altitude knob after selecting a lower altitude.

2. Pull the speed knob when the aircraft is in EXPEDITE mode.

3. Select a lower altitude while in ALTITUDE ACQUISITION mode.

It was the third condition that is suspected to have occurred. The pilot must not have been aware the aircraft was within 200 feet of the previously entered target altitude, which triggers ALTITUDE ACQUISITION mode. He therefore may not have expected selection of a lower altitude at that time to result in a mode transition and did not closely monitor his mode annunciations during this high workload time. He discovered what happened ten seconds before impact, but that was too late to recover with the engines at idle [182].

Other factors contributed to his not discovering the problem until too late, one of which is the problem in maintaining consistent process models when there are multiple controllers as discussed in the next section. The pilot flying (PF) had disengaged his flight director[1] during approach and was assuming the pilot not flying (PNF) would do the same. The result would have been a mode configuration in which airspeed is automatically controlled by the autothrottle (the SPEED mode), which is the recommended procedure for the approach phase of flight. The PNF never turned off his flight director, however, and the OPEN DESCENT mode became active when a lower altitude was selected. This indirect mode change led to the hazardous state and eventually the accident, as noted earlier. But a complicating factor was that each pilot only received an indication of the status of his own flight

---

1. The flight director is automation that gives visual cues to the pilot via an easily interpreted display of the aircraft's flight path. The preprogrammed path, automatically computed, furnishes the steering commands necessary to obtain and hold a desired path.

director and not all the information necessary to determine whether the desired mode would be engaged. The lack of feedback and resulting incomplete knowledge of the aircraft state (incorrect aircraft process model) contributed to the pilots not detecting the unsafe state in time to correct it.

Indirect mode transitions can be identified in software designs. What to do in response to identifying them or deciding not to include them in the first place is more problematic and the tradeoffs and mitigating design features must be considered for each particular system. The decision is just one of the many involving the benefits of complexity in system design versus the hazards that can result.

Coordination of Multiple Controller Process Models

When multiple controllers are engaging in coordinated control of a process, inconsistency between their process models can lead to hazardous control actions. Careful design of communication channels and coordinated activity is required. In aircraft, this coordination, called crew resource management, is accomplished through careful design of the roles of each controller to enhance communication and to ensure consistency among their process models.

A special case of this problem occurs when one human controller takes over for another. The handoff of information about both the state of the controlled process and any automation being supervised by the human must be carefully designed.

Thomas describes an incident involving loss of communication for an extended time between ground air traffic control and an aircraft [199]. In this incident, a ground controller had taken over after a controller shift change. Aircraft are passed from one air traffic control sector to another through a carefully designed set of exchanges, called a *handoff*, during which the aircraft is told to switch to the radio frequency for the new sector. When, after a shift change the new controller gave an instruction to a particular aircraft and received no acknowledgment, the controller decided to take no further action; she assumed that the lack of acknowledgment was an indication that the aircraft had already switched to the new sector and was talking to the next controller.

Process model coordination during shift changes is partially controlled in a *position relief briefing*. This briefing normally covers all aircraft that are currently on the correct radio frequency or have not checked in yet. When the particular flight in question was not mentioned in the briefing, the new controller interpreted that as meaning that the aircraft was no longer being controlled by this station. She did not call the next controller to verify this status because the aircraft had not been mentioned in the briefing.

The design of the air traffic control system includes redundancy to try to avoid errors—if the aircraft does not check in with the next controller, then that controller would call her. When she saw the aircraft (on her display) leave her airspace and no such call was received, she interpreted that as another indication that the aircraft was indeed talking to the next controller.

A final factor implicated in the loss of communication was that when the new controller took over, there was little traffic at the aircraft's altitude and no danger of collision. Common practice for controllers in this situation is to initiate an early handoff to the next controller. So although the aircraft was only halfway through her sector, the new controller assumed an early handoff had occurred.

An additional causal factor in this incident involves the way controllers track which aircraft have checked in and which have already been handed off to the next controller. The old system was based on printed flight progress strips and included a requirement to mark the strip when an aircraft had checked in. The new system uses electronic flight progress strips to display the same information, but there is no standard method to indicate the check-in has occurred. Instead, each individual controller develops his or her own personal method to keep track of this status. In this particular loss of communication case, the controller involved would type a symbol in a comment area to mark any aircraft that she had already handed off to the next sector. The controller that was relieved reported that he usually relied on his memory or checked a box to indicate which aircraft he was communicating with.

That a carefully designed and coordinated process such as air traffic control can suffer such problems with coordinating multiple controller process models (and procedures) attests to the difficulty of this design problem and the necessity for careful design and analysis.

9.4.8 Providing Information and Feedback

Designing feedback in general was covered in section 9.3.2. This section covers feedback design principles specific to human controllers. Important problems in designing feedback include what information should be provided, how to make the feedback process more robust, and how the information should be presented to human controllers.

Types of Feedback

Hazard analysis using STPA will provide information about the types of feedback needed and when. Some additional guidance can be provided to the designer, once again, using general safety design principles.

Two basic types of feedback are needed:

1. *The state of the controlled process:* This information is used to (1) update the controllers' process models and (2) to detect faults and failures in the other parts of the control loop, system, and environment.

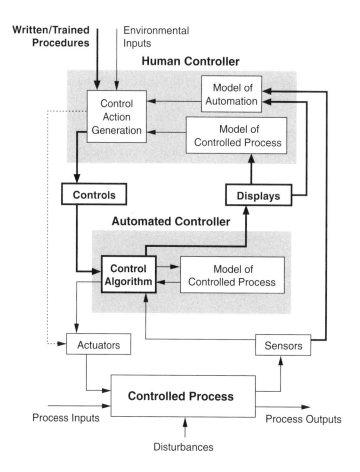

2. *The effect of the controllers' actions:*   This feedback is used to detect human errors. As discussed in the section on design for error tolerance, the key to making errors observable—and therefore remediable—is to provide feedback about them. This feedback may be in the form of information about the effects of controller actions, or it may simply be information about the action itself on the chance that it was inadvertent.

Updating Process Models

Updating process models requires feedback about the current state of the system and any changes that occur. In a system where rapid response by operators is necessary, timing requirements must be placed on the feedback information that the controller uses to make decisions. In addition, when task performance requires or implies need for the controller to assess timeliness of information, the feedback display should include time and date information associated with data.

When a human controller is supervising or monitoring automation, the automation should provide an indication to the controller and to bystanders that it is functioning. The addition of a light to the power interlock example in chapter 8 is a simple example of this type of feedback. For robot systems, bystanders should be signaled when the machine is powered up or warning provided when a hazardous zone is entered. An assumption should not be made that humans will not have to enter the robot's area. In one fully automated plant, an assumption was made that the robots would be so reliable that the human controllers would not have to enter the plant often and, therefore, the entire plant could be powered down when entry was required. The designers did not provide the usual safety features such as elevated walkways for the humans and alerts, such as aural warnings, when a robot was moving or about the move. After plant startup, the robots turned out to be so unreliable that the controllers had to enter the plant and bail them out several times during a shift. Because powering down the entire plant had such a negative impact on productivity, the humans got into the habit of entering the automated area of the plant without powering everything down. The inevitable occurred and someone was killed [72].

The automation should provide information about its internal state (such as the state of sensors and actuators), its control actions, its assumptions about the state of the system, and any anomalies that might have occurred. Processing requiring several seconds should provide a status indicator so human controllers can distinguish automated system processing from failure. In one nuclear power plant, the analog component that provided alarm annunciation to the operators was replaced with a digital component performing the same function. An argument was made that a safety analysis was not required because the replacement was "like for like." Nobody considered, however, that while the functional behavior might be the same, the failure behavior could be different. When the previous analog alarm annunciator failed, the screens went blank and the failure was immediately obvious to the human operators. When the new digital system failed, however, the screens froze, which was not immediately apparent to the operators, delaying critical feedback that the alarm system was not operating.

While the detection of nonevents is relatively simple for automated controllers—for instance, watchdog timers can be used—such detection is very difficult for humans. The *absence* of a signal, reading, or key piece of information is not usually immediately obvious to humans and they may not be able to recognize that a missing signal can indicate a change in the process state. In the Turkish Airlines flight TK 1951 accident at Amsterdam's Schiphol Airport in 2009, for example, the pilots did not notice the absence of a critical mode shift [52]. The design must ensure that lack of important signals will be registered and noticed by humans.

While safety interlocks are being overridden for test or maintenance, their status should be displayed to the operators and testers. Before allowing resumption of normal operations, the design should require confirmation that the interlocks have been restored. In one launch control system being designed by NASA, the operator could turn off alarms temporarily. There was no indication on the display, however, that the alarms had been disabled. If a shift change occurred and another operator took over the position, the new operator would have no way of knowing that alarms were not being annunciated.

If the information an operator needs to efficiently and safety control the process is not readily available, controllers will use experimentation to test their hypotheses about the state of the controlled system. If this kind of testing can be hazardous, then a safe way for operators to test their hypotheses should be provided rather than simply forbidding it. Such facilities will have additional benefits in handling emergencies.

The problem of feedback in emergencies is complicated by the fact that disturbances may lead to failure of sensors. The information available to the controllers (or to an automated system) becomes increasingly unreliable as the disturbance progresses. Alternative means should be provided to check safety-critical information as well as ways for human controllers to get additional information the designer did not foresee would be needed in a particular situation.

Decision aids need to be designed carefully. With the goal of providing assistance to the human controller, automated systems may provide feedforward (as well as feedback) information. Predictor displays show the operator one or more future states of the process parameters, as well as their present state or value, through a fast-time simulation, a mathematical model, or other analytic method that projects forward the effects of a particular control action or the progression of a disturbance if nothing is done about it.

Incorrect feedforward information can lead to process upsets and accidents. Humans can become dependent on automated assistance and stop checking whether the advice is reasonable if few errors occur. At the same time, if the process (control algorithm) truly can be accurately predetermined along with all future states of the system, then it should be automated. Humans are usually kept in systems when automation is introduced because they can vary their process models and control algorithms when conditions change or errors are detected in the original models and algorithms. Automated assistance such as predictor displays may lead to overconfidence and complacency and therefore overreliance by the operator. Humans may stop performing their own mental predictions and checks if few discrepancies are found over time. The operator then will begin to rely on the decision aid.

If decision aids are used, they need to be designed to reduce overdependence and to support operator skills and motivation rather than to take over functions in the name of support. Decision aids should provide assistance only when requested and their use should not become routine. People need to practice making decisions if we expect them to do so in emergencies or to detect erroneous decisions by automation.

Detecting Faults and Failures

A second use of feedback is to detect faults and failures in the controlled system, including the physical process and any computer controllers and displays. If the operator is expected to monitor a computer or automated decision making, then the computer must make decisions in a manner and at a rate that operators can follow. Otherwise they will not be able to detect faults and failures reliably in the system being supervised. In addition, the loss of confidence in the automation may lead the supervisor to disconnect it, perhaps under conditions where that could be hazardous, such as during critical points in the automatic landing of an airplane. When human supervisors can observe on the displays that proper corrections are being made by the automated system, they are less likely to intervene inappropriately, even in the presence of disturbances that cause large control actions.

For operators to anticipate or detect hazardous states, they need to be continuously updated about the process state so that the system progress and dynamic state can be monitored. Because of the poor ability of humans to perform monitoring over extended periods of time, they will need to be involved in the task in some way, as discussed earlier. If possible, the system should be designed to fail obviously or to make graceful degradation obvious to the supervisor.

The status of safety-critical components or state variables should be highlighted and presented unambiguously and completely to the controller. If an unsafe condition is detected by an automated system being supervised by a human controller, then the human controller should be told what anomaly was detected, what action was taken, and the current system configuration. Overrides of potentially hazardous failures or any clearing of the status data should not be permitted until all of the data has been displayed and probably not until the operator has acknowledged seeing it. A system may have a series of faults that can be overridden safely if they occur singly, but multiple faults could result in a hazard. In this case, the supervisor should be made aware of all safety-critical faults prior to issuing an override command or resetting a status display.

Alarms are used to alert controllers to events or conditions in the process that they might not otherwise notice. They are particularly important for low-probability events. The overuse of alarms, however, can lead to management by exception, overload and the incredulity response.

Designing a system that encourages or forces an operator to adopt a management-by-exception strategy, where the operator waits for alarm signals before taking

action, can be dangerous. This strategy does not allow operators to prevent disturbances by looking for early warning signals and trends in the process state.

The use of computers, which can check a large number of system variables in a
short amount of time, has made it easy to add alarms and to install large numbers
of them. In such plants, it is common for alarms to occur frequently, often five to
seven times an hour [196]. Having to acknowledge a large number of alarms may
leave operators with little time to do anything else, particularly in an emergency
[196]. A shift supervisor at the Three Mile Island (TMI) hearings testified that the
control room never had less than 52 alarms lit [98]. During the TMI incident, more
than a hundred alarm lights were lit on the control board, each signaling a different
malfunction, but providing little information about sequencing or timing. So many
alarms occurred at TMI that the computer printouts were running hours behind the
events and, at one point jammed, losing valuable information. Brooks claims that
operators commonly suppress alarms in order to destroy historical information
when they need real-time alarm information for current decisions [26]. Too many
alarms can cause confusion and a lack of confidence and can elicit exactly the wrong
response, interfering with the operator's ability to rectify the problems causing
the alarms.

Another phenomenon associated with alarms is the incredulity response, which
leads to not believing and ignoring alarms after many false alarms have occurred.
The problem is that in order to issue alarms early enough to avoid drastic countermeasures, the alarm limits must be set close to the desired operating point. This goal
is difficult to achieve for some dynamic processes that have fairly wide operating
ranges, leading to the problem of spurious alarms. Statistical and measurement
errors may add to the problem.

A great deal has been written about alarm management, particularly in the
nuclear power arena, and sophisticated disturbance and alarm analysis systems have
been developed. Those designing alarm systems should be familiar with current
knowledge about such systems. The following are just a few simple guidelines:

- *Keep spurious alarms to a minimum:*   This guideline will reduce overload and
  the incredulity response.

- *Provide checks to distinguish correct from faulty instruments:*   When response
  time is not critical, most operators will attempt to check the validity of the alarm
  [209]. Providing information in a form where this validity check can be made
  quickly and accurately, and not become a source of distraction, increases the
  probability of the operator acting properly.

- *Provide checks on alarm system itself:*   The operator has to know whether the
  problem is in the alarm or in the system. Analog devices can have simple checks
  such as "press to test" for smoke detectors or buttons to test the bulbs in a

lighted gauge. Computer-displayed alarms are more difficult to check; checking usually requires some additional hardware or redundant information that does not come through the computer. One complication comes in the form of alarm analysis systems that check alarms and display a prime cause along with associated effects. Operators may not be able to perform validity checks on the complex logic necessarily involved in these systems, leading to overreliance [209]. Weiner and Curry also worry that the priorities might not always be appropriate in automated alarm analysis and that operators may not recognize this fact.

- *Distinguish between routine and safety-critical alarms:*   The form of the alarm, such as auditory cues or message highlighting, should indicate degree or urgency. Alarms should be categorized as to which are the highest priority.

- *Provide temporal information about events and state changes:*   Proper decision making often requires knowledge about the timing and sequencing of events. Because of system complexity and built-in time delays due to sampling intervals, however, information about conditions or events is not always timely or even presented in the sequence in which the events actually occurred. Complex systems are often designed to sample monitored variables at different frequencies: some variables may be sampled every few seconds while, for others, the intervals may be measured in minutes. Changes that are negated within the sampling period may not be recorded at all. Events may become separated from their circumstances, both in sequence and time [26].

- *Require corrective action when necessary:*   When faced with a lot of undigested and sometimes conflicting information, humans will first try to figure out what is going wrong. They may become so involved in attempts to save the system that they wait too long to abandon the recovery efforts. Alternatively, they may ignore alarms they do not understand or they think are not safety critical. The system design may need to ensure that the operator cannot clear a safety-critical alert without taking corrective action or without performing subsequent actions required to complete an interrupted operation. The Therac-25, a linear accelerator that massively overdosed multiple patients, allowed operators to proceed with treatment five times after an error message appeared simply by pressing one key [115]. No distinction was made between errors that could be safety-critical and those that were not.

- *Indicate which condition is responsible for the alarm:*   System designs with more than one mode or where more than one condition can trigger the alarm for a mode, must clearly indicate which condition is responsible for the alarm. In the Therac-25, one message meant that the dosage given was either too low or too high, without providing information to the operator about which of these errors had occurred. In general, determining the cause of an alarm may be difficult. In complex, tightly coupled plants, the point where the alarm is first triggered may be far away from where the fault actually occurred.

- *Minimize the use of alarms when they may lead to management by exception:* After studying thousands of near accidents reported voluntarily by aircraft crews and ground support personnel, one U.S. government report recommended that the altitude alert signal (an aural sound) be disabled for all but a few long-distance flights [141]. Investigators found that this signal had caused decreased altitude awareness in the flight crew, resulting in more frequent overshoots—instead of leveling off at 10,000 feet, for example, the aircraft continues to climb or descend until the alarm sounds. A study of such overshoots noted that they rarely occur in bad weather, when the crew is most attentive.

Robustness of the Feedback Process

Because feedback is so important to safety, robustness must be designed into feedback channels. The problem of feedback in emergencies is complicated by the fact that disturbances may lead to failure of sensors. The information available to the controllers (or to an automated system) becomes increasingly unreliable as the disturbance progresses.

One way to prepare for failures is to provide alternative sources of information and alternative means to check safety-critical information. It is also useful for the operators to get additional information the designers did not foresee would be needed in a particular situation. The emergency may have occurred because the designers made incorrect assumptions about the operation of the controlled system, the environment in which it would operate, or the information needs of the controller.

If automated controllers provide the only information about the controlled system state, the human controller supervising the automation can provide little oversight. The human supervisor must have access to independent sources of information to detect faults and failures, except in the case of a few failure modes such as total inactivity. Several incidents involving the command and control warning system at NORAD headquarters in Cheyenne Mountain involved situations where the computer had bad information and thought the United States was under nuclear attack. Human supervisors were able to ascertain that the computer was incorrect through direct contact with the warning sensors (satellites and radars). This direct contact showed the sensors were operating and had received no evidence of incoming missiles [180]. The error detection would not have been possible if the humans could only get information about the sensors from the computer, which had the wrong information. Many of these direct sensor inputs are being removed in the mistaken belief that only computer displays are required.

The main point is that human supervisors of automation cannot monitor its performance if the information used in monitoring is not independent from the thing being monitored. There needs to be provision made for failure of computer displays or incorrect process models in the software by providing alternate sources of information. Of course, any instrumentation to deal with a malfunction must not be disabled by the malfunction, that is, common-cause failures must be eliminated or controlled. As an example of the latter, an engine and pylon came off the wing of a DC-10, severing the cables that controlled the leading edge flaps and also four hydraulic lines. These failures disabled several warning signals, including a flap mismatch signal and a stall warning light [155]. If the crew had known the slats were retracted and had been warned of a potential stall, they might have been able to save the plane.

Displaying Feedback to Human Controllers

Computer displays are now ubiquitous in providing feedback information to human controllers, as are complaints about their design.

Many computer displays are criticized for providing too much data (data overload) where the human controller has to sort through large amounts of data to find the pieces needed. Then the information located in different locations may need to be integrated. Bainbridge suggests that operators should not have to page between displays to obtain information about abnormal states in the parts of the process other than the one they are currently thinking about; neither should they have to page between displays that provide information needed for a single decision process.

These design problems are difficult to eliminate, but performing a task analysis coupled with a hazard analysis can assist in better design as will making all the information needed for a single decision process visible at the same time, placing frequently used displays centrally, and grouping displays of information using the information obtained in the task analysis. It may also be helpful to provide alternative ways to display information or easy ways to request what is needed.

Much has been written about how to design computer displays, although a surprisingly large number of displays still seem to be poorly designed. The difficulty of such design is increased by the problem that, once again, conflicts can exist. For example, intuition seems to support providing information to users in a form that can be quickly and easily interpreted. This assumption is true if rapid reactions are required. Some psychological research, however, suggests that cognitive processing for meaning leads to better information retention: A display that requires little thought and work on the part of the operator may not support acquisition of the knowledge and thinking skills needed in abnormal conditions [168].

Once again, the designer needs to understand the tasks the user of the display is performing. To increase safety, the displays should reflect what is known about how the information is used and what kinds of displays are likely to cause human error. Even slight changes in the way information is presented can have dramatic effects on performance.

This rest of this section concentrates only on a few design guidelines that are especially important for safety. The reader is referred to the standard literature on display design for more information.

Safety-related information should be distinguished from non-safety-related information and highlighted. In addition, when safety interlocks are being overridden, their status should be displayed. Similarly, if safety-related alarms are temporarily inhibited, which may be reasonable to allow so that the operator can deal with the problem without being continually interrupted by additional alarms, the inhibit status should be shown on the display. Make warning displays brief and simple.

A common mistake is to make all the information displays digital simply because the computer is a digital device. Analog displays have tremendous advantages for processing by humans. For example, humans are excellent at pattern recognition, so providing scannable displays that allow operators to process feedback and diagnose problems using pattern recognition will enhance human performance. A great deal of information can be absorbed relatively easily when it is presented in the form of patterns.

Avoid displaying absolute values unless the human requires the absolute values. It is hard to notice changes such as events and trends when digital values are going up and down. A related guideline is to provide references for judgment. Often, for example, the user of the display does not need the absolute value but only the fact that it is over or under a limit. Showing the value on an analog dial with references to show the limits will minimize the required amount of extra and error-prone processing by the user. The overall goal is to minimize the need for extra mental processing to get the information the users of the display need for decision making or for updating their process models.

Another typical problem occurs when computer displays must be requested and accessed sequentially by the user, which makes greater memory demands upon the operator, negatively affecting difficult decision-making tasks [14]. With conventional instrumentation, all process information is constantly available to the operator: an overall view of the process state can be obtained by a glance at the console. Detailed readings may be needed only if some deviation from normal conditions is detected.

The alternative, a process overview display on a computer console, is more time consuming to process: To obtain additional information about a limited part of the process, the operator has to select consciously among displays.

In a study of computer displays in the process industry, Swaanenburg and colleagues found that most operators considered a computer display more difficult to work with than conventional parallel interfaces, especially with respect to getting an overview of the process state. In addition, operators felt the computer overview displays were of limited use in keeping them updated on task changes; instead, operators tended to rely to a large extent on group displays for their supervisory tasks. The researchers conclude that a group display, showing different process variables in reasonable detail (such as measured value, setpoint, and valve position), clearly provided the type of data operators preferred. Keeping track of the progress of a disturbance is very difficult with sequentially presented information [196]. One general lesson to be learned here is that the operators of the system need to be involved in display design decisions: The designers should not just do what is easiest to implement or satisfies their aesthetic senses.

Whenever possible, software designers should try to copy the standard displays with which operators have become familiar, and which were often developed for good psychological reasons, instead of trying to be creative or unique. For example, icons with a standard interpretation should be used. Researchers have found that icons often pleased system designers but irritated users [92]. Air traffic controllers, for example, found the arrow icons for directions on a new display useless and preferred numbers. Once again, including experienced operators in the design process and understanding why the current analog displays have developed as they have will help to avoid these basic types of design errors.

An excellent way to enhance human interpretation and processing is to design the control panel to mimic the physical layout of the plant or system. For example, graphical displays allow the status of valves to be shown within the context of piping diagrams and even the flow of materials. Plots of variables can be shown, highlighting important relationships.

The graphical capabilities of computer displays provides exciting potential for improving on traditional instrumentation, but the designs need to be based on psychological principles and not just on what appeals to the designer, who may never have operated a complex process. As Lees has suggested, the starting point should be consideration of the operator's tasks and problems; the display should evolve as a solution to these [110].

Operator inputs to the design process as well as extensive simulation and testing will assist in designing usable computer displays. Remember that the overall goal is to reduce the mental workload of the human in updating their process models and to reduce human error in interpreting feedback.

9.5   Summary

A process for safety-guided design using STPA and some basic principles for safe design have been described in this chapter. The topic is an important one and more still needs to be learned, particularly with respect to safe system design for human controllers. Including skilled and experienced operators in the design process from the beginning will help as will performing sophisticated human task analyses rather than relying primarily on operators interacting with computer simulations.

The next chapter describes how to integrate the disparate information and techniques provided so far in part III into a system-engineering process that integrates safety into the design process from the beginning, as suggested in chapter 6.

10   Integrating Safety into System Engineering

Previous chapters have provided the individual pieces of the solution to engineering a safer world. This chapter demonstrates how to put these pieces together to integrate safety into a system engineering process. No one process is being proposed: Safety must be part of any system engineering process.

The glue that integrates the activities of engineering and operating complex systems is specifications and the safety information system. Communication is critical in handling any emergent property in a complex system. Our systems today are designed and built by hundreds and often thousands of engineers and then operated by thousands and even tens of thousands more people. Enforcing safety constraints on system behavior requires that the information needed for decision making is available to the right people at the right time, whether during system development, operations, maintenance, or reengineering.

This chapter starts with a discussion of the role of specifications and how systems theory can be used as the foundation for the specification of complex systems. Then an example of how to put the components together in system design and development is presented. Chapters 11 and 12 cover how to maximize learning from accidents and incidents and how to enforce safety constraints during operations. The design of safety information systems is discussed in chapter 13.

10.1   The Role of Specifications and the Safety Information System

While engineers may have been able to get away with minimal specifications during development of the simpler electromechanical systems of the past, specifications are critical to the successful engineering of systems of the size and complexity we are attempting to build today. Specifications are no longer simply a means of archiving information; they need to play an active role in the system engineering process. They are a critical tool in stretching our intellectual capabilities to deal with increasing complexity.

Our specifications must reflect and support the system safety engineering process and the safe operation, evolution and change of the system over time. Specifications should support the use of notations and techniques for reasoning about hazards and safety, designing the system to eliminate or control hazards, and validating—at each step, starting from the very beginning of system development—that the evolving system has the desired safety level. Later, specifications must support operations and change over time.

Specification languages can help (or hinder) human performance of the various problem-solving activities involved in system requirements analysis, hazard analysis, design, review, verification and validation, debugging, operational use, and maintenance and evolution (sustainment). They do this by including notations and tools that enhance our ability to: (1) reason about particular properties, (2) construct the system and the software in it to achieve them, and (3) validate—at each step, starting from the very beginning of system development—that the evolving system has the desired qualities. In addition, systems and particularly the software components are continually changing and evolving; they must be designed to be changeable and the specifications must support evolution without compromising the confidence in the properties that were initially verified.

Documenting and tracking hazards and their resolution are basic requirements for any effective safety program. But simply having the safety engineer track them and maintain a hazard log is not enough—information must be derived from the hazards to inform the system engineering process and that information needs to be specified and recorded in a way that has an impact on the decisions made during system design and operations. To have such an impact, the safety-related information required by the engineers needs to be *integrated into* the environment in which safety-related engineering decisions are made. Engineers are unlikely to be able to read through volumes of hazard analysis information and relate it easily to the specific component upon which they are working. The information the system safety engineer has generated must be presented to the system designers, implementers, maintainers, and operators in such a way that they can easily find what they need to make safer decisions.

Safety information is not only important during system design; it also needs to be presented in a form that people can learn from, apply to their daily jobs, and use throughout the life cycle of projects. Too often, preventable accidents have occurred due to changes that were made after the initial design period. Accidents are frequently the result of safe designs becoming unsafe over time when changes in the system itself or in its environment violate the basic assumptions of the original hazard analysis. Clearly, these assumptions must be recorded and easily retrievable when changes occur. Good documentation is the most important in complex systems where nobody is able to keep all the information necessary to make safe decisions in their head.

What types of specifications are needed to support humans in system safety engineering and operations? Design decisions at each stage must be mapped into the goals and constraints they are derived to satisfy, with earlier decisions mapped or traced to later stages of the process. The result should be a seamless and gapless record of the progression from high-level requirements down to component requirements and designs or operational procedures. The rationale behind the design decisions needs to be recorded in a way that is easily retrievable by those reviewing or changing the system design. The specifications must also support the various types of formal and informal analysis used to decide between alternative designs and to verify the results of the design process. Finally, specifications must assist in the coordinated design of the component functions and the interfaces between them.

The notations used in specification languages must be easily readable and learnable. Usability is enhanced by using notations and models that are close to the mental models created by the users of the specification and the standard notations in their fields of expertise.

The structure of the specification is also important for usability. The structure will enhance or limit the ability to retrieve needed information at the appropriate times.

Finally, specifications should not limit the problem-solving strategies of the users of the specification. Not only do different people prefer different strategies for solving problems, but the most effective problem solvers have been found to change strategies frequently [167, 58]. Experts switch problem-solving strategy when they run into difficulties following a particular strategy and as new information is obtained that changes the objectives or subgoals or the mental workload needed to use a particular strategy. Tools often limit the strategies that can be used, usually implementing the favorite strategy of the tool designer, and therefore limiting the problem solving strategies supported by the specification.

One way to implement these principles is to use *intent specifications* [120].

10.2   Intent Specifications

Intent specifications are based on systems theory, system engineering principles, and psychological research on human problem solving and how to enhance it. The goal is to assist humans in dealing with complexity. While commercial tools exist that implement intent specifications directly, any specification languages and tools can be used that allow implementing the properties of an intent specification.

An intent specification differs from a standard specification primarily in its structure, not its content: no extra information is involved that is not commonly found in detailed specifications—the information is simply organized in a way that has been found to assist in its location and use. Most complex systems have voluminous documentation, much of it redundant or inconsistent, and it degrades quickly as changes are made over time. Sometimes important information is missing, particularly information about *why* something was done the way it was—the intent or design rationale. Trying to determine whether a change might have a negative impact on safety, if possible at all, is usually enormously expensive and often involves regenerating analyses and work that was already done but either not recorded or not easily located when needed. Intent specifications were designed to help with these problems: Design rationale, safety analysis results, and the assumptions upon which the system design and validation are based are integrated directly into the system specification and its structure, rather than stored in separate documents, so the information is at hand when needed for decision making.

The structure of an intent specification is based on the fundamental concept of hierarchy in systems theory (see chapter 3) where complex systems are modeled in terms of a hierarchy of levels of organization, each level imposing constraints on the degree of freedom of the components at the lower level. Different description languages may be appropriate at the different levels. Figure 10.1 shows the seven levels of an intent specification.

Intent specifications are organized along three dimensions: intent abstraction, part-whole abstraction, and refinement. These dimensions constitute the problem space in which the human navigates. Part-whole abstraction (along the horizontal dimension) and refinement (within each level) allow users to change their focus of attention to more or less detailed views within each level or model. The vertical dimension specifies the level of intent at which the problem is being considered.

Each intent level contains information about the characteristics of the environment, human operators or users, the physical and functional system components, and requirements for and results of verification and validation activities for that level. The safety information is embedded in each level, instead of being maintained in a separate safety log, but linked together so that it can easily be located and reviewed.

The vertical intent dimension has seven levels. Each level represents a different model of the system from a different perspective and supports a different type of reasoning about it. Refinement and decomposition occurs within each level of the specification, rather than between levels. Each level provides information not just about *what* and *how*, but *why*, that is, the design rationale and reasons behind the design decisions, including safety considerations.

Figure 10.2 shows an example of the information that might be contained in each level of the intent specification.

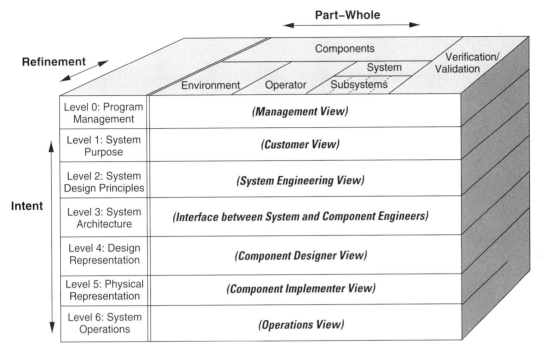

Figure 10.1
The structure of an intent specification.

The top level (level 0) provides a project management view and insight into the relationship between the plans and the project development status through links to the other parts of the intent specification. This level might contain the project management plans, the safety plan, status information, and so on.

Level 1 is the customer view and assists system engineers and customers in agreeing on what should be built and, later, whether that has been accomplished. It includes goals, high-level requirements and constraints (both physical and operator), environmental assumptions, definitions of accidents, hazard information, and system limitations.

Level 2 is the system engineering view and helps system engineers record and reason about the system in terms of the physical principles and system-level design principles upon which the system design is based.

Level 3 specifies the system architecture and serves as an unambiguous interface between system engineers and component engineers or contractors. At level 3, the system functions defined at level 2 are decomposed, allocated to components, and specified rigorously and completely. Black-box behavioral component models may be used to specify and reason about the logical design of the system as a whole and

	Environment	Operator	System and components	V&V
Level 0 Prog. Mgmt.	Project management plans, status information, safety plan, etc.			
Level 1 System Purpose	Assumptions Constraints	Responsibilities Requirements I/F requirements	System goals, high-level requirements, design constraints, limitations	Preliminary Hazard Analysis, Review s
Level 2 System Principles	External interfaces	Task analyses Task allocation Controls, displays	Logic principles, control laws, functional decomposition and allocation	Validation plan and results, System Hazard Analysis
Level 3 Blackbox Models	Environment models	Operator Task models HCI models	Blackbox functional models Interface specifications	Analysis plans and results, Subsystem Hazard Analysis
Level 4 Design Rep.		HCI design	Software and hardware design specs	Test plans and results
Level 5 Physical Rep.		GUI design, physical controls design	Software code, hardware assembly instructions	Test plans and results
Level 6 Operations	Audit procedures	Operator manuals Maintenance Training materials	Error reports, change requests, etc.	Performance monitoring and audits

Figure 10.2
An example of the information in an intent specification.

the interactions among individual system components without being distracted by implementation details.

If the language used at level 3 is formal (rigorously defined), then it can play an important role in system validation. For example, the models can be executed in system simulation environments to identify system requirements and design errors early in development. They can also be used to automate the generation of system and component test data, various types of mathematical analyses, and so forth. It is important, however, that the black-box (that is, transfer function) models be easily reviewed by domain experts—most of the safety-related errors in specifications will be found by expert review, not by automated tools or formal proofs.

A readable but formal and executable black-box requirements specification language was developed by the author and her students while helping the FAA specify the TCAS (Traffic Alert and Collision Avoidance System) requirements [123]. Reviewers can learn to read the specifications with a few minutes of instruction about the notation. Improvements have been made over the years, and it is being used successfully on real systems. This language provides an existence case that a readable and easily learnable but formal specification language is possible. Other languages with the same properties, of course, can also be used effectively.

The next two levels, Design Representation and Physical Representation, provide the information necessary to reason about individual component design and implementation issues. Some parts of level 4 may not be needed if at least portions of the physical design can be generated automatically from the models at level 3.

The final level, Operations, provides a view of the operational system and acts as the interface between development and operations. It assists in designing and performing system safety activities during system operations. It may contain required or suggested operational audit procedures, user manuals, training materials, maintenance requirements, error reports and change requests, historical usage information, and so on.

Each level of an intent specification supports a different type of reasoning about the system, with the highest level assisting systems engineers in their reasoning about system-level goals, constraints, priorities, and tradeoffs. The second level, System Design Principles, allows engineers to reason about the system in terms of the physical principles and laws upon which the design is based. The Architecture level enhances reasoning about the logical design of the system as a whole, the interactions between the components, and the functions computed by the components without being distracted by implementation issues. The lowest two levels provide the information necessary to reason about individual component design and implementation issues. The mappings between levels provide the relational information that allows reasoning across hierarchical levels and traceability of requirements to design.

Hyperlinks are used to provide the relational information that allows reasoning within and across levels, including the tracing from high-level requirements down to implementation and vice versa. Examples can be found in the rest of this chapter.

The structure of an intent specification does not imply that the development must proceed from the top levels down to the bottom levels in that order, only that at the end of the development process, all levels are complete. Almost all development involves work at all of the levels at the same time.

When the system changes, the environment in which the system operates changes, or components are reused in a different system, a new or updated safety analysis is required. Intent specifications can make that process feasible and practical.

Examples of intent specifications are available [121, 151] as are commercial tools to support them. But most of the principles can be implemented without special tools beyond a text editor and hyperlinking facilities. The rest of this chapter assumes only these very limited facilities are available.

10.3   An Integrated System and Safety Engineering Process

There is no agreed upon best system engineering process and probably cannot be one—the process needs to match the specific problem and environment in which it is being used. What is described in this section is how to integrate safety engineering into *any* reasonable system engineering process.

The system engineering process provides a logical structure for problem solving. Briefly, first a need or problem is specified in terms of objectives that the system must satisfy and criteria that can be used to rank alternative designs. Then a process of system synthesis takes place that usually involves considering alternative designs. Each of the alternatives is analyzed and evaluated in terms of the stated objectives and design criteria, and one alternative is selected. In practice, the process is highly iterative: The results from later stages are fed back to early stages to modify objectives, criteria, design decisions, and so on.

Design alternatives are generated through a process of system architecture development and analysis. The system engineers first develop requirements and design constraints for the system as a whole and then break the system into subsystems and design the subsystem interfaces and the subsystem interface topology. System functions and constraints are refined and allocated to the individual subsystems. The emerging design is analyzed with respect to desired system performance characteristics and constraints, and the process is iterated until an acceptable system design results.

The difference in safety-guided design is that hazard analysis is used throughout the process to generate the safety constraints that are factored into the design decisions as they are made. The preliminary design at the end of this process must be described in sufficient detail that subsystem implementation can proceed independently. The subsystem requirements and design processes are subsets of the larger system engineering process.

This general system engineering process has some particularly important aspects. One of these is the focus on interfaces. System engineering views each system as an integrated whole even though it is composed of diverse, specialized components, which may be physical, logical (software), or human. The objective is to design subsystems that when integrated into the whole provide the most effective system possible to achieve the overall objectives. The most challenging problems in building complex systems today arise in the interfaces between components. One example is the new highly automated aircraft where most incidents and accidents have been blamed on human error, but more properly reflect difficulties in the collateral design of the aircraft, the avionics systems, the cockpit displays and controls, and the demands placed on the pilots.

A second critical factor is the integration of humans and nonhuman system components. As with safety, a separate group traditionally does human factors design and analysis. Building safety-critical systems requires integrating both system safety and human factors into the basic system engineering process, which in turn has important implications for engineering education. Unfortunately, neither safety nor human factors plays an important role in most engineering education today.

During program and project planning, a system safety plan, standards, and project development safety control structure need to be designed including policies, procedures, the safety management and control structure, and communication channels. More about safety management plans can be found in chapters 12 and 13.

Figure 10.3 shows the types of activities that need to be performed in such an integrated process and the system safety and human factors inputs and products. Standard validation and verification activities are not shown, since they should be included throughout the entire process.

The rest of this chapter provides an example using TCAS II. Other examples are interspersed where TCAS is not appropriate or does not provide an interesting enough example.

10.3.1 Establishing the Goals for the System

The first step in any system engineering process is to identify the goals of the effort. Without agreeing on where you are going, it is not possible to determine how to get there or when you have arrived.

TCAS II is a box required on most commercial and some general aviation aircraft that assists in avoiding midair collisions. The goals for TCAS II are to:

G1: *Provide affordable and compatible collision avoidance system options for a broad spectrum of National Airspace System users.*

G2: *Detect potential midair collisions with other aircraft in all meteorological conditions; throughout navigable airspace, including airspace not covered by ATC primary or secondary radar systems; and in the absence of ground equipment.*

TCAS was intended to be an independent backup to the normal Air Traffic Control (ATC) system and the pilot's "see and avoid" responsibilities. It interrogates air traffic control transponders on aircraft in its vicinity and listens for the transponder replies. By analyzing these replies with respect to slant range and relative altitude, TCAS determines which aircraft represent potential collision threats and provides appropriate display indications, called advisories, to the flight crew to assure proper

Agree on system goals

Identify constraints on how goals can be achieved
   • Define accidents (unacceptable losses)
   • Identify hazards
   • Formulate system-level safety and non-safety constraints

Select a system architecture
   • Architectural trade analysis
   • Preliminary hazard analysis

Identify environmental assumptions
Create a concept of operations Perform a preliminary operator task analysis
Refine goals into testable and achievable system-level functional requirements

Refine safety constraints and functional requirements
   • Identify preliminary safety control structure
   • Perform STPA

Perform safety-driven system design and analysis
   • Make system-level design decisions to satisfy functional requirements
     and safety constraints
   • Define component responsibilities
   • Identify potentially unsafe control actions and restate as constraints
     on system and component behavior

Implementation (construction and manufacturing)
Document system limitations
Perform final safety assessment
Safety certification
Field testing, installation, and training

Operations, including maintenance and upgrades
   • Change analysis
   • Incident and accident analysis
   • Performance monitoring
   • Periodic audits

Decommissioning

Figure 10.3
System safety and human factors integrated into the set of typical system engineering tasks. Standard verification and validation activities are not shown as they are assumed to be performed throughout the whole process, not just at the end where they are often concentrated.

separation. Two types of advisories can be issued. *Resolution advisories* (RAs) provide instructions to the pilots to ensure safe separation from nearby traffic in the vertical plane.[1] *Traffic advisories* (TAs) indicate the positions of intruding aircraft that may later cause resolution advisories to be displayed.

TCAS is an example of a system created to directly impact safety where the goals are all directly related to safety. But system safety engineering and safety-driven design can be applied to systems where maintaining safety is not the only goal and, in fact, human safety is not even a factor. The example of an outer planets explorer spacecraft was shown in chapter 7. Another example is the air traffic control system, which has both safety and nonsafety (throughput) goals.

10.3.2 Defining Accidents

Before any safety-related activities can start, the definition of an accident needs to be agreed upon by the system customer and other stakeholders. This definition, in essence, establishes the goals for the safety effort.

Defining accidents in TCAS is straightforward—only one is relevant, a midair collision. Other more interesting examples are shown in chapter 7.

Basically, the criterion for specifying events as accidents is that the losses are so important that they need to play a central role in the design and tradeoff process. In the outer planets explorer example in chapter 7, some of the losses involve the mission goals themselves while others involve losses to other missions or a negative impact on our solar system ecology.

Priorities and evaluation criteria may be assigned to the accidents to indicate how conflicts are to be resolved, such as conflicts between safety goals or conflicts between mission goals and safety goals and to guide design choices at lower levels. The priorities are then inherited by the hazards related to each of the accidents and traced down to the safety-related design features.

10.3.3 Identifying the System Hazards

Once the set of accidents has been agreed upon, hazards can be derived from them. This process is part of what is called Preliminary Hazard Analysis (PHA) in System Safety. The hazard log is usually started as soon as the hazards to be considered are identified. While much of the information in the hazard log will be filled in later, some information is available at this time.

There is no right or wrong list of hazards—only an agreement by all involved on what hazards will be considered. Some hazards that were considered during the design of TCAS are listed in chapter 7 and are repeated here for convenience:

---

1. Horizontal advisories were originally planned for later versions of TCAS but have not yet been implemented.

1. TCAS causes or contributes to a near midair collision (NMAC), defined as a pair of controlled aircraft violating minimum separation standards.

2. TCAS causes or contributes to a controlled maneuver into the ground.

3. TCAS causes or contributes to the pilot losing control over the aircraft.

4. TCAS interferes with other safety-related aircraft systems (for example, ground proximity warning).

5. TCAS interferes with the ground-based air traffic control system (e.g., transponder transmissions to the ground or radar or radio services).

6. TCAS interferes with an ATC advisory that is safety-related (e.g., avoiding a restricted area or adverse weather conditions).

Once accidents and hazards have been identified, early concept formation (sometimes called high-level architecture development) can be started for the integrated system and safety engineering process.

10.3.4 Integrating Safety into Architecture Selection and System Trade Studies

An early activity in the system engineering of complex systems is the selection of an overall architecture for the system, or as it is sometimes called, system concept formation. For example, an architecture for manned space exploration might include a transportation system with parameters and options for each possible architectural feature related to technology, policy, and operations. Decisions will need to be made early, for example, about the number and type of vehicles and modules, the destinations for the vehicles, the roles and activities for each vehicle including dockings and undockings, trajectories, assembly of the vehicles (in space or on Earth), discarding of vehicles, prepositioning of vehicles in orbit and on the planet surface, and so on. Technology options include type of propulsion, level of autonomy, support systems (water and oxygen if the vehicle is used to transport humans), and many others. Policy and operational options may include crew size, level of international investment, types of missions and their duration, landing sites, and so on. Decisions about these overall system concepts clearly must precede the actual implementation of the system.

How are these decisions made? The selection process usually involves extensive tradeoff analysis that compares the different feasible architectures with respect to some important system property or properties. Cost, not surprisingly, usually plays a large role in the selection process while other properties, including system safety, are usually left as a problem to be addressed later in the development lifecycle. Many of the early architectural decisions, however, have a significant and lasting impact on safety and may not be reversible after the basic architectural decisions have been made. For example, the decision not to include a crew escape system on the Space Shuttle was an early architectural decision and has been impacting Shuttle safety for more than thirty years [74, 136]. After the *Challenger* accident and again after the *Columbia* loss, the idea resurfaced, but there was no cost-effective way to add crew escape at that time.

The primary reason why safety is rarely factored in during the early architectural tradeoff process, except perhaps informally, is that practical methods for analyzing safety, that is, hazard analysis methods that can be applied at that time, do not exist. But if information about safety were available early, it could be used in the selection process and hazards could be eliminated by the selection of appropriate architectural options or mitigated early when the cost of doing so is much less than later in the system lifecycle. Making basic design changes downstream becomes increasingly costly and disruptive as development progresses and, often, compromises in safety must be accepted that could have been eliminated if safety had been considered in the early architectural evaluation process.

While it is relatively easy to identify hazards at system conception, performing a hazard or risk assessment before a design is available is more problematic. At best, only a very rough estimate is possible. Risk is usually defined as a combination of severity and likelihood. Because these two different qualities (severity and likelihood) cannot be combined mathematically, they are commonly qualitatively combined using a risk matrix. Figure 10.4 shows a fairly standard form for such a matrix.

SEVERITY

	I Catastrophic	II Critical	III Marginal	IV Negligible
A Frequent	I–A	II–A	III–A	IV–A
B Moderate	I–B	II–B	III–B	IV–B
C Occasional	I–C	II–C	III–C	IV–C
D Remote	I–D	II–D	III–D	IV–D
E Unlikely	I–E	II–E	III–E	IV–E
F Impossible	I–F	II–F	III–F	IV–F

LIKELIHOOD

Figure 10.4
A standard risk matrix.

High-level hazards are first identified and, for each identified hazard, a qualitative evaluation is performed by classifying the hazard according to its severity and likelihood.

While severity can usually be evaluated using the worst possible consequences of that hazard, likelihood is almost always unknown and, arguably, unknowable for complex systems before any system design decisions have been made. The problem is even worse before a system architecture has been selected. Some probabilistic information is usually available about physical events, of course, and historical information may theoretically be available. But new systems are usually being created because existing systems and designs are not adequate to achieve the system goals, and the new systems will probably use new technology and design features that limit the accuracy of historical information. For example, historical information about the likelihood of propulsion-related losses may not be accurate for new spacecraft designs using nuclear propulsion. Similarly, historical information about the errors air traffic controllers make has no relevance for new air traffic control systems, where the type of errors may change dramatically.

The increasing use of software in most complex systems complicates the situation further. Much or even most of the software in the system will be new and have no historical usage information. In addition, statistical techniques that assume randomness are not applicable to software design flaws. Software and digital systems also introduce new ways for hazards to occur, including new types of component interaction accidents. Safety is a system property, and, as argued in part I, combining the probability of failure of the system components to be used has little or no relationship to the safety of the system as a whole.

There are no known or accepted rigorous or scientific ways to obtain probabilistic or even subjective likelihood information using historical data or analysis in the case of non-random failures and system design errors, including unsafe software behavior. When forced to come up with such evaluations, engineering judgment is usually used, which in most cases amounts to pulling numbers out of the air, often influenced by political and other nontechnical factors. Selection of a system architecture and early architectural trade evaluations on such a basis is questionable and perhaps one reason why risk usually does not play a primary role in the early architectural trade process.

Alternatives to the standard risk matrix are possible, but they tend to be application specific and so must be constructed for each new system. For many systems, the use of severity alone is often adequate to categorize the hazards in trade studies. Two examples of other alternatives are presented here, one created for augmented air traffic control technology and the other created and used in the early architectural trade study of NASA's Project Constellation, the program to return to the moon and later go on to Mars. The reader is encouraged to come up with their own methods appropriate for their particular application. The examples are not meant to be definitive, but simply illustrative of what is possible.

Example 1: A Human-Intensive System: Air Traffic Control Enhancements

Enhancements to the air traffic control (ATC) system are unique in that the problem is not to create a new or safer system but to maintain the very high level of safety built into the current system: The goal is to not degrade safety. The risk likelihood estimate can be restated, in this case, as the likelihood that safety will be degraded by the proposed changes and new tools. To tackle this problem, we created a set of criteria to be used in the evaluation of likelihood.[2] The criteria ranked various high-level architectural design features of the proposed set of ATC tools on a variety of factors related to risk in these systems. The ranking was qualitative and most criteria were ranked as having low, medium, or high impact on the likelihood of safety being degraded from the current level. For the majority of factors, "low" meant insignificant or no change in safety with respect to that factor in the new versus the current system, "medium" denoted the potential for a minor change, and "high" signified potential for a significant change in safety. Many of the criteria involve human-automation interaction, since ATC is a very human-intensive system and the new features being proposed involved primarily new automation to assist human air traffic controllers. Here are examples of the likelihood level criteria used:

- *Safety margins:* Does the new feature have the potential for (1) an insignificant or no change to the existing safety margins, (2) a minor change, or (3) a significant change.

- *Situation awareness:* What is the level of change in the potential for reducing situation awareness.

- *Skills currently used and those necessary to backup and monitor the new decision-support tools:* Is there an insignificant or no change in the controller skills, a minor change, or a significant change.

- *Introduction of new failure modes and hazard causes:* Do the new tools have the same function and failure modes as the system components they are replacing, are new failure modes and hazards introduced but well understood and effective mitigation measures can be designed, or are the new failure modes and hazard causes difficult to control.

- *Effect of the new software functions on the current system hazard mitigation measures:* Can the new features render the current safety measures ineffective or are they unrelated to current safety features.

---

2. These criteria were developed for a NASA contract by the author and have not been published previously.

• *Need for new system hazard mitigation measures:*   Will the proposed changes
require new hazard mitigation measures.

These criteria and others were converted into a numerical scheme so they could be
combined and used in an early risk assessment of the changes being contemplated
and their potential likelihood for introducing significant new risk into the system.
The criteria were weighted to reflect their relative importance in the risk analysis.

Example 2:   Early Risk Analysis of Manned Space Exploration
A second example was created by Nicolas Dulac and others as part of an MIT and
Draper Labs contract with NASA to perform an architectural tradeoff analysis for
future human space exploration [59]. The system engineers wanted to include safety
along with the usual factors, such as mass, to evaluate the candidate architectures,
but once again little information was available at this early stage of system engineering. It was not possible to evaluate likelihood using historical information; all of the
potential architectures involved new technology, new missions, and significant
amounts of software.

In the procedure developed to achieve the goal, the hazards were first identified
as shown in figure 10.5. As is the case at the beginning of any project, identifying
system hazards involved ten percent creativity and ninety percent experience.
Hazards were identified for each mission phase by domain experts under the guidance of the safety experts. Some hazards, such as fire, explosion, or loss of life-
support span multiple (if not all) mission phases and were grouped as *General
Hazards.* The control strategies used to mitigate them, however, may depend on the
mission phase in which they occur.

Once the hazards were identified, the severity of each hazard was evaluated by
considering the worst-case loss associated with the hazard. In the example, the losses
are evaluated for each of three categories: humans (H), mission (M), and equipment
(E). Initially, potential damage to the Earth and planet surface environment was
included in the hazard log. In the end, the environment component was left out of
the analysis because project managers decided to replace the analysis with mandatory compliance with NASA's planetary protection standards. A risk analysis can be
replaced by a customer policy on how the hazards are to be treated. A more complete example, however, for a different system would normally include environmental hazards.

A severity scale was created to account for the losses associated with each of the
three categories. The scale used is shown in figure 10.6, but obviously a different
scale could easily be created to match the specific policies or standard practice in
different industries and companies.

As usual, severity was relatively easy to handle but the likelihood of the potential
hazard occurring was unknowable at this early stage of system engineering. In

ID#	Phase	Hazard	Severity H M E
G1	General	Flammable substance in presence of ignition source (Fire)	4 4 4
G2	General	Flammable substance in presence of ignition source in confined space (explosion)	4 4 4
G3	General	Loss of life support (includes power, temperature, oxygen, air pressure, food, water, ...)	4 4 4
G4	General	Crew injury or illness	4 4 1
G5	General	Solar or nuclear radiation exceeding safe levels	3 3 2
G6	General	Collision (with micrometeroids, debris, during rendezvous or separation maneuver, ...)	4 4 4
G7	General	Loss of attitude control	4 4 4
G8	General	Engines do not ignite	4 4 2
PL1	Pre Launch	Damage to payload	2 3 3
PL2	Pre Launch	Launch delay (due to weather, pre launch test failues, etc.)	1 4 1
L1	Launch	Incorrect propulsion/trajectory/control during ascent	4 4 4
L2	Launch	Loss of structural integrity (due to aerodynamic loads, vibrations, ...)	4 4 4
L3	Launch	Incorrect stage separation	4 4 4
E1	EVA in Space	Astronaut lost in space	4 4 1
AS1	Assembly	Incorrect propulsion/control during rendezvous	4 4 4
AS2	Assembly	Inability to dock	1 4 3
AS3	Assembly	Inability to achieve airlock during docking	1 4 3
AS4	Assembly	Inability to undock	4 4 3
T1	Course Change	Incorrect propulsion/trajectory/control during course change burn	4 4 3
D1	Descent	Inability to undock	4 4 3
D2	Descent	Incorrect propulsion/trajectory/control during descent	4 4 4
D3	Descent	Loss of structural integrity (due to inadequate thermal control, aerodynamic loads, vibrations, ...)	4 4 4
AC1	Ascent	Incorrect stage separation (including ascent module disconnecting from descent stage)	4 3 3
AC2	Ascent	Incorrect propulsion/trajectory/control during ascent	4 3 3
AC3	Ascent	Loss of structural integrity (due to aerodynamic loads, vibrations, ...)	4 3 3
S1	Surface Ops	Crew members stranded on Moon/Mars surface during EVA (Extra Vehicle Activity)	4 3 3
S2	Surface Ops	Crew members lost on Moon/Mars surface during EVA	4 3 3
S3	Surface Ops	Equipment damage (includingdamage related to lunar dust)	2 3 3
NP1	Nuclear Power	Nuclear fuel released on Earth surface	4 4 2
NP2	Nuclear Power	Insufficient power generation (reactor does not work)	4 3 3
NP3	Nuclear Power	Insufficient reactor cooling (leading to reactor meltdown)	4 3 3
RE1	Reentry	Inability to undock	4 3 3
RE2	Reentry	Incorrect propulsion/trajectory/control during descent	4 3 3
RE3	Reentry	Loss of structural integrity (due to inadequate thermal control, aerodynamic loads, vibrations, ...)	4 3 4

Figure 10.5
System-level hazards and associated severities.

Severity Level	Human	Mission	Equipment
4	Loss of Life	Mission abort or mission loss	System loss
3	Severe Injury or Illness	Major mission objectives incomplete	Major system damage
2	Minor Injury or illness	Minor mission objectives incomplete	Minor system damage
1	No /insignificant injury or illness	All mission objectives completed	No/insignificant damage

Figure 10.6
Custom severity scale for the candidate architectures analysis.

addition, space exploration is the polar opposite of the ATC example above as the system did not already exist and the architectures and missions would involve things never attempted before, which created a need for a different approach to estimating likelihood.

We decided to use the *mitigation potential* of the hazard in the candidate architecture as an estimator of, or surrogate for, likelihood. Hazards that are more easily mitigated in the design and operations are less likely to lead to accidents. Similarly, hazards that have been eliminated during system design, and thus are not part of that candidate architecture or can easily be eliminated in the detailed design process, cannot lead to an accident.

The safety goal of the architectural analysis process was to assist in selecting the architecture with the fewest serious hazards and highest mitigation potential for those hazards that were not eliminated. Not all hazards will be eliminated even if they can be. One reason for not eliminating hazards might be that it would reduce the potential for achieving other important system goals or constraints. Obviously, safety is not the only consideration in the architecture selection process, but it is important enough in this case to be a criterion in the selection process.

Mitigation potential was chosen as a surrogate for likelihood for two reasons: (1) the potential for eliminating or controlling the hazard in the design or operations has a direct and important bearing on the likelihood of the hazard occurring (whether traditional or new designs and technology are used) and (2) mitigatibility of the hazard can be determined before an architecture or design is selected—indeed, it assists in the selection process.

Figure 10.7 shows an example from the hazard log created during the PHA effort. The example hazard shown is *nuclear reactor overheating*. Nuclear power generation and use, particularly during planetary surface operations, was considered to be an important option in the architectural tradeoffs. The potential accident and its effects are described in the hazard log as:

> Nuclear core meltdown would cause loss of power, and possibly radiation exposure. Surface operations must abort mission and evacuate. If abort is unsuccessful or unavailable at the time, the crew and surface equipment could be lost. There would be no environmental impact on Earth.

The hazard is defined as the nuclear reactor operating at temperatures above the design limits.

Although some causal factors can be hypothesized early, a hazard analysis using STPA can be used to generate a more complete list of causal factors later in the development process to guide the design process after an architecture is chosen.

Like severity, mitigatibility was evaluated by domain experts under the guidance of safety experts. Both the cost of the potential mitigation strategy and its

Hazard Name:	**Nuclear reactor overheating**										

Mission Phase:
(Circle all appropriate)

Pre–Launch　To–Space Launch　In–Space Assembly　To–M Transfer　To–M Descent　**(Surface Exploring)**　From M Ascent　To–E Transfer　In–E Orbit Arriving　On–E Landing　On–E Recovery

Operation/Event:	Ex: docking, lift off, etc. **Power Generation for surface exploration activities**

Vehicle(s)/Systems Affected:	Ex: CEV, DAV, rover, etc. **Surface nuclear power generation, and all systems used on M surface (HAB, DAV, rovers, powered equipment)**

Subsystem(s) Affected:	Ex: engine, heat shield, etc. **Nuclear Reactor, cooling subsystem**

Severity (1–4):

Human	Mission	Equipment	Environment
4	**4**	**3**	**1**

Accident/Effect Description:	What potential losses could reslt from the hazard occurrence? What are the worst potential effects, assuming no mitigation strategies are implemented? What damage could result? Explain the severity ratings provided above. **Nuclear reactor core meltdown would cause loss of power, and possibly radiation exposure. Surface operations must abort mission and evacuate. If abort is unsuccessful or unavailable at the time, the crew could be lost. All surface equipment is lost. No environmental impact on Earth.**

Hazard Description:	Describe the hazard as a system state. What other environmental conditions could influence the effect of the hazard occurrence? **Nuclear reactor operating at temperature above design limits.**

Causal Factors/ Assumptions:	What conditions allow the hazard to occur? **TBD. Possible causes include: thermal control system malfunction, solar radiation protection inadequate, insufficient radiator heat rejection.**

Mitigation Strategy:		Cost/Difficulty (L,M,H)	Mitigation Priority (1–4)
	1. Surface power generation does not rely on nuclear technology	M	4
	2. Backup power generation sytem is available for surface operations	H	1

Figure 10.7
A sample from the hazard log generated during the preliminary hazard analysis for the space architecture candidate tradeoff analysis.

Level	General Description	Detailed Description
4	Eliminate	Complete elimination of the hazard from the design
3	Prevent	Reduction of the likelihood that the hazard will occur
2	Control	Reduction of the likelihood that the hazard results in an accident
1	Reduce Damage	Reduction of damage if an accident does occur

Figure 10.8
A sample hazard-mitigation priority scale.

effectiveness were evaluated. For the nuclear power example, two strategies were identified: the first is not to use nuclear power generation at all. The cost of this option was evaluated as medium (on a low, medium, high scale). But the mitigation potential was rated as high because it eliminates the hazard completely. The mitigation priority scale used is shown in figure 10.8. The second mitigation potential identified by the engineers was to provide a backup power generation system for surface operations. The difficulty and cost was rated high and the mitigation rating was 1, which was the lowest possible level, because at best it would only reduce the damage if an accident occurred but potential serious losses would still occur. Other mitigation strategies are also possible but have been omitted from the sample hazard log entry shown.

None of the effort expended here is wasted. The information included in the hazard log about the mitigation strategies will be useful later in the design process if the final architecture selected uses surface nuclear power generation. NASA might also be able to use the information in future projects and the creation of such early risk analysis information might be common to companies or industries and not have to be created for each project. As new technologies are introduced to an industry, new hazards or mitigation possibilities could be added to the previously stored information.

The final step in the process is to create safety risk metrics for each candidate architecture. Because the system engineers on the project created hundreds of feasible architectures, the evaluation process was automated. The actual details of the mathematical procedures used are of limited general interest and are available elsewhere [59]. Weighted averages were used to combine mitigation factors and severity factors to come up with a final *Overall Residual Safety-Risk Metric*. This metric was then used in the evaluation and ranking of the potential manned space exploration architectures.

By selecting and deselecting options in the architecture description, it was also possible to perform a first-order assessment of the relative importance of each architectural option in determining the Overall Residual Safety-Risk Metric.

While hundreds of parameters were considered in the risk analysis, the process allowed the identification of major contributors to the hazard mitigation potential of selected architectures and thus informed the architecture selection process and the tradeoff analysis. For example, important contributors to increased safety were determined to include the use of heavy module and equipment prepositioning on the surface of Mars and the use of minimal rendezvous and docking maneuvers. Prepositioning modules allows for pretesting and mitigates the hazards associated with loss of life support, equipment damage, and so on. On the other hand, prepositioning modules increases the reliance on precision landing to ensure that all landed modules are within range of each other. Consequently, using heavy prepositioning may require additional mitigation strategies and technology development to reduce the risk associated with landing in the wrong location. All of this information must be considered in selecting the best architecture. As another example, on one hand, a transportation architecture requiring no docking at Mars orbit or upon return to Earth inherently mitigates hazards associated with collisions or failed rendezvous and docking maneuvers. On the other hand, having the capability to dock during an emergency, even though it is not required during nominal operations, provides additional mitigation potential for loss of life support, especially in Earth orbit.

Reducing these considerations to a number is clearly not ideal, but with hundreds of potential architectures it was necessary in this case in order to pare down the choices to a smaller number. More careful tradeoff analysis is then possible on the reduced set of choices.

While mitigatibility is widely applicable as a surrogate for likelihood in many types of domains, the actual process used above is just one example of how it might be used. Engineers will need to adapt the scales and other features of the process to the customary practices in their own industry. Other types of surrogates or ways to handle likelihood estimates in early phases of projects are possible beyond the two examples provided in this section. While none of these approaches is ideal, they are much better than ignoring safety in decision making or selecting likelihood estimates based solely on wishful thinking or the politics that often surround the preliminary hazard analysis process.

After a conceptual design is chosen, development begins.

10.3.5 Documenting Environmental Assumptions

An important part of the system development process is to determine and document the assumptions under which the system requirements and design features are derived and upon which the hazard analysis is based. Assumptions will be identified and specified throughout the system engineering process and the engineering specifications to explain decisions or to record fundamental information upon which the design is based. If the assumptions change over time or the system changes and the assumptions are no longer true, then the requirements and the safety constraints and design features based on those assumptions need to be revisited to ensure safety has not been compromised by the change.

Because operational safety depends on the accuracy of the assumptions and models underlying the design and hazard analysis processes, the operational system should be monitored to ensure that:

1. The system is constructed, operated, and maintained in the manner assumed by the designers.

2. The models and assumptions used during initial decision making and design are correct.

3. The models and assumptions are not violated by changes in the system, such as workarounds or unauthorized changes in procedures, or by changes in the environment.

Operational feedback on trends, incidents, and accidents should trigger reanalysis when appropriate. Linking the assumptions throughout the document with the parts of the hazard analysis based on that assumption will assist in performing safety maintenance activities.

Several types of assumptions are relevant. One is the assumptions under which the system will be used and the environment in which the system will operate. Not only will these assumptions play an important role in system development, but they also provide part of the basis for creating the operational safety control structure and other operational safety controls such as creating feedback loops to ensure the assumptions underlying the system design and the safety analyses are not violated during operations as the system and its environment change over time.

While many of the assumptions that originate in the existing environment into which the new system will be integrated can be identified at the beginning of development, additional assumptions will be identified as the design process continues and new requirements and design decisions and features are identified. In addition, assumptions that the emerging system design imposes on the surrounding environment will become clear only after detailed decisions are made in the design and safety analyses.

Examples of important environment assumptions for TCAS II are that:

EA1: *High-integrity communications exist between aircraft.*

EA2: *The TCAS-equipped aircraft carries a Mode-S air traffic control transponder.*[3]

---

3. An aircraft *transponder* sends information to help air traffic control maintain aircraft separation. Primary radar generally provides bearing and range position information, but lacks altitude information. Mode A transponders transmit only an identification signal, while Mode C and Mode S transponders also report pressure altitude. Mode S is newer and has more capabilities than Mode C, some of which are required for the collision avoidance functions in TCAS.

EA3: *All aircraft have operating transponders.j*

EA4: *All aircraft have legal identification numbers.*

EA5: *Altitude information is available from intruding targets with a minimum precision of 100 feet.*

EA6: *The altimetry system that provides own aircraft pressure altitude to the TCAS equipment will satisfy the requirements in RTCA Standard . . .*

EA7: *Threat aircraft will not make an abrupt maneuver that thwarts the TCAS escape maneuver.*

As noted, these assumptions must be enforced in the overall safety control structure. With respect to assumption EA4, for example, identification numbers are usually provided by the aviation authorities in each country, and that requirement will need to be ensured by international agreement or by some international agency. The assumption that aircraft have operating transponders (EA3) may be enforced by the airspace rules in a particular country and, again, must be ensured by some group. Clearly, these assumptions play an important role in the construction of the safety control structure and assignments of responsibilities for the final system. For TCAS, some of these assumptions will already be imposed by the existing air transportation safety control structure while others may need to be added to the responsibilities of some group(s) in the control structure. The last assumption, EA7, imposes constraints on pilots and the air traffic control system.

Environment requirements and constraints may lead to restrictions on the use of the new system (in this case, TCAS) or may indicate the need for system safety and other analyses to determine the constraints that must be imposed on the system being created (TCAS again) or the larger encompassing system to ensure safety. The requirements for the integration of the new subsystem safely into the larger system must be determined early. Examples for TCAS include:

E1: *The behavior or interaction of non-TCAS equipment with TCAS must not degrade the performance of the TCAS equipment or the performance of the equipment with which TCAS interacts.*

E2: *Among the aircraft environmental alerts, the hierarchy shall be: Windshear has first priority, then the Ground Proximity Warning System (GPWS), then TCAS.*

E3: *The TCAS alerts and advisories must be independent of those using the master caution and warning system.*

10.3.6 System-Level Requirements Generation

Once the goals and hazards have been identified and a conceptual system architecture has been selected, system-level requirements generation can begin. Usually, in the early stages of a project, goals are stated in very general terms, as shown in G1 and G2. One of the first steps in the design process is to refine the goals into testable and achievable high-level requirements (the "shall" statements). Examples of high-level functional requirements implementing the goals for TCAS are:

1.18: *TCAS shall provide collision avoidance protection for any two aircraft closing horizontally at any rate up to 1200 knots and vertically up to 10,000 feet per minute.*

> **Assumption:** *This requirement is derived from the assumption that commercial aircraft can operate up to 600 knots and 5000 fpm during vertical climb or controlled descent (and therefore two planes can close horizontally up to 1200 knots and vertically up to 10,000 fpm).*

1.19.1: *TCAS shall operate in enroute and terminal areas with traffic densities up to 0.3 aircraft per square nautical miles (i.e., 24 aircraft within 5 nmi).*

> **Assumption:** *Traffic density may increase to this level by 1990, and this will be the maximum density over the next 20 years.*

As stated earlier, *assumptions* should continue to be specified when appropriate to explain a decision or to record fundamental information on which the design is based. Assumptions are an important component of the documentation of design rationale and form the basis for safety audits during operations. Consider the above requirement labeled 1.18, for example. In the future, if aircraft performance limits change or there are proposed changes in airspace management, the origin of the specific numbers in the requirement (1,200 and 10,000) can be determined and evaluated for their continued relevance. In the absence of the documentation of such assumptions and how they impact the detailed design decisions, numbers tend to become "gospel," and everyone is afraid to change them.

Requirements (and constraints) must also be included for the human operator and for the human–computer interface. These requirements will in part be derived from the *concept of operations*, which should in turn include a *human task analysis* [48, 47], to determine how TCAS is expected to be used by pilots (which, again, should be checked in safety audits during operations). These analyses use information about the goals of the system, the constraints on how the goals are achieved, including safety constraints, how the automation will be used, how humans now control the system and work in the system without automation, and the tasks humans need to perform and how the automation will support them in performing these tasks. The task analysis must also consider workload and its impact on operator performance. Note that a low workload may be more dangerous than a high one.

Requirements on the operator (in this case, the pilot) are used to guide the design of the TCAS-pilot interface, the design of the automation logic, flight-crew tasks and procedures, aircraft flight manuals, and training plans and program. Traceability links should be provided to show the relationships. Links should also be provided to the parts of the hazard analysis from which safety-related requirements are derived. Examples of TCAS II operator safety requirements and constraints are:

OP.4: *After the threat is resolved, the pilot shall return promptly and smoothly to his/her previously assigned fight path (→ HA-560, ↓3.3).*

OP.9: *The pilot must not maneuver on the basis of a Traffic Advisory only (→ HA-630, ↓2.71.3).*

The requirements and constraints include links to the hazard analysis that produced the information and to design documents and decisions to show where the requirements are applied. These two examples have links to the parts of the hazard analysis from which they were derived, links to the system design and operator procedures where they are enforced, and links to the user manuals (in this case, the pilot manuals) to explain why certain activities or behaviors are required.

The links not only provide traceability from requirements to implementation and vice versa to assist in review activities, but they also embed the design rationale information into the specification. If changes need to be made to the system, it is easy to follow the links and determine why and how particular design decisions were made.

10.3.7 Identifying High-Level Design and Safety Constraints

Design constraints are restrictions on how the system can achieve its purpose. For example, TCAS is not allowed to interfere with the ground-level air traffic control system while it is trying to maintain adequate separation between aircraft. Avoiding interference is not a goal or purpose of TCAS—the best way to achieve the goal is not to build the system at all. It is instead a constraint on how the system can achieve its purpose, that is, a constraint on the potential system designs. Because of the need to evaluate and clarify tradeoffs among alternative designs, separating these two types of intent information (goals and design constraints) is important.

For safety-critical systems, constraints should be further separated into safety-related and not safety-related. One nonsafety constraint identified for TCAS, for example, was that requirements for new hardware and equipment on the aircraft be minimized or the airlines would not be able to afford this new collision avoidance system. Examples of nonsafety constraints for TCAS II are:

C.1: *The system must use the transponders routinely carried by aircraft for ground ATC purposes (↓2.3, 2.6).*

Rationale: *To be acceptable to airlines, TCAS must minimize the amount of new hardware needed.*

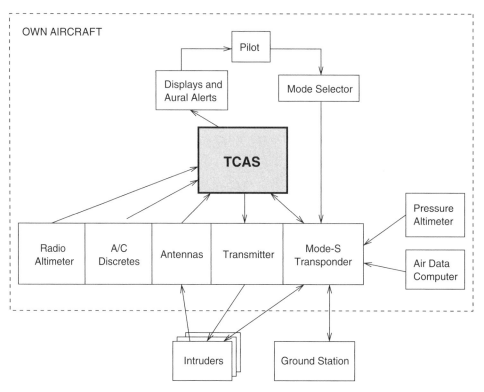

Figure 10.9
The system interface topology for TCAS.

> **C.4:** *TCAS must comply with all applicable FAA and FCC policies, rules, and philosophies (↓2.30, 2.79).*

The physical environment with which TCAS interacts is shown in figure 10.9. The constraints imposed by these existing environmental components must also be identified before system design can begin.

Safety-related constraints should have two-way links to the system hazard log and to any analysis results that led to that constraint being identified as well as links to the design features (usually level 2) included to eliminate or control them. Hazard analyses are linked to level 1 requirements and constraints, to design features on level 2, and to system limitations (or accepted risks). An example of a level 1 safety constraint derived to prevent hazards is:

> **SC.3:** *TCAS must generate advisories that require as little deviation as possible from ATC clearances (→ H6, HA-550, ↓2.30).*

The link in SC.3 to <u>2.30</u> points to the level 2 system design feature that implements this safety constraint. The other links provide traceability to the hazard (H6) from which the constraint was derived and to the parts of the hazard analysis involved, in this case the part of the hazard analysis labeled HA-550.

The following is another example of a safety constraint for TCAS II and some constraints refined from it, all of which stem from a high-level environmental constraint derived from safety considerations in the encompassing system into which TCAS will be integrated. The refinement will occur as safety-related decisions are made and guided by an STPA hazard analysis:

SC.2: *TCAS must not interfere with the ground ATC system or other aircraft transmissions to the ground ATC system* ($\to$ <u>H5</u>).

> **SC.2.1:** *The system design must limit interference with ground-based secondary surveillance radar, distance-measuring equipment channels, and with other radio services that operate in the 1030/1090 MHz frequency band* ($\downarrow$<u>2.5.1</u>).

>> **SC.2.1.1:** *The design of the Mode S waveforms used by TCAS must provide compatibility with Modes A and C of the ground-based secondary surveillance radar system* ($\downarrow$<u>2.6</u>).

>> **SC.2.1.2:** *The frequency spectrum of Mode S transmissions must be controlled to protect adjacent distance-measuring equipment channels* ($\downarrow$<u>2.13</u>).

>> **SC.2.1.3:** *The design must ensure electromagnetic compatibility between TCAS and [...]* [$\downarrow$<u>21.4</u>].

> **SC.2.2:** *Multiple TCAS units within detection range of one another (approximately 30 nmi) must be designed to limit their own transmissions. As the number of such TCAS units within this region increases, the interrogation rate and power allocation for each of them must decrease in order to prevent undesired interference with ATC* ($\downarrow$<u>2.13</u>).

Assumptions are also associated with safety constraints. As an example of such an assumption, consider:

SC.6: *TCAS must not disrupt the pilot and ATC operations during critical phases of flight nor disrupt aircraft operation* ($\to$ <u>H3</u>, $\downarrow$<u>2.2.3</u>, <u>2.19</u>, <u>2.24.2</u>).

> **SC.6.1:** *The pilot of a TCAS-equipped aircraft must have the option to switch to the Traffic-Advisory-Only mode where TAs are displayed but display of resolution advisories is inhibited* ($\downarrow$ <u>2.2.3</u>).

> **Assumption:** *This feature will be used during final approach to parallel runways, when two aircraft are projected to come close to each other and TCAS would call for an evasive maneuver (↓ 6.17).*

The specified assumption is critical for evaluating safety during operations. Humans tend to change their behavior over time and use automation in different ways than originally intended by the designers. Sometimes, these new uses are dangerous. The hyperlink at the end of the assumption (↓ 6.17) points to the required auditing procedures for safety during operations and to where the procedures for auditing this assumption are specified.

Where do these safety constraints come from? Is the system engineer required to simply make them up? While domain knowledge and expertise is always going to be required, there are procedures that can be used to guide this process.

The highest-level safety constraints come directly from the identified hazards for the system. For example, TCAS must not cause or contribute to a near miss (H1), TCAS must not cause or contribute to a controlled maneuver into the ground (H2), and TCAS must not interfere with the ground-based ATC system. STPA can be used to refine these high-level design constraints into more detailed design constraints as described in chapter 8.

The first step in STPA is to create the high-level TCAS operational safety control structure. For TCAS, this structure is shown in figure 10.10. For simplicity, much of the structure above ATC operations management has been omitted and the roles and responsibilities have been simplified here. In a real design project, roles and responsibilities will be augmented and refined as development proceeds, analyses are performed, and design decisions are made. Early in the system concept formation, specific roles may not all have been determined, and more will be added as the design concepts are refined. One thing to note is that there are three groups with potential responsibilities over the pilot's response to a potential NMAC: TCAS, the ground ATC, and the airline operations center which provides the airline procedures for responding to TCAS alerts. Clearly any potential conflicts and coordination problems between these three controllers will need to be resolved in the overall air traffic management system design. In the case of TCAS, the designers decided that because there was no practical way, at that time, to downlink information to the ground controllers about any TCAS advisories that might have been issued for the crew, the pilot was to immediately implement the TCAS advisory and the co-pilot would transmit the TCAS alert information by radio to ground ATC. The airline would provide the appropriate procedures and training to implement this protocol.

Part of defining this control structure involves identifying the responsibilities of each of the components related to the goal of the system, in this case collision avoidance. For TCAS, these responsibilities include:

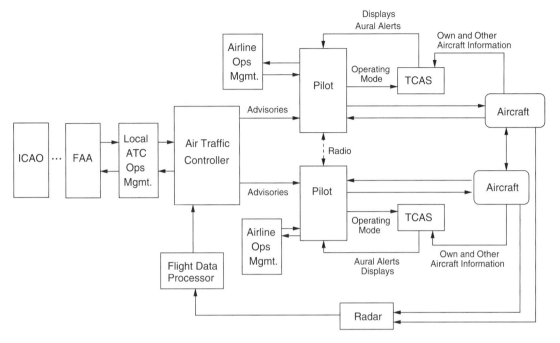

Figure 10.10
The high-level operational TCAS control structure.

- *Aircraft Components* (e.g., transponders, antennas): Execute control maneuvers, read and send messages to other aircraft, etc.

- *TCAS:* Receive information about its own and other aircraft, analyze the information received and provide the pilot with (1) information about where other aircraft in the vicinity are located and (2) an escape maneuver to avoid potential NMAC threats.

- *Aircraft Components* (e.g., transponders, antennas): Execute pilot-generated TCAS control maneuvers, read and send messages to and from other aircraft, etc.

- *Pilot:* Maintain separation between own and other aircraft, monitor the TCAS displays, and implement TCAS escape maneuvers. The pilot must also follow ATC advisories.

- *Air Traffic Control:* Maintain separation between aircraft in the controlled airspace by providing advisories (control actions) for the pilot to follow. TCAS is designed to be independent of and a backup for the air traffic controller so ATC does not have a direct role in the TCAS safety control structure but clearly has an indirect one.

- *Airline Operations Management:* Provide procedures for using TCAS and following TCAS advisories, train pilots, and audit pilot performance.
- *ATC Operations Management:* Provide procedures, train controllers, audit performance of controllers and of the overall collision avoidance system.
- *ICAO:* Provide worldwide procedures and policies for the use of TCAS and provide oversight that each country is implementing them.

After the general control structure has been defined (or alternative candidate control structures identified), the next step is to determine how the controlled system (the two aircraft) can get into a hazardous state. That information will be used to generate safety constraints for the designers. STAMP assumes that hazardous states (states that violate the safety constraints) are the result of ineffective control. Step 1 of STPA is to identify the potentially inadequate control actions.

Control actions in TCAS are called resolution advisories or RAs. An RA is an aircraft escape maneuver created by TCAS for the pilots to follow. Example resolution advisories are DESCEND, INCREASE RATE OF CLIMB TO 2500 FMP, and DON'T DESCEND. Consider the TCAS component of the control structure (see figure 10.10) and the NMAC hazard. The four types of control flaws for this example translate into:

1. The aircraft are on a near collision course, and TCAS does not provide an RA that avoids it (that is, does not provide an RA, or provides an RA that does not avoid the NMAC).
2. The aircraft are in close proximity and TCAS provides an RA that degrades vertical separation (causes an NMAC).
3. The aircraft are on a near collision course and TCAS provides a maneuver too late to avoid an NMAC.
4. TCAS removes an RA too soon.

These inadequate control actions can be restated as high-level constraints on the behavior of TCAS:

1. TCAS must provide resolution advisories that avoid near midair collisions.
2. TCAS must not provide resolution advisories that degrade vertical separation between two aircraft (that is, cause an NMAC).
3. TCAS must provide the resolution advisory while enough time remains for the pilot to avoid an NMAC. (A human factors and aerodynamic analysis should be performed at this point to determine exactly how much time that implies.)
4. TCAS must not remove the resolution advisory before the NMAC is resolved.

Similarly, for the pilot, the inadequate control actions are:

1. The pilot does not provide a control action to avoid a near midair collision.

2. The pilot provides a control action that does not avoid the NMAC.

3. The pilot provides a control action that causes an NMAC that would not otherwise have occurred.

4. The pilot provides a control action that could have avoided the NMAC but it was too late.

5. The pilot starts a control action to avoid an NMAC but stops it too soon.

Again, these inadequate pilot control actions can be restated as safety constraints that can be used to generate pilot procedures. Similar hazardous control actions and constraints must be identified for each of the other system components. In addition, inadequate control actions must be identified for the other functions provided by TCAS (beyond RAs) such as traffic advisories.

Once the high-level design constraints have been identified, they must be refined into more detailed design constraints to guide the system design and then augmented with new constraints as design decisions are made, creating a seamless integrated and iterative process of system design and hazard analysis.

Refinement of the constraints involves determining how they could be violated. The refined constraints will be used to guide attempts to eliminate or control the hazards in the system design or, if that is not possible, to prevent or control them in the system or component design. This process of scenario development is exactly the goal of hazard analysis and STPA. As an example of how the results of the analysis are used to refine the high-level safety constraints, consider the second high-level TCAS constraint: that TCAS must not provide resolution advisories that degrade vertical separation between two aircraft (cause an NMAC):

SC.7: *TCAS must not create near misses (result in a hazardous level of vertical separation that would not have occurred had the aircraft not carried TCAS) ($\rightarrow$ H1).*

> **SC.7.1:** *Crossing Maneuvers must be avoided if possible ($\downarrow$ 2.36, $\downarrow$ 2.38, $\downarrow$ 2.48, $\downarrow$ 2.49.2).*

> **SC.7.2:** *The reversal of a displayed advisory must be extremely rare[4] ($\downarrow$ 2.51, $\downarrow$ 2.56.3, $\downarrow$ 2.65.3, $\downarrow$ 2.66).*

> **SC.7.3:** *TCAS must not reverse an advisory if the pilot will have insufficient time to respond to the RA before the closest point of approach (four seconds*

---

4. This requirement is clearly vague and untestable. Unfortunately, I could find no definition of "extremely rare" in any of the TCAS documentation to which I had access.

or less) or if own and intruder aircraft are separated by less than 200 feet vertically when ten seconds or less remain to closest point of approach (↓ 2.52).

Note again that pointers are used to trace these constraints into the design features used to implement them.

10.3.8 System Design and Analysis

Once the basic requirements and design constraints have been at least partially specified, the system design features that will be used to implement them must be created. A strict top-down design process is, of course, not usually feasible. As design decisions are made and the system behavior becomes better understood, additions and changes will likely be made in the requirements and constraints. The specification of assumptions and the inclusion of traceability links will assist in this process and in ensuring that safety is not compromised by later decisions and changes. It is surprising how quickly the rationale behind the decisions that were made earlier is forgotten.

Once the system design features are determined, (1) an internal control structure for the system itself is constructed along with the interfaces between the components and (2) functional requirements and design constraints, derived from the system-level requirements and constraints, are allocated to the individual system components.

System Design

What has been presented so far in this chapter would appear in level 1 of an intent specification. The second level of an intent specification contains *System Design Principles*—the basic system design and scientific and engineering principles needed to achieve the behavior specified in the top level, as well as any derived requirements and design features not related to the level 1 requirements.

While traditional design processes can be used, STAMP and STPA provide the potential for safety-driven design. In safety-driven design, the refinement of the high-level hazard analysis is intertwined with the refinement of the system design to guide the development of the system design and system architecture. STPA can be used to generate safe design alternatives or applied to the design alternatives generated in some other way to continually evaluate safety as the design progresses and to assist in eliminating or controlling hazards in the emerging design, as described in chapter 9.

For TCAS, this level of the intent specification includes such general principles as the basic *tau* concept, which is related to all the high-level alerting goals and constraints:

2.2: *Each TCAS-equipped aircraft is surrounded by a protected volume of airspace. The boundaries of this volume are shaped by the tau and DMOD criteria (↑1.20.3).*

2.2.1: *TAU: In collision avoidance, time-to-go to the closest point of approach (CPA) is more important than distance-to-go to the CPA. Tau is an approximation of the time in seconds to CPA. Tau equals 3600 times the slant range in nmi, divided by the closing speed in knots.*

2.2.2: *DMOD: If the rate of closure is very low, a target could slip in very close without crossing the tau boundaries and triggering an advisory. In order to provide added protection against a possible maneuver or speed change by either aircraft, the tau boundaries are modified (called DMOD). DMOD varies depending on own aircraft's altitude regime (→ 2.2.4).*

The principles are linked to the related higher-level requirements, constraints, assumptions, limitations, and hazard analysis as well as to lower-level system design and documentation and to other information at the same level. Assumptions used in the formulation of the design principles should also be specified at this level.

For example, design principle 2.51 (related to safety constraint SC-7.2 shown in the previous section) describes how sense[5] reversals are handled:

2.51: *Sense Reversals: (↓ Reversal-Provides-More-Separation) In most encounter situations, the resolution advisory will be maintained for the duration of an encounter with a threat aircraft (↑SC-7.2). However, under certain circumstances, it may be necessary for that sense to be reversed. For example, a conflict between two TCAS-equipped aircraft will, with very high probability, result in selection of complementary advisory senses because of the coordination protocol between the two aircraft. However, if coordination communication between the two aircraft is disrupted at a critical time of sense selection, both aircraft may choose their advisories independently (↑HA-130). This could possibly result in selection of incompatible senses (↑HA-395).*

2.51.1: *. . . [information about how incompatibilities are handled]*

Design principle 2.51 describes the conditions under which reversals of TCAS advisories can result in incompatible senses and lead to the creation of a hazard by TCAS. The pointer labeled HA-395 points to the part of the hazard analysis analyzing that problem. The hazard analysis portion labeled HA-395 would have a complementary pointer to section 2.51. The design decisions made to handle such

---

5. The *sense* is the direction of the advisory, such as descend or climb.

incompatibilities are described in 2.51.1, but that part of the specification is omitted here. 2.51 also contains a hyperlink (↓Reversal-Provides-More-Separation) to the detailed functional level 3 logic (component black-box requirements specification) used to implement the design decision.

Information about the allocation of these design decisions to individual system components and the logic involved is located in level 3, which in turn has links to the implementation of the logic in lower levels. If a change has to be made to a system component (such as a change to a software module), it is possible to trace the function computed by that module upward in the intent specification levels to determine whether the module is safety critical and if (and how) the change might affect system safety.

As another example, the TCAS design has a built-in bias against generating advisories that would result in the aircraft crossing paths (called *altitude crossing advisories*).

> **2.36.2:** *A bias against altitude crossing RAs is also used in situations involving intruder level-offs at least 600 feet above or below the TCAS aircraft (↑SC.7.1). In such a situation, an altitude-crossing advisory is deferred if an intruder aircraft that is projected to cross own aircraft's altitude is more than 600 feet away vertically (↓ Alt_Separation_Test).*
>
> **Assumption:** *In most cases, the intruder will begin a level-off maneuver when it is more than 600 feet away and so should have a greatly reduced vertical rate by the time it is within 200 feet of its altitude clearance (thereby either not requiring an RA if it levels off more than ZTHR[6] feet away or requiring a non-crossing advisory for level-offs begun after ZTHR is crossed but before the 600 foot threshold is reached).*

Again, the example above includes a pointer down to the part of the black box component requirements (functional) specification (*Alt_Separation_Test*) that embodies the design principle. Links could also be provided to detailed mathematical analyses used to support and validate the design decisions.

As another example of using links to embed design rationale in the specification and of specifying limitations (defined later) and potential hazardous behavior that could not be controlled in the design, consider the following. TCAS II advisories may need to be inhibited because of an inadequate climb performance for the particular aircraft on which TCAS is installed. The collision avoidance maneuvers posted as advisories (called RAs or resolution advisories) by TCAS assume an aircraft's ability to safely achieve them. If it is likely they are beyond the capability

---

6. The vertical dimension, called ZTHR, used to determine whether advisories should be issued varies from 750 to 950 feet, depending on the TCAS aircraft's altitude.

of the aircraft, then TCAS must know beforehand so it can change its strategy and issue an alternative advisory. The performance characteristics are provided to TCAS through the aircraft interface (via what are called *aircraft discretes*). In some cases, no feasible solutions to the problem could be found. An example design principle related to this problem found at level 2 of the TCAS intent specification is:

> **2.39:** *Because of the limited number of inputs to TCAS for aircraft, performance inhibits, in some instances where inhibiting RAs would be appropriate it is not possible to do so (↑L6). In these cases, TCAS may command maneuvers that may significantly reduce stall margins or result in stall warning (↑SC9.1). Conditions where this may occur include . . . The aircraft flight manual or flight manual supplement should provide information concerning this aspect of TCAS so that flight crews may take appropriate action (↓ [Pointers to pilot procedures on level 3 and Aircraft Flight Manual on level 6).*

Finally, design principles may reflect tradeoffs between higher-level goals and constraints. As examples:

> **2.2.3:** *Tradeoffs must be made between necessary protection (↑1.18) and unnecessary advisories (↑SC.5, SC.6). This is accomplished by controlling the sensitivity level, which controls the tau, and therefore the dimensions of the protected airspace around each TCAS-equipped aircraft. The greater the sensitivity level, the more protection is provided but the higher is the incidence of unnecessary alerts. Sensitivity level is determined by . . .*

> **2.38:** *The need to inhibit* CLIMB *RAs because of inadequate aircraft climb performance will increase the likelihood of TCAS II (a) issuing crossing maneuvers, which in turn increases the possibility that an RA may be thwarted by the intruder maneuvering (↑SC7.1, HA-115), (b) causing an increase in* DESCEND *RAs at low altitude (↑SC8.1), and (c) providing no RAs if below the descend inhibit level (1200 feet above ground level on takeoff and 1000 feet above ground level on approach).*

Architectural Design, Functional Allocation, and Component Implementation (Level 3)

Once the general system design concepts are agreed upon, the next step usually involves developing the design architecture and allocating behavioral requirements and constraints to the subsystems and components. Once again, two-way tracing should exist between the component requirements and the system design principles and requirements. These links will be available to the subsystem developers to be used in their implementation and development activities and in verification (testing and reviews). Finally, during field testing and operations, the links and recorded assumptions and design rationale can be used in safety change analysis, incident and accident analysis, periodic audits, and performance monitoring as required to ensure that the operational system is and remains safe.

Level 3 of an intent specification contains the system architecture, that is, the allocation of functions to components and the designed communication paths among those components (including human operators). At this point, a black-box functional requirements specification language becomes useful, particularly a formal language that is executable. SpecTRM-RL is used as the example specification language in this section [85, 86]. An early version of the language was developed in 1990 to specify the requirements for TCAS II and has been refined and improved since that time. SpecTRM-RL is part of a larger specification management system called SpecTRM (Specification Tools and Requirements Methodology). Other languages, of course, can be used.

One of the first steps in low-level architectural design is to break the system into a set of components. For TCAS, only three components were used: surveillance, collision avoidance, and performance monitoring.

The environment description at level 3 includes the assumed behavior of the external components (such as the altimeters and transponders for TCAS), including perhaps failure behavior, upon which the correctness of the system design is predicated, along with a description of the interfaces between the TCAS system and its environment. Figure 10.11 shows part of a SpecTRM-RL description of an environment component, in this case an altimeter.

RADIO ALTIMETER

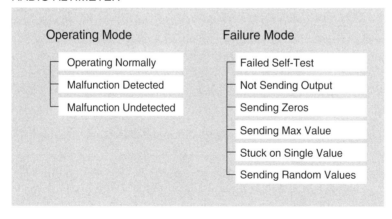

Figure 10.11
Part of the SpecTRM-RL description of an environment component (a radio altimeter). Modeling failure behavior is especially important for safety analyses. In this example, (1) the altimeter may be operating correctly, (2) it may have failed in a way that the failure can be detected by TCAS II (that is, it fails a self-test and sends a status message to TCAS or it is not sending any output at all), or (3) the malfunctioning is undetected and it sends an incorrect radio altitude.

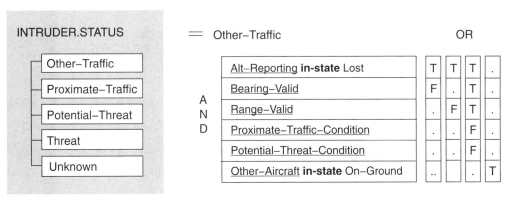

Description: A threat is reclassified as other traffic if its altitude reporting has been lost (↟2.13) and either the bearing or range inputs are invalid; if its altitude reporting has been lost and both the range and bearing are valid but neither the proximate nor potential threat classification criteria are satisfied; or the aircraft is on the ground (↟2.12).

Mapping to Level 2: ↟2.23, ↟2.29

Mapping to Level 4: ↟4.7.1, Traffic–Advisory

Figure 10.12
Example from the level 3 SpecTRM-RL model of the collision avoidance logic. It defines the criteria for downgrading the status of an intruder (into our protected volume) from being labeled a threat to being considered simply as other traffic. Intruders can be classified in decreasing order of importance as a threat, a potential threat, proximate traffic, and other traffic. In the example, the criterion for taking the transition from state *Threat* to state *Other Traffic* is represented by an AND/OR table, which evaluates to TRUE if any of its columns evaluates to TRUE. A column is TRUE if all of its rows that have a "T" are TRUE and all of its rows with an "F" are FALSE. Rows containing a dot represent "don't care" conditions.

A system is an abstraction and the system boundaries can be set anywhere convenient for the purposes of the specifier. In this example, the environment includes any component that was already on the aircraft or in the airspace control system and was not newly designed or built as part of the TCAS effort.

All communications between the system and external components need to be described in detail, including the designed interfaces. The black-box behavior of each component also needs to be specified. This specification serves as the functional requirements for the components. What is included in the component specification will depend on whether the component is part of the environment or part of the system being constructed. Figure 10.12 shows part of the SpecTRM-RL description of the behavior of the CAS (collision avoidance system) subcomponent. SpecTRM-RL specifications are intended to be both easily readable with minimum instruction and formally analyzable. They are also executable and can be used in a system simulation environment. Readability was a primary goal in the design of SpecTRM-RL, as was completeness with regard to safety. Most of the requirements completeness criteria described in *Safeware* and rewritten as functional design principles in chapter 9 of this book are included in the syntax of the language to assist in system safety reviews of the requirements.

SpecTRM-RL explicitly shows the process model used by the controller and describes the required behavior in terms of this model. A state machine model is used to describe the system component's process model, in this case the state of the aircraft and the air space around it, and the ways the process model can change state.

Logical behavior is specified in SpecTRM-RL using AND/OR tables. Figure 10.12 shows a small part of the specification of the TCAS collision avoidance logic. For TCAS, an important state variable is the status of the other aircraft around the TCAS aircraft, called *intruders*. Intruders are classified into four groups: Other Traffic, Potential Threat, and Threat. The figure shows the logic for classifying an intruder as Other Traffic using an AND/OR table. The information in the tables can be visualized in additional ways.

The rows of the table represent AND relationships, while the columns represent OR. The state variable takes the specified value (in this case, *Other Traffic*) if any of the columns evaluate to TRUE. A column evaluates to TRUE if all the rows have the value specified for that row in the column. A dot in the table indicates that the value for the row is irrelevant. Underlined variables represent hyperlinks. For example, clicking on "Alt Reporting" would show how the Alt Reporting variable is defined: In our TCAS intent specification[7] [121], the altitude report for an aircraft is defined as *Lost* if no valid altitude report has been received in the past six seconds. Bearing Valid, Range Valid, Proximate Traffic Condition, and Proximate Threat Condition are *macros*, which simply means that they are defined using separate logic tables. The additional logic for the macros could have been inserted here, but sometimes the logic gets very complex and it is easier for specifiers and reviewers if, in those cases, the tables are broken up into smaller pieces (a form of refinement abstraction). This decision is, of course, up to the creator of the table.

The behavioral descriptions at this level are purely black-box: They describe the inputs and outputs of each component and their relationships *only* in terms of externally visible behavior. Essentially it represents the transfer function across the component. Any of these components (except the humans, of course) could be implemented either in hardware or software. Some of the TCAS surveillance

---

7. A SpecTRM-RL model of TCAS was created by the author and her students Jon Reese, Mats Heimdahl, and Holly Hildreth to assist in the certification of TCAS II. Later, as an experiment to show the feasibility of creating intent specifications, the author created the level 1 and level 2 intent specification for TCAS. Jon Reese rewrote the level 3 collision avoidance system logic from the early version of the language into SpecTRM-RL.

functions are, in fact, implemented using analog devices by some vendors and digital by others. Decisions about physical implementation, software design, internal variables, and so on are limited to levels of the specification below this one. Thus, this level serves as a rugged interface between the system designers and the component designers and implementers (including subcontractors).

Software need not be treated any differently than the other parts of the system. Most safety-related software problems stem from requirements flaws. The system requirements and system hazard analysis should be used to determine the behavioral safety constraints that must be enforced on software behavior and that the software must enforce on the controlled system. Once that is accomplished, those requirements and constraints are passed to the software developers (through the black-box requirements specifications), and they use them to generate and validate their designs just as the hardware developers do.

Other information at this level might include flight crew requirements such as description of tasks and operational procedures, interface requirements, and the testing requirements for the functionality described on this level. If the black-box requirements specification is executable, system testing can be performed early to validate requirements using system and environment simulators or hardware-in-the-loop simulation. Including a visual operator task-modeling language permits integrated simulation and analysis of the entire system, including human–computer interactions [15, 177].

Models at this level are reusable, and we have found that these models provide the best place to provide component reuse and build component libraries [119]. Reuse of application software at the code level has been problematic at best, contributing to a surprising number of accidents [116]. Level 3 black-box behavioral specifications provide a way to make the changes almost always necessary to reuse software in a format that is both reviewable and verifiable. In addition, the black-box models can be used to maintain the system and to specify and validate changes before they are made in the various manufacturers' products. Once the changed level 3 specifications have been validated, the links to the modules implementing the modeled behavior can be used to determine which modules need to be changed and how. Libraries of component models can also be developed and used in a plug-and-play fashion, making changes as required, in order to develop product families [211].

The rest of the development process, involving the implementation of the component requirements and constraints and documented at levels 4 and 5 of intent specifications, is straightforward and differs little from what is normally done today.

10.3.9 Documenting System Limitations

When the system is completed, the system limitations need to be identified and documented. Some of the identification will, of course, be done throughout the development. This information is used by management and stakeholders to determine whether the system is adequately safe to use, along with information about each of the identified hazards and how they were handled.

Limitations should be included in level 1 of the intent specification, because they properly belong in the customer view of the system and will affect both acceptance and certification.

Some limitations may be related to the basic functional requirements, such as these:

L4: *TCAS does not currently indicate horizontal escape maneuvers and therefore does not (and is not intended to) increase horizontal separation.*

Limitations may also relate to environment assumptions. For example:

L1: *TCAS provides no protection against aircraft without transponders or with nonoperational transponders (→EA3, HA-430).*

L6: *Aircraft, performance limitations constrain the magnitude of the escape maneuver that the flight crew can safely execute in response to a resolution advisory. It is possible for these limitations to preclude a successful resolution of the conflict (→H3, ↓2.38, 2.39).*

L4: *TCAS is dependent on the accuracy of the threat aircraft's reported altitude. Separation assurance may be degraded by errors in intruder pressure altitude as reported by the transponder of the intruder aircraft (→EA5).*

   Assumption: *This limitation holds for the airspace existing at the time of the initial TCAS deployment, where many aircraft use pressure altimeters rather than GPS. As more aircraft install GPS systems with greater accuracy than current pressure altimeters, this limitation will be reduced or eliminated.*

Limitations are often associated with hazards or hazard causal factors that could not be completely eliminated or controlled in the design. Thus they represent accepted risks. For example,

L3: *TCAS will not issue an advisory if it is turned on or enabled to issue resolution advisories in the middle of a conflict (→ HA-405).*

L5: *If only one of two aircraft is TCAS equipped while the other has only ATCRBS altitude-reporting capability, the assurance of safe separation may be reduced (→ HA-290).*

In the specification, both of these system limitations would have pointers to the relevant parts of the hazard analysis along with an explanation of why they could not be eliminated or adequately controlled in the system design. Decisions about deployment and certification of the system will need to be based partially on these limitations and their impact on the safety analysis and safety assumptions of the encompassing system, which, in the case of TCAS, is the overall air traffic system.

A final type of limitation is related to problems encountered or tradeoffs made during system design. For example, TCAS has a high-level performance-monitoring requirement that led to the inclusion of a self-test function in the system design to determine whether TCAS is operating correctly. The following system limitation relates to this self-test facility:

L9: *Use by the pilot of the self-test function in flight will inhibit TCAS operation for up to 20 seconds depending upon the number of targets being tracked. The ATC transponder will not function during some portion of the self-test sequence (↓6.52).*

These limitations should be linked to the relevant parts of the development and, most important, operational specifications. For example, L9 may be linked to the pilot operations manual.

10.3.10   System Certification, Maintenance, and Evolution

At this point in development, the safety requirements and constraints are documented and traced to the design features used to implement them. A hazard log contains the hazard information (or links to it) generated during the development process and the results of the hazard analysis performed. The log will contain embedded links to the resolution of each hazard, such as functional requirements, design constraints, system design features, operational procedures, and system limitations. The information documented should be easy to collect into a form that can be used for the final safety assessment and certification of the system.

Whenever changes are made in safety-critical systems or software (during development or during maintenance and evolution), the safety of the change needs to be reevaluated. This process can be difficult and expensive if it has to start from scratch each time. By providing links throughout the specification, it should be easy to assess whether a particular design decision or piece of code was based on the original safety analysis or safety-related design constraint and only that part of the safety analysis process repeated or reevaluated.

11 Analyzing Accidents and Incidents (CAST)

The causality model used in accident or incident analysis determines what we look for, how we go about looking for "facts," and what we see as relevant. In our experience using STAMP-based accident analysis, we find that even if we use only the information presented in an existing accident report, we come up with a very different view of the accident and its causes.

Most accident reports are written from the perspective of an event-based model. They almost always clearly describe the events and usually one or several of these events is chosen as the "root cause(s)." Sometimes "contributory causes" are identified. But the analysis of why those events occurred is usually incomplete: The analysis frequently stops after finding someone to blame—usually a human operator—and the opportunity to learn important lessons is lost.

An accident analysis technique should provide a framework or process to assist in understanding the entire accident process and identifying the most important systemic causal factors involved. This chapter describes an approach to accident analysis, based on STAMP, called CAST (Causal Analysis based on STAMP). CAST can be used to identify the questions that need to be answered to fully understand why the accident occurred. It provides the basis for maximizing learning from the events.

The use of CAST does not lead to identifying single causal factors or variables. Instead it provides the ability to examine the entire sociotechnical system design to identify the weaknesses in the existing safety control structure and to identify changes that will not simply eliminate symptoms but potentially all the causal factors, including the systemic ones.

One goal of CAST is to get away from assigning blame and instead to shift the focus to *why* the accident occurred and how to prevent similar losses in the future. To accomplish this goal, it is necessary to minimize hindsight bias and instead to determine why people behaved the way they did, given the information they had at the time.

An example of the results of an accident analysis using CAST is presented in chapter 5. Additional examples are in appendixes B and C. This chapter describes the steps to go through in producing such an analysis. An accident at a fictional chemical plant called Citichem [174] is used to demonstrate the process.[1] The accident scenario was developed by Risk Management Pro to train accident investigators and describes a realistic accident process similar to many accidents that have occurred in chemical plants. While the loss involves release of a toxic chemical, the analysis serves as an example of how to do an accident or incident analysis for any industry.

An accident investigation process is not being specified here, but only a way to document and analyze the results of such a process. Accident investigation is a much larger topic that goes beyond the goals of this book. This chapter only considers how to analyze the data once it has been collected and organized. The accident analysis process described in this chapter does, however, contribute to determining what questions should be asked during the investigation. When attempting to apply STAMP-based analysis to existing accident reports, it often becomes apparent that crucial information was not obtained, or at least not included in the report, that is needed to fully understand why the loss occurred and how to prevent future occurrences.

11.1   The General Process of Applying STAMP to Accident Analysis

In STAMP, an accident is regarded as involving a complex process, not just individual events. Accident analysis in CAST then entails understanding the dynamic process that led to the loss. That accident process is documented by showing the sociotechnical safety control structure for the system involved and the safety constraints that were violated at each level of this control structure and why. The analysis results in multiple views of the accident, depending on the perspective and level from which the loss is being viewed.

Although the process is described in terms of steps or parts, no implication is being made that the analysis process is linear or that one step must be completed before the next one is started. The first three steps are the same ones that form the basis of all the STAMP-based techniques described so far.

1. Identify the system(s) and hazard(s) involved in the loss.

2. Identify the system safety constraints and system requirements associated with that hazard.

3. Document the safety control structure in place to control the hazard and enforce the safety constraints. This structure includes the roles and responsi-

1. Maggie Stringfellow and John Thomas, two MIT graduate students, contributed to the CAST analysis of the fictional accident used in this chapter.

bilities of each component in the structure as well as the controls provided or created to execute their responsibilities and the relevant feedback provided to them to help them do this. This structure may be completed in parallel with the later steps.

4. Determine the proximate events leading to the loss.

5. Analyze the loss at the physical system level. Identify the contribution of each of the following to the events: physical and operational controls, physical failures, dysfunctional interactions, communication and coordination flaws, and unhandled disturbances. Determine why the physical controls in place were ineffective in preventing the hazard.

6. Moving up the levels of the safety control structure, determine how and *why* each successive higher level allowed or contributed to the inadequate control at the current level. For each system safety constraint, either the responsibility for enforcing it was never assigned to a component in the safety control structure or a component or components did not exercise adequate control to ensure their assigned responsibilities (safety constraints) were enforced in the components below them. Any human decisions or flawed control actions need to be understood in terms of (at least): the information available to the decision maker as well as any required information that was *not* available, the behavior-shaping mechanisms (the context and influences on the decision-making process), the value structures underlying the decision, and any flaws in the process models of those making the decisions and why those flaws existed.

7. Examine overall coordination and communication contributors to the loss.

8. Determine the dynamics and changes in the system and the safety control structure relating to the loss and any weakening of the safety control structure over time.

9. Generate recommendations.

In general, the description of the role of each component in the control structure will include the following:

• Safety Requirements and Constraints
• Controls
• Context
  – Roles and responsibilities
  – Environmental and behavior-shaping factors
• Dysfunctional interactions, failures, and flawed decisions leading to erroneous control actions

- Reasons for the flawed control actions and dysfunctional interactions
  - Control algorithm flaws
  - Incorrect process or interface models.
  - Inadequate coordination or communication among multiple controllers
  - Reference channel flaws
  - Feedback flaws

The next sections detail the steps in the analysis process, using Citichem as a running example.

11.2   Creating the Proximal Event Chain

While the event chain does not provide the most important causality information, the basic events related to the loss do need to be identified so that the physical process involved in the loss can be understood.

For Citichem, the physical process events are relatively simple: A chemical reaction occurred in storage tanks 701 and 702 of the Citichem plant when the chemical contained in the tanks, K34, came in contact with water. K34 is made up of some extremely toxic and dangerous chemicals that react violently to water and thus need to be kept away from it. The runaway reaction led to the release of a toxic cloud of tetrachloric cyanide (TCC) gas, which is flammable, corrosive, and volatile. The TCC blew toward a nearby park and housing development, in a city called Oakbridge, killing more than four hundred people.

The direct events leading to the release and deaths are:

1. Rain gets into tank 701 (and presumably 702), both of which are in Unit 7 of the Citichem Oakbridge plant. Unit 7 was shut down at the time due to lowered demand for K34.

2. Unit 7 is restarted when a large order for K34 is received.

3. A small amount of water is found in tank 701 and an order is issued to make sure the tank is dry before startup.

4. T34 transfer is started at unit 7.

5. The level gauge transmitter in the 701 storage tank shows more than it should.

6. A request is sent to maintenance to put in a new level transmitter.

7. The level transmitter from tank 702 is moved to tank 701. (Tank 702 is used as a spare tank for overflow from tank 701 in case there is a problem.)

8. Pressure in Unit 7 reads as too high.

9. The backup cooling compressor is activated.

10. Tank 701 temperature exceeds 12 degrees Celsius.

11. A sample is run, an operator is sent to check tank pressure, and the plant manager is called.

12. Vibration is detected in tank 701.

13. The temperature and pressure in tank 701 continue to increase.

14. Water is found in the sample that was taken (see event 11).

15. Tank 701 is dumped into the spare tank 702

16. A runaway reaction occurs in tank 702.

17. The emergency relief valve jams and runoff is not diverted into the backup scrubber.

18. An uncontrolled gas release occurs.

19. An alarm sounds in the plant.

20. Nonessential personnel are ordered into units 2 and 3, which have positive pressure and filtered air.

21. People faint outside the plant fence.

22. Police evacuate a nearby school.

23. The engineering manager calls the local hospital, gives them the chemical name and a hotline phone number to learn more about the chemical.

24. The public road becomes jammed and emergency crews cannot get into the surrounding community.

25. Hospital personnel cannot keep up with steady stream of victims.

26. Emergency medical teams are airlifted in.

These events are presented as one list here, but separation into separate interacting component event chains may be useful sometimes in understanding what happened, as shown in the friendly fire event description in chapter 5.

The Citichem event chain here provides a superficial analysis of what happened. A deep understanding of why the events occurred requires much more information. Remember that the goal of a STAMP-based analysis is to determine why the events occurred—*not* who to blame for them—and to identify the changes that could prevent them and similar events in the future.

11.3   Defining the System(s) and Hazards Involved in the Loss

Citichem has two relevant physical processes being controlled: the physical plant and public health. Because separate and independent controllers were controlling

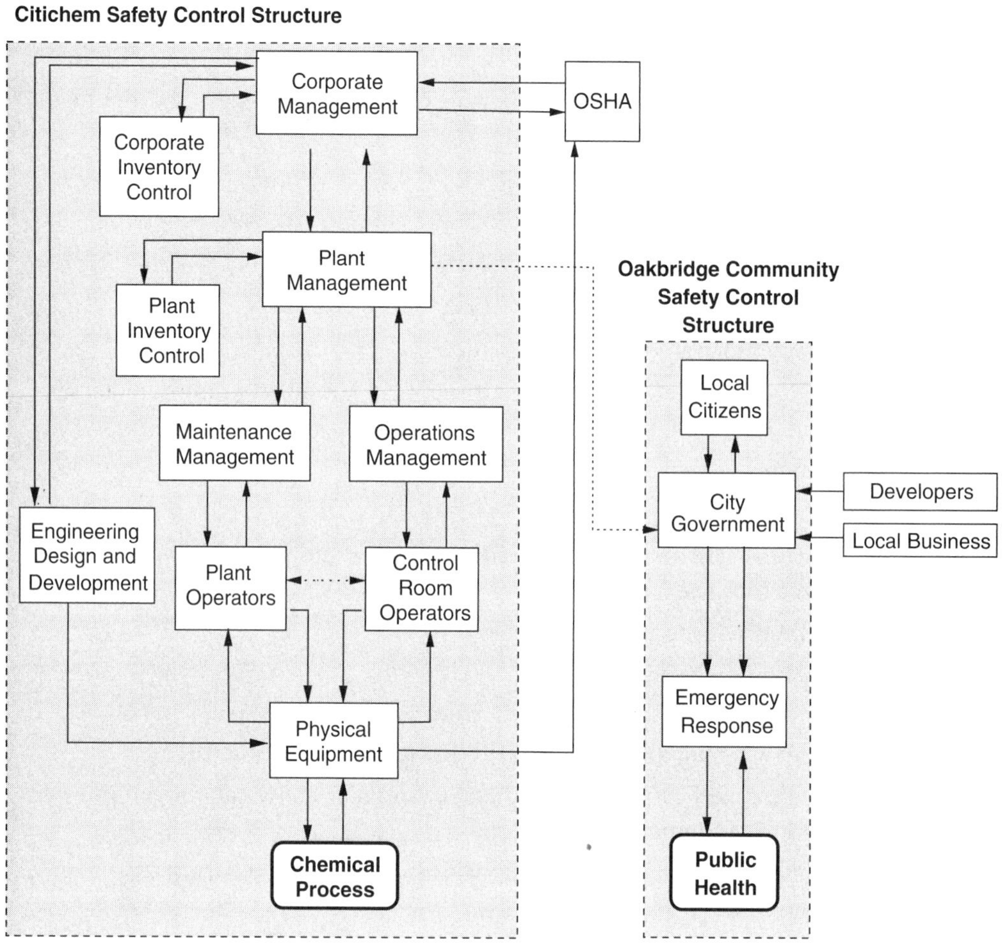

Citichem Safety Control Structure

Figure 11.1
The two safety control structures most relevant to the Citichem accident analysis.

these two processes, it makes sense to consider them as two interacting but independent systems: (1) the chemical company, which controls the chemical process,
and (2) the public political structure, which has responsibilities for public health.
Figure 11.1 shows the major components of the two safety control structures and
interactions between them. Only the major structures are shown in the figure;
the details will be added throughout this chapter.[2] No information was provided

---

2. OSHA, the Occupational Safety and Health Administration, is part of a third larger governmental
control structure, which has many other components. For simplicity, only OSHA is shown and considered
in the example analysis.

about the design and engineering process for the Citichem plant in the accident description, so details about it are omitted. A more complete example of a development control structure and analysis of its role can be found in appendix B.

The analyst(s) also needs to identify the hazard(s) being avoided and the safety constraint(s) to be enforced. An accident or loss event for the combined chemical plant and public health structure can be defined as death, illness, or injury due to exposure to toxic chemicals.

The hazards being controlled by the two control structures are related but different. The public health structure hazard is *exposure of the public to toxic chemicals*. The system-level safety constraints for the public health control system are that:

1. The public must not be exposed to toxic chemicals.

2. Measures must be taken to reduce exposure if it occurs.

3. Means must be available, effective, and used to treat exposed individuals outside the plant.

The hazard for the chemical plant process is *uncontrolled release of toxic chemicals*. Accordingly, the system-level constraints are that:

1. Chemicals must be under positive control at all times.

2. Measures must be taken to reduce exposure if inadvertent release occurs.

3. Warnings and other measures must be available to protect workers in the plant and minimize losses to the outside community.

4. Means must be available, effective, and used to treat exposed individuals inside the plant.

Hazards and safety-constraints must be within the design space of those who developed the system and within the operational space of those who operate it. For example, the chemical plant designers cannot be responsible for those things outside the boundaries of the chemical plant over which they have no control, although they may have some influence over them. Control over the environment of a plant is usually the responsibility of the community and various levels of government. As another example, while the operators of the plant may cooperate with local officials in providing public health and emergency response facilities, responsibility for this function normally lies in the public domain. Similarly, while the community and local government may have some influence on the design of the chemical plant, the company engineers and managers control detailed design and operations.

Once the goals and constraints are determined, the controls in place to enforce them must be identified.

11.4   Documenting the Safety Control Structure

If STAMP has been used as the basis for previous safety activities, such as the original engineering process or the investigation and analysis of previous incidents and accidents, a model of the safety-control structure may already exist. If not, it must be created although it can be reused in the future. Chapters 12 and 13 provide information about the design of safety-control structures.

The components of the structure as well as each component's responsibility with respect to enforcing the system safety constraints must be identified. Determining what these are (or what they should be) can start from system safety requirements. The following are some example system safety requirements that might be appropriate for the Citichem chemical plant example:

1. Chemicals must be stored in their safest form.

2. The amount of toxic chemicals stored should be minimized.

3. Release of toxic chemicals and contamination of the environment must be prevented.

4. Safety devices must be operable and properly maintained at all times when potentially toxic chemicals are being processed or stored.

5. Safety equipment and emergency procedures (including warning devices) must be provided to reduce exposure in the event of an inadvertent chemical release.

6. Emergency procedures and equipment must be available and operable to treat exposed individuals.

7. All areas of the plant must be accessible to emergency personnel and equipment during emergencies. Delays in providing emergency treatment must be minimized.

8. Employees must be trained to

   a. Perform their jobs safely and understand proper use of safety equipment

   b. Understand their responsibilities with regards to safety and the hazards related to their job

   c. Respond appropriately in an emergency

9. Those responsible for safety in the surrounding community must be educated about potential hazards from the plant and provided with information about how to respond appropriately.

A similar list of safety-related requirements and responsibilities might be generated for the community safety control structure.

These general system requirements must be enforced somewhere in the safety control structure. As the accident analysis proceeds, they are used as the starting point for generating more specific constraints, such as constraints for the specific chemicals being handled. For example, requirement 4, when instantiated for TCC, might generate a requirement to prevent contact of the chemical with water. As the accident analysis proceeds, the identified responsibilities of the components can be mapped to the system safety requirements—the opposite of the forward tracing used in safety-guided design. If STPA was used in the design or analysis of the system, then the safety control structure documentation should already exist.

In some cases, general requirements and policies for an industry are established by the government or by professional associations. These can be used during an accident analysis to assist in comparing the actual safety control structure (both in the plant and in the community) at the time of the accidents with the standards or best practices of the industry and country. Accident analyses can in this way be made less arbitrary and more guidance provided to the analysts as to what should be considered to be inadequate controls.

The specific designed controls need not all be identified before the rest of the analysis starts. Additional controls will be identified as the analysts go through the next steps of the process, but a good start can usually be made early in the analysis process.

11.5   Analyzing the Physical Process

Analysis starts with the physical process, identifying the physical and operational controls and any potential physical failures, dysfunctional interactions and communication, or unhandled external disturbances that contributed to the events. The goal is to determine why the physical controls in place were ineffective in preventing the hazard. Most accident analyses do a good job of identifying the physical contributors to the events.

Figure 11.2 shows the requirements and controls at the Citichem physical plant level as well as failures and inadequate controls. The physical contextual factors contributing to the events are included.

The most likely reason for water getting into tanks 701 and 702 were inadequate controls provided to keep water out during a recent rainstorm (an unhandled external disturbance to the system in figure 4.8), but there is no way to determine that for sure.

Accident investigations, when the events and physical causes are not obvious, often make use of a hazard analysis technique, such as fault trees, to create scenarios to consider. STPA can be used for this purpose. Using control diagrams of the physical system, scenarios can be generated that could lead to the lack of enforcement

Physical Plant Safety Controls

Safety Requirements and Constraints Violated:
- Prevent runaway reactions
- Prevent inadvertent release of toxic chemicals or explosion
- Convert released chemicals into a nonhazardous of less hazardous form
- Provide indicators (alarms) of the existence of hazardous conditions
- Provide protection against human or environmental exposure after a release
- Provide emergency equipment to treat exposed individuals

Emergency and Safety Equipment (Controls): Partial list
- Air monitors
- Windsock to determine which way wind is blowing
- Automatic temperature controls to prevent overheating
- Pressure relief system to deal with excessive pressure
- Gauges and indicators to provide information about the state of the process
- Flares and scrubbers to burn off or neutralize released gas
- Positive pressure and filtered air in some units to protect employees
- Spare tank for runoff
- Emergency showers
- Eyewash fountain
- Protective equipment for employees
- Sirens

Failures and Inadequate Controls:
- Inadequate protection against water getting into tanks
- Inadequate monitoring of chemical process: Gauges were missing or inoperable
- Inadequate emergency relief system
    - Emergency relief valve jammed (could not send excess gas to scrubber)
    - Pop-up pressure relief valves in Units 7 and 9 were too small: Small amounts of corrosion in valves could prevent venting if non-gas material is present
    - Relief valve lines too small to relieve pressure fast enough: This is in effect a single point of failure for the emergency relief system

Physical Contextual Factors:
- The plant was built in a remote location 30 years ago so it would have a buffer area around it, but the city grew closer over the years
- The only access to the plant is a two-lane narrow road. There was a plan to widen the road in the future, but that never happened
- Approximately 24 different chemical products are manufactured at Oakbridge, most of which are toxic to humans and some very toxic
- The plant manufactures K34, which contains Tetra Chloric Cyanide (TCC). TCC is flammable, corrosive and volatile. It is extremely toxic and dangerous and reacts violently with water
- Unit 7 was previously used to manufacture pesticide, but production was moved to Mexico because it was cheaper to make there. At the time of the start of the accident proximal events, Unit 7 was shut down and was not being used. It was restarted to provide extra K34
- The plant operates 24 hours a day, with three different shifts
- The plant already was operating at capacity before the decision to increase production of K34

Figure 11.2
STAMP analysis at the Citichem physical plant level.

of the safety constraint(s) at the physical level. The safety design principles in chapter 9 can provide assistance in identifying design flaws.

As is common in the process industry, the physical plant safety equipment (controls) at Citichem were designed as a series of barriers to satisfy the system safety constraints identified earlier, that is, to protect against runaway reactions, protect against inadvertent release of toxic chemicals or an explosion (uncontrolled energy), convert any released chemicals into a non-hazardous or less hazardous form, provide protection against human or environmental exposure after release, and provide emergency equipment to treat exposed individuals. Citichem had the standard types of safety equipment installed, including gauges and other indicators of the physical system state. In addition, it had an emergency relief system and devices to minimize the danger from released chemicals such as a scrubber to reduce the toxicity of any released chemicals and a flare tower to burn off gas before it gets into the atmosphere.

A CAST accident analysis examines the controls to determine which ones did not work adequately and why. While there was a reasonable amount of physical safety controls provided at Citichem, much of this equipment was inadequate or not operational—a common finding after chemical plant accidents.

In particular, rainwater got into the tank, which implies the tanks were not adequately protected against rain despite the serious hazard created by the mixing of TCC with water. While the inadequate protection against rainwater should be investigated, no information was provided in the Citichem accident description. Did the hazard analysis process, which in the process industry often involves HAZOP, identify this hazard? If not, then the hazard analysis process used by the company needs to be examined to determine why an important factor was omitted. If it was not omitted, then the flaw lies in the translation of the hazard analysis results into protection against the hazard in the design and operations. Were controls to protect against water getting into the tank provided? If not, why not? If so, why were they ineffective?

Critical gauges and monitoring equipment were missing or inoperable at the time of the runaway reaction. As one important example, the plant at the time of the accident had no operational level indicator on tank 702 despite the fact that this equipment provided safety-critical information. One task for the accident analysis, then, is to determine whether the indicator was designated as safety-critical, which would (or should) trigger more controls at the higher levels, such as higher priority in maintenance activities. The inoperable level indicator also indicates a need to look at higher levels of the control structure that are responsible for providing and maintaining safety-critical equipment.

As a final example, the design of the emergency relief system was inadequate: The emergency relief valve jammed and excess gas could not be sent to the scrubber.

The pop-up relief valves in Unit 7 (and Unit 9) at the plant were too small to allow the venting of the gas if non-gas material was present. The relief valve lines were also too small to relieve the pressure fast enough, in effect providing a single point of failure for the emergency relief system. Why an inadequate design existed also needs to be examined in the higher-level control structure. What group was responsible for the design and why did a flawed design result? Or was the design originally adequate but conditions changed over time?

The physical contextual factors identified in figure 11.2 play a role in the accident causal analysis, such as the limited access to the plant, but their importance becomes obvious only at higher levels of the control structure.

At this point of the analysis, several recommendations are reasonable: add protection against rainwater getting into the tanks, change the design of the valves and vent pipes in the emergency relief system, put a level indicator on Tank 702, and so on. Accident investigations often stop here with the physical process analysis or go one step higher to determine what the operators (the direct controllers of the physical process) did wrong.

The other physical process being controlled here, public health, must be examined in the same way. There were very few controls over public health instituted in Oakbridge, the community surrounding the plant, and the ones that did exist were inadequate. The public had no training in what to do in case of an emergency, the emergency response system was woefully inadequate, and unsafe development was allowed, such as the creation of a children's park right outside the walls of the plant. The reasons for these inadequacies, as well as the inadequacies of the controls on the physical plant process, are considered in the next section.

11.6 Analyzing the Higher Levels of the Safety Control Structure

While the physical control inadequacies are relatively easy to identify in the analysis and are usually handled well in any accident analysis, understanding why those physical failures or design inadequacies existed requires examining the higher levels of safety control: Fully understanding the behavior at any level of the sociotechnical safety control structure requires understanding how and why the control at the next higher level allowed or contributed to the inadequate control at the current level. Most accident reports include some of the higher-level factors, but usually incompletely and inconsistently, and they focus on finding someone or something to blame.

Each relevant component of the safety control structure, starting with the lowest physical controls and progressing upward to the social and political controls, needs to be examined. How are the components to be examined determined? Considering everything is not practical or cost effective. By starting at the bottom, the relevant components to consider can be identified. At each level, the flawed behavior or inadequate controls are examined to determine why the behavior occurred and why the controls at higher levels were not effective at preventing that behavior. For example, in the STAMP-based analysis of an accident where an aircraft took off from the wrong runway during construction at the airport, it was discovered that the airport maps provided to the pilot were out of date [142]. That led to examining the procedures at the company that provided the maps and the FAA procedures for ensuring that maps are up-to-date.

Stopping after identifying inadequate control actions by the lower levels of the safety control structure is common in accident investigation. The result is that the cause is attributed to "operator error," which does not provide enough information to prevent accidents in the future. It also does not overcome the problems of hindsight bias. In hindsight, it is always possible to see that a different behavior would have been safer. But the information necessary to identify that safer behavior is usually only available after the fact. To improve safety, we need to understand the reasons people acted the way they did. Then we can determine if and how to change conditions so that better decisions can be made in the future.

The analyst should start from the assumption that most people have good intentions and do not purposely cause accidents. The goal then is to understand *why* people did not or could not act differently. People acted the way they did for very good reasons; we need to understand why the behavior of the people involved made sense to them at the time [51].

Identifying these reasons requires examining the context and behavior-shaping factors in the safety control structure that influenced that behavior. What contextual factors should be considered? Usually the important contextual and behavior-shaping factors become obvious in the process of explaining why people acted the way they did. Stringfellow has suggested a set of general factors to consider [195]:

- *History:* Experiences, education, cultural norms, behavioral patterns: how the historical context of a controller or organization may impact their ability to exercise adequate control.

- *Resources:* Staff, finances, time.

- *Tools and Interfaces:* Quality, availability, design, and accuracy of tools. Tools may include such things as risk assessments, checklists, and instruments as well as the design of interfaces such as displays, control levers, and automated tools.

- *Training:* Quality, frequency, and availability of formal and informal training.

- *Human Cognition Characteristics:* Person–task compatibility, individual tolerance of risk, control role, innate human limitations.

- *Pressures:* Time, schedule, resource, production, incentive, compensation, political. Pressures can include any positive or negative force that can influence behavior.

- *Safety Culture:* Values and expectations around such things as incident reporting, workarounds, and safety management procedures.

- *Communication:* How the communication techniques, form, styles, or content impacted behavior.

- *Human Physiology:* Intoxication, sleep deprivation, and the like.

We also need to look at the process models used in the decision making. What information did the decision makers have or did they need related to the inadequate control actions? What other information could they have had that would have changed their behavior? If the analysis determines that the person was truly incompetent (not usually the case), then the focus shifts to ask why an incompetent person was hired to do this job and why they were retained in their position. A useful method to assist in understanding human behavior is to show the process model of the human controller at each important event in which he or she participated, that is, what information they had about the controlled process when they made their decisions.

Let's follow some of the physical plant inadequacies up the safety control structure at Citichem. Three examples of STAMP-based analyses of the inadequate control at Citichem are shown in figure 11.3: a maintenance worker, the maintenance manager, and the operations manager.

During the investigation, it was discovered that a maintenance worker had found water in tank 701. He was told to check the Unit 7 tanks to ensure they were ready for the T34 production startup. Unit 7 had been shut down previously (see "Physical Plant Context"). The startup was scheduled for 10 days after the decision to produce additional K34 was made. The worker found a small amount of water in tank 701, reported it to the maintenance manager, and was told to make sure the tank was "bone dry." However, water was found in the sample taken from tank 701 right before the uncontrolled reaction. It is unknown (and probably unknowable) whether the worker did not get all the water out or more water entered later through the same path it entered previously or via a different path. We do know he was fatigued and working a fourteen-hour day, and he may not have had time to do the job properly. He also believed that the tank's residual water was from condensation, not rain. No independent check was made to determine whether all the water was removed.

Some potential recommendations from what has been described so far include establishing procedures for quality control and checking safety-critical activities. Any existence of a hazardous condition—such as finding water in a tank that is to

Maintenance Manager

Safety-Related Responsibilities:
- Maintain plant equpment in a safe condition
- Report safety-related problems found

Context
- Has been with company a long time
- Has an inadequate workforce (understaffed)
- Workers are tired, lots of overtime
- Under extreme schedule pressures, unrealistic schedule
- No organization responsible safety analyses and risk assessments during operations

Unsafe Decisions and Control Actions:
- Directed worker to dry Tank 701 but does not tell him to check other tanks. No check made to determine whether tank 701 is really dry?
- Did not follow up finding water in tank with investigation of how it got there
- Did not inform plant manager that water had been found in tank 701
- Did not overhaul the pumps in Unit 7. Decided to test them instead due to time pressures
- Did not notify the plant manager that pumps were not overhauled
- Agrees to delay needed maintenance for 10 days
- Made all these decisions without an analysis of hazards involved

Process Model Flaws:
- Believed the tank's residual water was from condensation, not rain getting into the tank (?)

Informal
Communication
◁ - - - - - - - - ▷

Operations Manager

Safety-Related Responsibilities:
- Develop operating procedures that adequately control hazards
- Provide operator training on plant hazards and safe operating procedures. Audit to ensure training is effective
- Oversee operations to ensure that (safety-related) policies and procedures are being followed

Context
- Under same performance pressures as everyone else
- No organization responsible for safety analyses and risk assessments
- Understaffed

Unsafe Decisions and Control Actions:
- Decides to take level guage from tank 702 and put it on 701 Runs unit 7 without a level guage on tank 702. Ignores concerns by operators about operating a tank with no gauge
- Agrees to or makes changes without thoroughly hanalyzing hazards involved
- Agrees to start unit 7 in ten days knowing he does not have the personnel to do a thorough inspection and adequate startup activities

Process Model Flaws:
- Thinks tank 702 is empty. Does not know that water was found by maintenance in tank 701
- Inaccurate assessment of likelihood of having to use Tank 702
- Like the others, most likely does not understand the limitatinos of the design of the safety equipment

Maintenance Worker

Safety-Related Responsibilities:
- Maintain plant equipment in a safe condition as directed by the maintenance manager
- Report any problems found

Context
- Fatigued, working 14 hour days

Unsafe Decisions and Control Actions:
- Inadequate removal of water from Tank 701?

Process Model Flaws:
- Believed tank's residual water was from condensation, not a rain leak

· · ·

Figure 11.3
Middle-management-level analysis at Citichem.

be used to produce a chemical that is highly reactive to water—should trigger an in-depth investigation of why it occurred before any dangerous operations are started or restarted. In addition, procedures should be instituted to ensure that those performing safety-critical operations have the appropriate skills, knowledge, and physical resources, which, in this case, include adequate rest. Independent checks of critical activities also seem to be needed.

The maintenance worker was just following the orders of the maintenance manager, so the role of maintenance management in the safety-control structure also needs to be investigated. The runaway reaction was the result of TCC coming in contact with water. The operator who worked for the maintenance manager told him about finding water in tank 701 after the rain and was directed to remove it. The maintenance manager does not tell him to check the spare tank 702 for water and does not appear to have made any other attempts to perform that check. He apparently accepted the explanation of condensation as the source of the water and did not, therefore, investigate the leak further.

Why did the maintenance manager, a long-time employee who had always been safety conscious in the past, not investigate further? The maintenance manager was working under extreme time pressure and with inadequate staff to perform the jobs that were necessary. There was no reporting channel to someone with specified responsibility for investigating hazardous events, such as finding water in a tank used for a toxic chemical that should never contact water. Normally an investigation would not be the responsibility of the maintenance manager but would fall under the purview of the engineering or safety engineering staff. There did not appear to be anyone at Citichem with the responsibility to perform the type of investigation and risk analysis required to understand the reason for water being in the tank. Such events should be investigated thoroughly by a group with designated responsibility for process safety, which presumes, of course, such a group exists.

The maintenance manager did protest (to the plant manager) about the unsafe orders he was given and the inadequate time and resources he had to do his job adequately. At the same time, he did not tell the plant manager about some of the things that had occurred. For example, he did not inform the plant manager about finding water in tank 701. If the plant manager had known these things, he might have acted differently. There was no problem-reporting system in this plant for such information to be reliably communicated to decision makers: Communication relied on chance meetings and informal channels.

Lots of recommendations for changes could be generated from this part of the analysis, such as providing rigorous procedures for hazard analysis when a hazardous condition is detected and training and assigning personnel to do such an analysis. Better communication channels are also indicated, particularly problem reporting channels.

The operations manager (figure 11.3) also played a role in the accident process. He too was under extreme pressure to get Unit 7 operational. He was unaware that the maintenance group had found water in tank 701 and thought 702 was empty. During the effort to get Unit 7 online, the level indicator on tank 701 was found to be not working. When it was determined that there were no spare level indicators at the plant and that delivery would require two weeks, he ordered the level indicator on 702 to be temporarily placed on tank 701—tank 702 was only used for overflow in case of an emergency, and he assessed the risk of such an emergency as low. This flawed decision clearly needs to be carefully analyzed. What types of risk and safety analyses were performed at Citichem? What training was provided on the hazards? What policies were in place with respect to disabling safety-critical equipment? Additional analysis also seems warranted for the inventory control procedures at the plant and determining why safety-critical replacement parts were out of stock.

Clearly, safety margins were reduced at Citichem when operations continued despite serious failures of safety devices. Nobody noticed the degradation in safety. Any change of the sort that occurred here—startup of operations in a previously shut down unit and temporary removal of safety-critical equipment—should have triggered a hazard analysis and a management of change (MOC) process. Lots of accidents in the chemical industry (and others) involve unsafe workarounds. The causal analysis so far should trigger additional investigation to determine whether adequate management of change and control of work procedures had been provided but not enforced or were not provided at all. The first step in such an analysis is to determine who was responsible (if anyone) for creating such procedures and who was responsible for ensuring they were followed. The goal again is not to find someone to blame but simply to identify the flaws in the process for running Citichem so they can be fixed.

At this point, it appears that decision making by higher-level management (above the maintenance and operations manager) and management controls were inadequate at Citichem. Figures 11.4 and 11.5 show example STAMP-based analysis results for the Citichem plant manager and Citichem corporate management. The plant manager made many unsafe decisions and issued unsafe control actions that directly contributed to the accident or did not initiate control actions necessary for safety (as shown in figure 11.4). At the same time, it is clear that he was under extreme pressure to increase production and was missing information necessary to make better decisions. An appropriate safety control structure at the plant had not been established leading to unsafe operational practices and inaccurate risk assessment by most of the managers, especially those higher in the control structure. Some of the lower level employees tried to warn against the high-risk practices, but appropriate communication channels had not been established to express these concerns.

Citichem Oakbridge Plant Manager

Safety-Related Responsibilities:
- Ensure safe operation of the plant
  - Establish a safety organization and ensure it has adequate resources, appropriate expertise, and communication channels to all parts of the plant
  - Seek and use appropriate inputs from the safety organization when making safety-critical decisions
  - Establish appropriate responsibilitiy, accountability, and authority for safety-related decision making and activities at all plant management levels
  - Provide oversight to ensure compliance with company safety policies and standards at the plant
  - Create and oversee communication channels for safety-related information
- Ensure appropriate emergency preparedness and response within the plant
- Ensure that adequate emergency preparedness information is provided to the community

Context:
- Under pressure to manufacture a large amount of K34 in a short time to satisfy company sales orders. If unsuccessful, corporate could close the plant and move operations to Mexico. The need for a turnaround (major maintenance) on Unit 9 increases the pressure even more
- The plant is already stretched to capacity. The additional resources needed for increased production are not available and no budget to add more employees so must increase production without additional workers
- Highly skilled and very experienced (has been working for company for over 20 years). He has strong ties to the community and wants to ensure that Oakbridge remains a key revenue source for Citichem corporate so the employees can keep their jobs
- The Citichem Oakbridge plant as had very few acidents in the past 30 years
- The plant passes several OSHA inspections every year

Unsafe Decisions and Control Actions:
- Agrees to produce extra K34 without the resources to do it safely
- Initiates start up of Unit 7 under unsafe conditions (all safety-related equipment is not operational, and pumps are not overhauled)
- Delayed in responding to new information about inadequacy of emergency relief system design
- Does not use safety analysis information when making safety-related decisions. No management of change policies to evaluate hazards involved before changes are made
- Established inadequate inventory control policies and procedures to ensure safety-related equipment in stock at all times
- Has not set up and enforced a policy for thorough incident/accident investigation
- Has not established appropriate communication channels within the plant for safety-related information, including a problem-reporting system
- Did not warn community about dangers of development next to the plant
- Did not make sure community has information necessary for emergency preparedness activities and the handling of chemical emergencies

Process Model Flaws:
- Inaccurate risk assessment. Believes the "risks are acceptable, considering the benefits." Does not tie the recent incidents to decreasing safety margins
- Incorrectly believed the pumps had been overhauled
- Did not know that water had been found in Tank 701
- Does not know about lack of working indicator on Tank 702 and lack of spare parts

Figure 11.4
Citichem plant management analysis.

Citichem Corporate Management

Safety-Related Responsibilities:
* Ensure that Citichem plants are operated safely and that adequate equipment and resources are provided to accomplish this goal
* Ensure that communication with communities surrounding the plants is adequate and information exchanged to reduce risk of injury for those explosed to chemicals should a release or other hazardous event occur
* Provide leadership on safety issues, including the creation and enforcement of a company safety policy

Context:
* Price competition has increased. The British recently cut their K34 prices
* Chemical plants are cheaper to operate in Mexico (and many other countries) than in the U.S
* Production at Oakbridge has been increased in the past without an incident despite warnings of decreased safety margins

Unsafe Decisions and Control Actions:
* Long-term planning of production goals and creation of sales targets was performed with inadequate regard for safety. There was no hazard or risk analysis of increasing the production of K34 at Oakbridge given the current resources there
* Inadequate allocation of resources to Oakbridge for increased production and unrealistic schedule
* Ignored feedback from Oakbridge Plant Manager that increasing production without increasing resources would require cutting safety margins to inadequate levels
* Inadequate oversight and enforcement of maintenance schedules and other plant operations related to safety
* Inadequate inventory control policies for safety-critical components and parts
* Implemented a policy of not disclosing what chemicals are used and the products they make to surrounding communities because of business competition reasons (which hindered community emergency response)
* Did not require in-depth analysis of incidents and accidents

Process Model Flaws:
* Inaccurate assessment of risk of increased production
* Belief that the only way to eliminate the risks was to eliminate the industry — that risk cannot be reduced without reducing profits or productivity
* Belief that recent incidents were not indicative of true high risk in the system and resulted simply from the employees own errors and negligence

Figure 11.5
Corporate-level Citichem management analysis.

Safety controls were almost nonexistent at the corporate management level. The upper levels of management provided inadequate leadership, oversight and management of safety. There was either no adequate company safety policy or it was not followed, either of which would lead to further causal analysis. A proper process safety management system clearly did not exist at Citichem. Management was under great competitive pressures, which may have led to ignoring corporate safety controls or adequate controls may never have been established. Everyone had very flawed mental models of the risks of increasing production without taking the proper precautions. The recommendations should include consideration of what kinds of changes might be made to provide better information about risks to management decision makers and about the state of plant operations with respect to safety.

Like any major accident, when analyzed thoroughly, the process leading to the loss is complex and multi-faceted. A complete analysis of this accident is not needed here. But a look at some of the factors involved in the plant's environment, including the control of public health, is instructive.

Figure 11.6 shows the STAMP-based analysis of the Oakbridge city emergency-response system. Planning was totally inadequate or out of date. The fire department did not have the proper equipment and training for a chemical emergency, the hospital also did not have adequate emergency resources or a backup plan, and the evacuation plan was ten years out of date and inadequate for the current level of population.

Understanding why these inadequate controls existed requires understanding the context and process model flaws. For example, the police chief had asked for resources to update equipment and plans, but the city had turned him down. Plans had been made to widen the road to Oakbridge so that emergency equipment could be brought in, but those plans were never implemented and the planners never went back to their plans to see if they were realistic for the current conditions. Citichem had a policy against disclosing what chemicals they produce and use, justifying this policy by the need for secrecy from their competitors, making it impossible for the hospital to stockpile the supplies and provide the training required for emergencies, all of which contributed to the fatalities in the accident. The government had no disclosure laws requiring chemical companies to provide such information to emergency responders.

Clear recommendations for changes result from this analysis, for example, updating evacuation plans and making changes to the planning process. But again, stopping at this level does not help to identify systemic changes that could improve community safety: The analysts should work their way up the control structure to understand the entire accident process. For example, why was an inadequate emergency response system allowed to exist?

Oakbridge Emergency Response

Safety-Related Responsibilities:

General: Provide appropriate emergency response such as fire fighting, evacuation, medical intervention

- *Fire Chief:*
  - Ensure there is adequate fire fighting equipment and emergency planning in case of a serious incident
  - Effectively communicate emergency needs to city council, mayor, and city manager (city government)
  - Learn about potential safety hazards posed by the plant (including the chemicals being manufactured and stored there)
  - Coordinate with medical facilities and other emergency responders

- *Fire Brigade:* Ensure there is adequate emergency equipment and training inside and outside the plant and drill those outside the boundaries of the plant

- *Doctors and Hospital (and other medical facilities in the area):*
  - Learn what chemicals and other dangerous products at Citichem could affect the health of the population surrounding the plant
  - Obtain adequate supplies and the information necessary to respond in an emergency as well as plan for obtaining additional human resources if required
  - Coordinate with other emergency responders (e.g., the fire department)
  - Conduct regular drills to assess and improve planning for emergency response

Context:

- Evacuation plan for city is 10 years out of date and hopelessly inadequate for the current population The police chief has asked for money several times to fund a study to update the plans, but each time is is turned down by the city
- The city government does not rank emergency preparedness as a high priority
- Citichem has a policy against disclosing what chemicals are used and the products they make. The state has no disclosure law to force them to provide this information
- Citichem is better equipped to fight chemical spills, has better equipment than they do, and knows more than the fire brigade about chemical spills
- The fire chief prefers that Citichem handle its own problems. This preference reinforces the lack of preparedness
- Plans to widen the road to Oakbridge were never implemented

Unsafe Decisions and Control Actions:

- Unless Citichem requests assistance (which they never have), the fire brigade stays "outside the fence." The fire brigade made no attempt to learn about potential Citichem hazards nor to ensure that adequate emergency equipment, training, and resources were available within Citichem
- Just about everyone outside Citichem made inadequate preparation for emergencies
- The hospital did not obtain adequate resources for an emergency and made no backup plan

Process Model Flaws:

- Hospital knows nothing about the health hazards of the plant
- Fire brigade does not know what chemicals are being used at the plant
- Everyone believed risk from the plant was low

Figure 11.6
STAMP analysis of the Oakbridge emergency response system.

The analysis in figure 11.7 helps to answer this question. For example, the members of the city government had inadequate knowledge of the hazards associated with the plant, and they did not try to obtain more information about them or about the impact of increased development close to the plant. At the same time, they turned down requests for the funding to upgrade the emergency response system as the population increased as well as attempts by city employees to provide emergency response pamphlets for the citizens and set up appropriate communication channels.

Why did they make what in retrospect look like such bad decisions? With inadequate knowledge about the risks, the benefits of increased development were ranked above the dangers from the plant in the priorities used by the city managers. A misunderstanding about the dangers involved in the chemical processing at the plant contributed also to the lack of planning and approval for emergency-preparedness activities.

The city government officials were subjected to pressures from local developers and local businesses that would benefit financially from increased development. The developer sold homes before the development was approved in order to increase pressure on the city council. He also campaigned against a proposed emergency response pamphlet for local residents because he was afraid it would reduce his sales. The city government was subjected to additional pressure from local businessmen who wanted more development in order to increase their business and profits. The residents did not provide opposing pressure to counteract the business influences and trusted that government would protect them: No community organizations existed to provide oversight of the local government safety controls and to ensure that government was adequately considering their health and safety needs (figure 11.8).

The city manager had the right instincts and concern for public safety, but she lacked the freedom to make decisions on her own and the clout to influence the mayor or city council. She was also subject to external pressures to back down on her demands and no structure to assist her in resisting those pressures.

In general, there are few requirements for serving on city councils. In the United States, they are often made up primarily of those with conflicts of interest, such as real estate agents and developers. Mayors of small communities are often not paid a full salary and must therefore have other sources of income, and city council members are likely to be paid even less, if at all.

If community-level management is unable to provide adequate controls, controls might be enforced by higher levels of government. A full analysis of this accident would consider what controls existed at the state and federal levels and why they were not effective in preventing the accident.

Oakbridge City Government

Safety-Related Responsibilities:
- Ensure that emergency preparedness planning is adequate and in place and provide necessary resources
- Ensure public safety. Approve only development that does not degrade public safety below acceptable levels

Context:
- Under pressure to create a hospitable environment for investment and development. Need support of business community and people who work at the plant and live in the community to be elected and to perform their duties
- Believe the plant is safe based on the fact that it has been there for 30 years and there have been no worse consequences than "bad smells." The plant passes several OSHA inspections every year
- The city manager worked for 18 years to get to where she is and does not want to lose her position Although she sees many problems, she feels she has no ability to change the system
- Oakbridge can use the extra tax base from additional development. Development brings jobs, more opportunities, increased tax revenues, better schools, better housing, and benefits for the local business community
- There was very little turnout for the public hearing on new development by the plant
- There is lots of pressure from developers and local businessmen to allow development

Unsafe Decisions and Control Actions:
- City council turned down funding for an emergency response pamphlet and never produced one
- City government did not ensure that adequate emergency preparedness was in place. The City Council turned down funding to update the emergency evacuation plan
- Allowed development without having an adequately sized road in place for emergency access. Argued road would be widened the next year, but then never ensured that that happened
- Allowed erosion of the physical safety buffer. Approved a children's park near the plant fence
- Ranked development and increasing the tax base over ensuring public safety
- Did not attempt to get a proper risk assessment of the increased development. Instead they took the Citichem plant manager's word that the risks were acceptable with respect to the benefits (jobs, revenues, etc.)
- The expressed concerns by the city manager were not heeded or considered adequately. Attempts by the city manager to get insight into the potential hazards and to set up formal communications between the plant and the city were thwarted

Process Model Flaws:
- Believed risk from plant was less than it really was. Assumed past perceived safety guarantees future safety
- Believed the two-lane, narrow road was not an issue because of plans to widen to four lanes "next year"

Figure 11.7
STAMP analysis of the Oakbridge city government's role in the accident.

Oakbridge Residents (Local Citizens)

Safety-Related Responsibilities:
- Ensure that elected officials are adequately executing their responsibilities with repect to public safety
- Inform themselves about potential community hazards, protection mechanisms, and emergency preparedness when moving into communities near chemical or other plants
- Understand what to do in case of an emergency

Context:
- People want to live near where they work
- Usually cheaper to live in communities near industrial plants (especially smelly ones)
- No information is available to the public about the hazards of the plant and there is often no way for them to obtain this information without the assistance of government and public disclosure laws
- Development brings jobs, more opportunities, better schools, better housing

Unsafe Decisions and Control Actions:
- Did not show up for hearings on the new development or display interest in any other way
- Did not ask about hazards or risks associated with the plant before or after moning to Oakbridge or about the state of emergency preparedness

Process Model Flaws:
- Do not know about or understand the hazards of the plant
- Do not know about the lack of emergency preparedness in their community
- Assume elected officials and local government are adequately looking out for their safety

Figure 11.8
Analysis of the role of the Oakbridge residents.

11.7   A Few Words about Hindsight Bias and Examples

One of the most common mistakes in accident analyses is the use of hindsight bias. Words such as "could have" or "should have" in accident reports are judgments that are almost always the result of such bias [50]. It is not the role of the accident analyst to render judgment in terms of what people did or did *not* do (although that needs to be recorded) but to understand *why* they acted the way they did.

Although hindsight bias is usually applied to the operators in an accident report, because most accident reports focus on the operators, it theoretically could be applied to people at any level of the organization: "The plant manager should have known ..."

The biggest problem with hindsight bias in accident reports is not that it is unfair (which it usually is), but that an opportunity to learn from the accident and prevent future occurrences is lost. It is always possible to identify a better decision in retrospect—or there would not have been a loss or near miss—but it may have been difficult or impossible to identify that the decision was flawed at the time it had to be made. To improve safety and to reduce errors, we need to understand why the decision made sense to the person at the time and redesign the system to help people make better decisions.

Accident investigation should start with the assumption that most people have good intentions and do not purposely cause accidents. The goal of the investigation, then, is to understand why they did the wrong thing in that particular situation. In particular, what were the contextual or systemic factors and flaws in the safety control structure that influenced their behavior? Often, the person had an inaccurate view of the state of the process and, given that view, did what appeared to be the right thing at the time but turned out to be wrong with respect to the actual state. The solution then is to redesign the system so that the controller has better information on which to make decisions.

As an example, consider a real accident report on a chemical overflow from a tank, which injured several workers in the vicinity [118]. The control room operator issued an instruction to open a valve to start the flow of liquid into the tank. The flow meter did not indicate a flow, so the control room operator asked an outside operator to check the manual valves near the tank to see if they were closed. The control room operator believed that the valves were normally left in an open position to facilitate conducting the operation remotely. The tank level at this time was 7.2 feet.

The outside operator checked and found the manual valves at the tank open. The outside operator also saw no indication of flow on the flow meter and made an effort to visually verify that there was no flow. He then began to open and close the valves manually to try to fix the problem. He reported to the control room operator that he heard a clunk that may have cleared an obstruction, and the control room operator tried opening the valve remotely again. Both operators still saw no flow on the flow meter. The outside operator at this time got a call to deal with a problem in a different part of the plant and left. He did not make another attempt to visually verify if there was flow. The control room operator left the valve in the closed position. In retrospect, it appears that the tank level at this time was approximately 7.7 feet.

Twelve minutes later, the high-level alarm on the tank sounded in the control room. The control room operator acknowledged the alarm and turned it off. In retrospect, it appears that the tank level at this time was approximately 8.5 feet, although there was no indication of the actual level on the control board. The control room operator got an alarm about an important condition in another part of the plant and turned his attention to dealing with that alarm. A few minutes later, the tank overflowed.

The accident report concluded, "The available evidence should have been sufficient to give the control room operator a clear indication that [the tank] was indeed filling and required immediate attention." This statement is a classic example of hindsight bias—note the use of the words "should have ..." The report does not identify what that evidence was. In fact, the majority of the evidence that both operators had at this time was that the tank was *not* filling.

To overcome hindsight bias, it is useful to examine exactly what evidence the operators had at time of each decision in the sequence of events. One way to do this is to draw the operator's process model and the values of each of the relevant variables in it. In this case, both operators thought the control valve was closed—the control room operator had closed it and the control panel indicated that it was closed, the flow meter showed no flow, and the outside operator had visually checked and there was no flow. The situation is complicated by the occurrence of other alarms that the operators had to attend to at the same time.

Why did the control board show the control valve was closed when it must have actually been open? It turns out that there is no way for the control room operator to get confirmation that the valve has actually closed after he commands it closed. The valve was not equipped with a valve stem position monitor, so the control room operator only knows that a signal has gone to the valve for it to close but not whether it has actually done so. The operators in many accidents, including Three Mile Island, have been confused about the actual position of valves due to similar designs.

An additional complication is that while there is an alarm in the tank that should sound when the liquid level reaches 7.5 feet, that alarm was not working at the time, and the operator did not know it was not working. So the operator had extra reason to believe the liquid level had not risen above 7.5 feet, given that he believed there was no flow into the tank and the 7.5-foot alarm had not sounded. The level transmitter (which provided the information to the 7.5-foot alarm) had been operating erratically for a year and a half, but a work order had not been written to repair it until the month before. It had supposedly been fixed two weeks earlier, but it clearly was not working at the time of the spill.

The investigators, in retrospect knowing that there indeed had to have been some flow, suggested that the control room operator "could have" called up trend data on the control board and detected the flow. But this suggestion is classic hindsight bias. The control room operator had no reason to perform this extra check and was busy taking care of critical alarms in other parts of the plant. Dekker notes the distinction between *data availability*, which is what can be shown to have been physically available somewhere in the situation, and *data observability*, which is what was observable given the features of the interface and the multiple interleaving tasks, goals, interests, and knowledge of the people looking at it [51]. The trend data were available to the control room operator, but they were not observable without taking special actions that did not seem necessary at the time.

While that explains why the operator did not know the tank was filling, it does not fully explain why he did not respond to the high-level alarm. The operator said that he thought the liquid was "tickling" the sensor and triggering a false alarm. The accident report concludes that the operator should have had sufficient evidence the tank was indeed filling and responded to the alarm. Not included in the official accident report was the fact that nuisance alarms were relatively common in this unit: they occurred for this alarm about once a month and were caused by sampling errors or other routine activities. This alarm had never previously signaled a serious problem. Given that all the observable evidence showed the tank was not filling and that the operator needed to respond to a serious alarm in another part of the plant at the time, the operator not responding immediately to the alarm does not seem unreasonable.

An additional alarm was involved in the sequence of events. This alarm was at the tank and denoted that a gas from the liquid in the tank was detected in the air outside the tank. The outside operator went to investigate. Both operators are faulted in the report for waiting thirty minutes to sound the evacuation horn after this alarm went off. The official report says:

> Interviews with operations personnel did not produce a clear reason why the response to the [gas] alarm took 31 minutes. The only explanation was that there was not a sense of urgency since, in their experience, previous [gas] alarms were attributed to minor releases that did not require a unit evacuation.

This statement is puzzling, because the statement itself provides a clear explanation for the behavior, that is, the previous experience. In addition, the alarm maxed out at 25 ppm, which is much lower than the actual amount in the air, but the control room operator had no way of knowing what the actual amount was. In addition, there are no established criteria in any written procedure for what level of this gas or what alarms constitute an emergency condition that should trigger sounding the evacuation alarm. Also, none of the alarms were designated as critical alarms, which the accident report does concede might have "elicited a higher degree of attention amongst the competing priorities" of the control room operator. Finally, there was no written procedure for responding to an alarm for this gas. The "standard response" was for an outside operator to conduct a field assessment of the situation, which he did.

While there is training information provided about the hazards of the particular gas that escaped, this information was not incorporated in standard operating or emergency procedures. The operators were apparently on their own to decide if an emergency existed and then were chastised for not responding (in hindsight) correctly. If there is a potential for operators to make poor decisions in safety-critical situations, then they need to be provided with the criteria to make such a decision. Expecting operators under stress and perhaps with limited information about the current system state and inadequate training to make such critical decisions based on their own judgment is unrealistic. It simply ensures that operators will be blamed when their decisions turn out, in hindsight, to be wrong.

One of the actions the operators were criticized for was trying to fix the problem rather than calling in emergency personnel immediately after the gas alarm sounded. In fact, this response is the *normal* one for humans (see chapter 9 and [115], as well as the following discussion): if it is not the desirable response, then procedures and training must be used to ensure that a different response is elicited. The accident report states that the safety policy for this company is:

> At units, any employee shall assess the situation and determine what level of evacuation and what equipment shutdown is necessary to ensure the safety of all personnel, mitigate the environmental impact and potential for equipment/property damage. When in doubt, evacuate.

There are two problems with such a policy.

The first problem is that evacuation responsibilities (or emergency procedures more generally) do not seem to be assigned to anyone but can be initiated by all employees. While this may seem like a good idea, it has a serious drawback because one consequence of such a lack of assigned control responsibility is that everyone may think that someone else will take the initiative—and the blame if the alarm is a false one. Although everyone should report problems and even sound an emergency alert when necessary, there must be someone who has the actual responsibility, authority, and accountability to do so. There should also be backup procedures for others to step in when that person does not execute his or her responsibility acceptably.

The second problem with this safety policy is that unless the procedures clearly say to execute emergency procedures, humans are very likely to try to diagnose the situation first. The same problem pops up in many accident reports—humans who are overwhelmed with information that they cannot digest quickly or do not understand, will first try to understand what is going on before sounding an alarm [115]. If management wants employees to sound alarms expeditiously and consistently, then the safety policy needs to specify exactly when alarms are required, not leave it up to personnel to "evaluate the situation" when they are probably confused and unsure as to what is going on (as in this case) and under pressure to make quick decisions under stressful situations. How many people, instead of dialing 911 immediately, try to put out a small kitchen fire themselves? That it often works simply reinforces the tendency to act in the same way during the next emergency. And it avoids the embarrassment of the firemen arriving for a non-emergency. As it turns out, the evacuation alert had been delayed in the past in this same plant, but nobody had investigated why that occurred.

The accident report concludes with a recommendation that "operator duty to respond to alarms needs to be reinforced with the work force." This recommendation is inadequate because it ignores *why* the operators did not respond to the alarms. More useful recommendations might have included designing more accurate and more observable feedback about the actual position of the control valve (rather than just the commanded position), about the state of flow into the tank, about the level of the liquid in the tank, and so on. The recommendation also ignores the ambiguous state of the company policy on responding to alarms.

Because the official report focused only on the role of the operators in the accident and did not even examine that in depth, a chance to detect flaws in the design and operation of the plant that could lead to future accidents was lost. To prevent future accidents, the report needed to explain such things as why the HAZOP performed on the unit did not identify any of the alarms in this unit as critical. Is there some deficiency in HAZOP or in the way it is being performed in this company? Why were there no procedures in place, or why were the ones in place ineffective, to respond to the emergency? Either the hazard was not identified, the company does not have a policy to create procedures for dealing with hazards, or it was an oversight and there was no procedure in place to check that there is a response for all identified hazards.

The report does recommend that a risk assessed procedure for filling this tank be created that defines critical operational parameters such as the sequence of steps required to initiate the filling process, the associated process control parameters, the safe level at which the tank is considered full, the sequence of steps necessary to conclude and secure the tank-filling process, and appropriate response to alarms. It does not say anything, however, about performing the same task for other processes in the plant. Either this tank and its safety-critical process are the only ones missing such procedures or the company is playing a sophisticated game of Whack-a-Mole (see chapter 13), in which only symptoms of the real problems are removed with each set of events investigated.

The official accident report concludes that the control room operator "did not demonstrate an awareness of risks associated with overflowing the tank and potential to generate high concentrations of [gas] if the [liquid in the tank] was spilled." No further investigation of why this was true was included in the report. Was there a deficiency in the training procedures about the hazards associated with his job responsibilities? Even if the explanation is that this particular operator is simply incompetent (probably not true) and although exposed to potentially effective training did not profit from it, then the question becomes why such an operator was allowed to continue in that job and why the evaluation of his training outcomes did not detect this deficiency. It seemed that the outside operator also had a poor understanding of the risks from this gas so there is clearly evidence that a systemic problem exists. An audit should have been performed to determine if a spill in this tank is the only hazard that is not understood and if these two operators are the only ones who are confused. Is this unit simply a poorly designed and managed one in the plant or do similar deficiencies exist in other units?

Other important causal factors and questions also were not addressed in the report such as why the level transmitter was not working so soon after it was supposedly fixed, why safety orders were so delayed (the average age of a safety-related work order in this plant was three months), why critical processes were allowed to operate with non-functioning or erratically functioning safety-related equipment, whether the plant management knew this was happening, and so on.

Hindsight bias and focusing only on the operator's role in accidents prevents us from fully learning from accidents and making significant progress in improving safety.

11.8   Coordination and Communication

The analysis so far has looked at each component separately. But coordination and communication between controllers are important sources of unsafe behavior.

Whenever a component has two or more controllers, coordination should be examined carefully. Each controller may have different responsibilities, but the control actions provided may conflict. The controllers may also control the same aspects of the controlled component's behavior, leading to confusion about who is responsible for providing control at any time. In the Walkerton *E. coli* water supply contamination example provided in appendix C, three control components were responsible for following up on inspection reports and ensuring the required changes were made: the Walkerton Public Utility Commission (WPUC), the Ministry of the Environment (MOE), and the Ministry of Health (MOH). The WPUC commissioners had no expertise in running a water utility and simply left the changes to the manager. The MOE and MOH both were responsible for performing the same oversight: The local MOH facility assumed that the MOE was performing this function, but the MOE's budget had been cut, and follow-ups were not done. In this case, each of the three responsible groups assumed the other two controllers were providing the needed oversight, a common finding after an accident.

A different type of coordination problem occurred in an aircraft collision near Überlingen, Germany, in 2002 [28, 212]. The two controllers—the automated on-board TCAS system and the ground air traffic controller—provided uncoordinated control instructions that conflicted and actually caused a collision. The loss would have been prevented if both pilots had followed their TCAS alerts or both had followed the ground ATC instructions.

In the friendly fire accident analyzed in chapter 5, the responsibility of the AWACS controllers had officially been disambiguated by assigning one to control aircraft within the no-fly zone and the other to monitor and control aircraft outside it. This partitioning of control broke down over time, however, with the result that neither controlled the Black Hawk helicopter on that fateful day. No performance

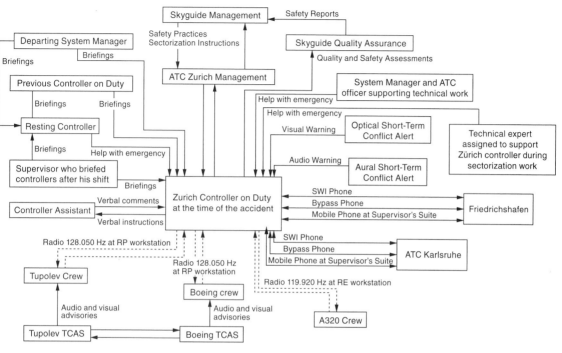

Figure 11.9
The communication links theoretically in place at the time of the Überlingen aircraft collision (adapted from [212]).

auditing occurred to ensure that the assumed and designed behavior of the safety control structure components was actually occurring.

Communication, both feedback and exchange of information, is also critical. All communication links should be examined to ensure they worked properly and, if they did not, the reasons for the inadequate communication must be determined. The Überlingen collision, between a Russian Tupolev aircraft and a DHL Boeing aircraft, provides a useful example. Wong used STAMP to analyze this accident and demonstrated how the communications breakdown on the night of the accident played an important role [212]. Figure 11.9 shows the components surrounding the controller at the Air Traffic Control Center in Zürich that was controlling both aircraft at the time and the feedback loops and communication links between the components. Dashed lines represent partial communication channels that are not available all the time. For example, only partial communication is available between the controller and multiple aircraft because only one party can transmit at one time when they are sharing a single radio frequency. In addition, the controller cannot directly receive information about TCAS advisories—the Pilot Not Flying (PNF) is supposed to report TCAS advisories to the controller over the radio. Finally, communicating all the time with all the aircraft requires the presence of two controllers at two different consoles, but only one controller was present at the time.

Nearly all the communication links were broken or ineffective at the time of the accident (see figure 11.10). A variety of conditions contributed to the lost links.

The first reason for the dysfunctional communication was unsafe practices such as inadequate briefings given to the two controllers scheduled to work the night shift, the second controller being in the break room (which was not officially allowed but was known and tolerated by management during times of low traffic), and the reluctance of the controller's assistant to speak up with ideas to assist in the situation due to feeling that he would be overstepping his bounds. The inadequate briefings were due to a lack of information as well as each party believing they were not responsible for conveying specific information, a result of poorly defined roles and responsibilities.

More links were broken due to maintenance work that was being done in the control room to reorganize the physical sectors. This work led to unavailability of the direct phone line used to communicate with adjacent ATC centers (including ATC Karlsruhe, which saw the impending collision and tried to call ATC Zurich) and the loss of an optical short-term conflict alert (STCA) on the console. The aural short-term conflict alert was theoretically working, but nobody in the control room heard it.

Unusual situations led to the loss of additional links. These include the failure of the bypass telephone system from adjacent ATC centers and the appearance of a delayed A320 aircraft landing at Friedrichshafen. To communicate with all three aircraft, the controller had to alternate between two consoles, changing all the aircraft–controller communication channels to partial links.

Finally, some links were unused because the controller did not realize they were available. These include possible help from the other staff present in the control room (but working on the resectorization) and a third telephone system that the controller did not know about. In addition, the link between the crew of the Tupolev aircraft and its TCAS unit was broken due to the crew ignoring the TCAS advisory.

Figure 11.10 shows the remaining links after all these losses. At the time of the accident, there were no complete feedback loops left in the system and the few remaining connections were partial ones. The exception was the connection between the TCAS units of the two aircraft, which were still communicating with each other. The TCAS unit can only provide information to the crew, however, so this remaining loop was unable to exert any control over the aircraft.

Another common type of communication failure is in the problem-reporting channels. In a large number of accidents, the investigators find that the problems were identified in time to prevent the loss but that the required problem-reporting

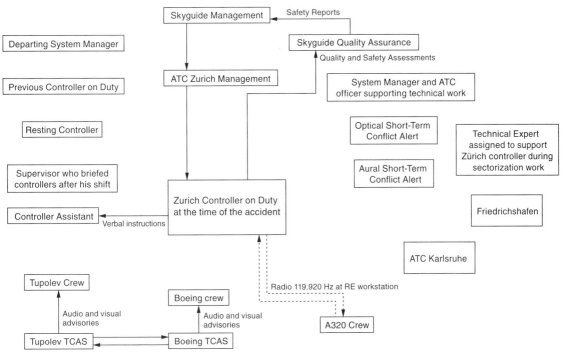

Figure 11.10
The actual state of the communication links and control loops at the time of the accident (adapted from [212]). Compare this figure with the designed communication links shown in figure 11.9.

channels were not used. Recommendations in the ensuing accident reports usually involve training people to use the reporting channels—based on an assumption that the lack of use reflected poor training—or attempting to enforce their use by reiterating the requirement that all problems be reported. These investigations, however, usually stop short of finding out why the reporting channels were not used. Often an examination and a few questions reveal that the formal reporting channels are difficult or awkward and time-consuming to use. Redesign of a poorly designed system will be more effective in ensuring future use than simply telling people they have to use a poorly designed system. Unless design changes are made, over time the poorly designed communication channels will again become underused.

At Citichem, all problems were reported orally to the control room operator, who was supposed to report them to someone above him. One conduit for information, of course, leads to a very fragile reporting system. At the same time, there were few formal communication and feedback channels established—communication was informal and ad hoc, both within Citichem and between Citichem and the local government.

11.9   Dynamics and Migration to a High-Risk State

As noted previously, most major accidents result from a migration of the system toward reduced safety margins over time. In the Citichem example, pressure from commercial competition was one cause of this degradation in safety. It is, of course, a very common one. Operational safety practices at Citichem had been better in the past, but the current market conditions led management to cut the safety margins and ignore established safety practices. Usually there are precursors signaling the increasing risks associated with these changes in the form of minor incidents and accidents, but in this case, as in so many others, these precursors were not recognized. Ironically, the death of the Citichem maintenance manager in an accident led the management to make changes in the way they were operating, but it was too late to prevent the toxic chemical release.

The corporate leaders pressured the Citichem plant manager to operate at higher levels of risk by threatening to move operations to Mexico, leaving the current workers without jobs. Without any way of maintaining an accurate model of the risk in current operations, the plant manager allowed the plant to move to a state of higher and higher risk.

Another change over time that affected safety in this system was the physical change in the separation of the population from the plant. Usually hazardous facilities are originally placed far from population centers, but the population shifts after the facility is created. People want to live near where they work and do not like long commutes. Land and housing may be cheaper near smelly, polluting plants. In third world countries, utilities (such as power and water) and transportation facilities may be more readily available near heavy industrial plants, as was the case at Bhopal.

At Citichem, an important change over time was the obsolescence of the emergency preparations as the population increased. Roads, hospital facilities, firefighting equipment, and other emergency resources became inadequate. Not only were there insufficient resources to handle the changes in population density and location, but financial and other pressures militated against those wanting to update the emergency resources and plans.

Considering the Oakbridge community dynamics, the city of Oakbridge contributed to the accident through the erosion of the safety controls due to the normal pressures facing any city government. Without any history of accidents, or risk assessments indicating otherwise, the plant was deemed safe, and officials allowed developers to build on previously restricted land. A contributing factor was the desire to increase city finances and business relationships that would assist in reelection of the city officials. The city moved toward a state where casualties would be massive when an accident did occur.

The goal of understanding the dynamics is to redesign the system and the safety control structure to make them more conducive to system safety. For example, behavior is influenced by recent accidents or incidents: As safety efforts are successfully employed, the feeling grows that accidents cannot occur, leading to reduction in the safety efforts, an accident, and then increased controls for a while until the system drifts back to an unsafe state and complacency again increases . . .

This complacency factor is so common that any system safety effort must include ways to deal with it. SUBSAFE, the U.S. nuclear submarine safety program, has been particularly successful at accomplishing this goal. The SUBSAFE program is described in chapter 14.

One way to combat this erosion of safety is to provide ways to maintain accurate risk assessments in the process models of the system controllers. The more and better information controllers have, the more accurate will be their process models and therefore their decisions.

In the Citichem example, the dynamics of the city migration toward higher risk might be improved by doing better hazard analyses, increasing communication between the city and the plant (e.g., learning about incidents that are occurring), and the formation of community citizen groups to provide counterbalancing pressures on city officials to maintain the emergency response system and the other public safety measures.

Finally, understanding the reason for such migration provides an opportunity to design the safety control structure to prevent it or to detect it when it occurs. Thorough investigation of incidents using CAST and the insight it provides can be used to redesign the system or to establish operational controls to stop the migration toward increasing risk before an accident occurs.

11.10   Generating Recommendations from the CAST Analysis

The goal of an accident analysis should not be just to address symptoms, to assign blame, or to determine which group or groups are more responsible than others.

Blame is difficult to eliminate, but, as discussed in section 2.7, blame is antithetical to improving safety. It hinders accident and incident investigations and the reporting of errors before a loss occurs, and it hinders finding the most important factors that need to be changed to prevent accidents in the future. Often, blame is assigned to the least politically powerful in the control hierarchy or to those people or physical components physically and operationally closest to the actual loss events. Understanding why inadequate control was provided and why it made sense for the controllers to act in the way they did helps to diffuse what seems to be a natural desire to assign blame for events. In addition, looking at how the entire safety control structure was flawed and conceptualizing accidents as complex processes rather than the result of independent events should reduce the finger pointing and arguments about others being more to blame that often arises when system components other than the operators are identified as being part of the accident process. "More to blame" is not a relevant concept in a systems approach to accident analysis and should be resisted and avoided. Each component in a system works together to obtain the results, and no part is more important than another.

The goal of the accident analysis should instead be to determine how to change or reengineer the entire safety-control structure in the most cost-effective and practical way to prevent similar accident processes in the future. Once the STAMP analysis has been completed, generating recommendations is relatively simple and follows directly from the analysis results.

One consequence of the completeness of a STAMP analysis is that many possible recommendations may result—in some cases, too many to be practical to include in the final accident report. A determination of the relative importance of the potential recommendations may be required in terms of having the greatest impact on the largest number of potential future accidents. There is no algorithm for identifying these recommendations, nor can there be. Political and situational factors will always be involved in such decisions. Understanding the entire accident process and the overall safety control structure should help with this identification, however.

Some sample recommendations for the Citichem example are shown throughout the chapter. A more complete list of the recommendations that might result from a STAMP-based Citichem accident analysis follows. The list is divided into four parts: physical equipment and design, corporate management, plant operations and management, and government and community.

Physical Equipment and Design

1. Add protection against rainwater getting into tanks.

2. Consider measures for preventing and detecting corrosion.

3. Change the design of the valves and vent pipes to respond to the two-phase flow problem (which was responsible for the valves and pipes being jammed).

4. Etc. (the rest of the physical plant factors are omitted)

Corporate Management

1. Establish a corporate safety policy that specifies:

   a. Responsibility, authority, accountability of everyone with respect to safety

   b. Criteria for evaluating decisions and for designing and implementing safety controls.

2. Establish a corporate process safety organization to provide oversight that is responsible for:

   a. Enforcing the safety policy

   b. Advising corporate management on safety-related decisions

   c. Performing risk analyses and overseeing safety in operations including performing audits and setting reporting requirements (to keep corporate process models accurate). A safety working group at the corporate level should be considered.

   d. Setting minimum requirements for safety engineering and operations at plants and overseeing the implementation of these requirements as well as management of change requirements for evaluating all changes for their impact on safety.

   e. Providing a conduit for safety-related information from below (a formal safety reporting system) as well as an independent feedback channel about process safety concerns by employees.

   f. Setting minimum physical and operational standards (including functioning equipment and backups) for operations involving dangerous chemicals.

   g. Establishing incident/accident investigation standards and ensuring recommendations are adequately implemented.

   h. Creating and maintaining a corporate process safety information system.

3. Improve process safety communication channels both within the corporate level as well as information and feedback channels from Citichem plants to corporate management.

4. Ensure that appropriate communication and coordination is occurring between the Citichem plants and the local communities in which they reside.

5. Strengthen or create an inventory control system for safety-critical parts at the corporate level. Ensure that safety-related equipment is in stock at all times.

Citichem Oakbridge Plant Management and Operations

1. Create a safety policy for the plant. Derive it from the corporate safety policy and make sure everyone understands it. Include minimum requirements for operations: for example, safety devices must be operational, and production should be shut down if they are not.

2. Establish a plant process safety organization and assign responsibility, authority, and accountability for this organization. Include a process safety manager whose primary responsibility is process safety. The responsibilities of this organization should include at least the following:

   a. Perform hazard and risk analysis.

    b. Advise plant management on safety-related decisions.

    c. Create and maintain a plant process safety information system.

    d. Perform or organize process safety audits and inspections using hazard analysis results as the preconditions for operations and maintenance.

    e. Investigate hazardous conditions, incidents, and accidents.

    f. Establish leading indicators of risk.

    g. Collect data to ensure process safety policies and procedures are being followed.

3. Ensure that everyone has appropriate training in process safety and the specific hazards associated with plant operations.

4. Regularize and improve communication channels. Create the operational feedback channels from controlled components to controllers necessary to maintain accurate process models to assist in safety-related decision making. If the channels exist but are not used, then the reason why they are unused should be determined and appropriate changes made.

5. Establish a formal problem reporting system along with channels for problem reporting that include management and rank and file workers. Avoid communication channels with a single point of failure for safety-related messages. Decisions on whether management is informed about hazardous operational events should be proceduralized. Any operational conditions found to exist that involve hazards should be reported and thoroughly investigated by those responsible for system safety.

6. Consider establishing employee safety committees with union representation (if there are unions at the plant). Consider also setting up a plant process safety working group.

7. Require that all changes affecting safety equipment be approved by the plant manager or by his or her designated representative for safety. Any outage of safety-critical equipment must be reported immediately.

8. Establish procedures for quality control and checking of safety-critical activities and follow-up investigation of safety excursions (hazardous conditions).

9. Ensure that those performing safety-critical operations have appropriate skills and physical resources (including adequate rest).

10. Improve inventory control procedures for safety-critical parts at the Oakbridge plant.

11. Review procedures for turnarounds, maintenance, changes, operations, etc. that involve potential hazards and ensure that these are being followed. Create an MOC procedure that includes hazard analysis on all planned changes.

12. Enforce maintenance schedules. If delays are unavoidable, a safety analysis should be performed to understand the risks involved.

13. Establish incident/accident investigation standards and ensure that they are being followed and recommendations are implemented.

14. Create a periodic audit system on the safety of operations and the state of the plant. Audit scope might be defined by such information as the hazard analysis, identified leading indicators of risk, and past incident/accident investigations.

15. Establish communication channels with the surrounding community and provide appropriate information for better decision making by community leaders and information to emergency responders and the medical establishment. Coordinate with the surrounding community to provide information and assistance in establishing effective emergency preparedness and response measures. These measures should include a warning siren or other notification of an emergency and citizen information about what to do in the case of an emergency.

Government and Community

1. Set policy with respect to safety and ensure that the policy is enforced.

2. Establish communication channels with hazardous industry in the community.

3. Establish and monitor information channels about the risks in the community. Collect and disseminate information on hazards, the measures citizens can take to protect themselves, and what to do in case of an emergency.

4. Encourage citizens to take responsibility for their own safety and to encourage local, state, and federal government to do the things necessary to protect them.

5. Encourage the establishment of a community safety committee and/or a safety ombudsman office that is not elected but represents the public in safety-related decision making.

6. Ensure that safety controls are in place before approving new development in hazardous areas, and if not (e.g., inadequate roads, communication channels, emergency response facilities), then perhaps make developers pay for them. Consider requiring developers to provide an analysis of the impact of new development on the safety of the community. Hire outside consultants to evaluate these impact analyses if such expertise is not available locally.

7. Establish an emergency preparedness plan and re-evaluate it periodically to determine if it is up to date. Include procedures for coordination among emergency responders.

8. Plan temporary measures for additional manpower in emergencies.

9. Acquire adequate equipment.

10. Provide drills and ensure alerting and communication channels exist and are operational.

11. Train emergency responders.

12. Ensure that transportation and other facilities exist for an emergency.

13. Set up formal communications between emergency responders (hospital staff, police, firefighters, Citichem). Establish emergency plans and means to periodically update them.

One thing to note from this example is that many of the recommendations are simply good safety management practices. While this particular example involved a system that was devoid of the standard safety practices common to most industries, many accident investigations conclude that standard safety management practices were not observed. This fact points to a great opportunity to prevent accidents simply by establishing standard safety controls using the techniques described in this book. While we want to learn as much as possible from each loss, preventing the losses in the first place is a much better strategy than waiting to learn from our mistakes.

These recommendations and those resulting from other thoroughly investigated accidents also provide an excellent resource to assist in generating the system safety requirements and constraints for similar types of systems and in designing improved safety control structures.

Just investigating the incident or accident is, of course, not enough. Recommendations must be implemented to be useful. Responsibility must be assigned for ensuring that changes are actually made. In addition, feedback channels should be established to determine whether the recommendations and changes were successful in reducing risk.

11.11 Experimental Comparisons of CAST with Traditional Accident Analysis

Although CAST is new, several evaluations have been done, mostly aviation-related.

Robert Arnold, in a master's thesis for Lund University, conducted a qualitative comparison of SOAM and STAMP in an Air Traffic Management (ATM) occurrence investigation. SOAM (Systemic Occurrence Analysis Methodology) is used by Eurocontrol to analyze ATM incidents. In Arnold's experiment, an incident was investigated using SOAM and STAMP and the usefulness of each in identifying systemic countermeasures was compared. The results showed that SOAM is a useful heuristic and a powerful communication device, but that it is weak with respect to emergent phenomena and nonlinear interactions. SOAM directs the investigator to consider the context in which the events occur, the barriers that failed, and the organizational factors involved, but not the processes that created them or how the entire system can migrate toward the boundaries of safe operation. In contrast, the author concludes,

> STAMP directs the investigator more deeply into the mechanism of the interactions between system components, and how systems adapt over time. STAMP helps identify the controls and constraints necessary to prevent undesirable interactions between system components. STAMP also directs the investigation through a structured analysis of the upper levels of the system's control structure, which helps to identify high level systemic countermeasures. The global ATM system is undergoing a period of rapid technological and political change. . . . The ATM is moving from centralized human controlled systems to semi-automated distributed decision making. . . . Detailed new systemic models like STAMP are now necessary to prevent undesirable interactions between normally functioning system components and to understand changes over time in increasingly complex ATM systems.

Paul Nelson, in another Lund University master's thesis, used STAMP and CAST to analyze the crash of Comair 5191 at Lexington, Kentucky, on August 27, 2006, when the pilots took off from the wrong runway [142]. The accident, of course, has been thoroughly investigated by the NTSB. Nelson concludes that the NTSB report narrowly targeted causes and potential solutions. No recommendations were put forth to correct the underlying safety control structure, which fostered process model inconsistencies, inadequate and dysfunctional control actions, and unenforced safety constraints. The CAST analysis, on the other hand, uncovered these useful levers for eliminating future loss.

Stringfellow compared the use of STAMP, augmented with guidewords for organizational and human error analysis, with the use of HFACS (Human Factors Analysis and Classification System) on the crash of a Predator-B unmanned aircraft near Nogales, Arizona [195]. HFACS, based on the Swiss Cheese Model (event-chain model), is an error-classification list that can be used to label types of errors, problems, or poor decisions made by humans and organizations [186]. Once again, although the analysis of the unmanned vehicle based on STAMP found all the factors found in the published analysis of the accident using HFACS [31, 195], the STAMP-based analysis identified additional factors, particularly those at higher levels of the safety control structure, for example, problems in the FAA's COA[3] approval process. Stringfellow concludes:

---

3. The COA or Certificate of Operation allows an air vehicle that does not nominally meet FAA safety standards access to the National Airspace System. The COA application process includes measures to mitigate risks, such as sectioning off the airspace to be used by the unmanned aircraft and preventing other aircraft from entering the space.

The organizational influences listed in HFACS . . . do not go far enough for engineers to create recommendations to address organizational problems. . . . Many of the factors cited in Swiss Cheese-based methods don't point to solutions; many are just another label for human error in disguise [195, p. 154].

In general, most accident analyses do a good job in describing *what* happened, but not *why*.

11.12 Summary

In this chapter, the process for performing accident analysis using STAMP as the basis is described and illustrated using a chemical plant accident as an example. Stopping the analysis at the lower levels of the safety-control structure, in this case at the physical controls and the plant operators, provides a distorted and incomplete view of the causative factors in the loss. Both a better understanding of why the accident occurred and how to prevent future ones are enhanced with a more complete analysis. As the entire accident process becomes better understood, individual mistakes and actions assume a much less important role in comparison to the role played by the environment and context in which their decisions and control actions take place. What may look like an error or even negligence by the low-level operators and controllers may appear much more reasonable given the full picture. In addition, changes at the lower levels of the safety-control structure often have much less ability to impact the causal factors in major accidents than those at higher levels.

At all levels, focusing on assessing blame for the accident does not provide the information necessary to prevent future accidents. Accidents are complex processes, and understanding the entire process is necessary to provide recommendations that are going to be effective in preventing a large number of accidents and not just preventing the symptoms implicit in a particular set of events. There is too much repetition of the same causes of accidents in most industries. We need to improve our ability to learn from the past.

Improving accident investigation may require training accident investigators in systems thinking and in the types of environmental and behavior shaping factors to consider during an analysis, some of which are discussed in later chapters. Tools to assist in the analysis, particularly graphical representations that illustrate interactions and causality, will help. But often the limitations of accident reports do not stem from the sincere efforts of the investigators but from political and other pressures to limit the causal factors identified to those at the lower levels of the management or political hierarchy. Combating these pressures is beyond the scope of this book. Removing blame from the process will help somewhat. Management also has to be educated to understand that safety pays and, in the longer term, costs less than the losses that result from weak safety programs and incomplete accident investigations.

12 Controlling Safety during Operations

In some industries, system safety is viewed as having its primary role in development and most of the activities occur before operations begin. Those concerned with safety may lose influence and resources after that time. As an example, one of the chapters in the *Challenger* accident report, titled "The Silent Safety Program," lamented:

> Following the successful completion of the orbital flight test phase of the Shuttle program, the system was declared to be operational. Subsequently, several safety, reliability, and quality assurance organizations found themselves with reduced and/or reorganized functional capabilities. . . . The apparent reason for such actions was a perception that less safety, reliability, and quality assurance activity would be required during "routine" Shuttle operations. This reasoning was faulty.

While safety-guided design eliminates some hazards and creates controls for others, hazards and losses may still occur in operations due to:

- Inadequate attempts to eliminate or control the hazards in the system design, perhaps due to inappropriate assumptions about operations.
- Inadequate implementation of the controls that designers assumed would exist during operations.
- Changes that occur over time, including violation of the assumptions underlying the design.
- Unidentified hazards, sometimes new ones that arise over time and were not anticipated during design and development.

Treating operational safety as a control problem requires facing and mitigating these potential reasons for losses.

A complete system safety program spans the entire life of the system and, in some ways, the safety program during operations is even more important than during development. System safety does not stop after development; it is just getting started. The focus now, however, shifts to the operations safety control structure.

This chapter describes the implications of STAMP on operations. Some topics that are relevant here are left to the next chapter on management: organizational design, safety culture and leadership, assignment of appropriate responsibilities throughout the safety control structure, the safety information system, and corporate safety policies. These topics span both development and operations and many of the same principles apply to each, so they have been put into a separate chapter. A final section of this chapter considers the application of STAMP and systems thinking principles to occupational safety.

12.1   Operations Based on STAMP

Applying the basic principles of STAMP to operations means that, like development, the goal during operations is enforcement of the safety constraints, this time on the operating system rather than in its design. Specific responsibilities and control actions required during operations are outlined in chapter 13.

Figure 12.1 shows the interactions between development and operations. At the end of the development process, the safety constraints, the results of the hazard analyses, as well as documentation of the safety-related design features and design rationale, should be passed on to those responsible for the maintenance and evolution of the system. This information forms the baseline for safe operations. For example, the identification of safety-critical items in the hazard analysis should be used as input to the maintenance process for prioritization of effort.

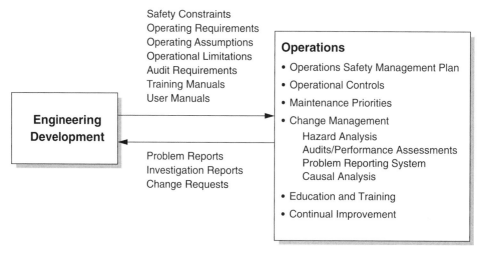

Figure 12.1
The relationship between development and operations.

At the same time, the accuracy and efficacy of the hazard analyses performed during development and the safety constraints identified need to be evaluated using the operational data and experience. Operational feedback on trends, incidents, and accidents should trigger reanalysis when appropriate. Linking the assumptions throughout the system specification with the parts of the hazard analysis based on that assumption will assist in performing safety maintenance activities. During field testing and operations, the links and recorded assumptions and design rationale can be used in safety change analysis, incident and accident analysis, periodic audits and performance monitoring as required to ensure that the operational system is and remains safe.

For example, consider the TCAS requirement that TCAS provide collision avoidance protection for any two aircraft closing horizontally at any rate up to 1,200 knots and vertically up to 10,000 feet per minute. As noted in the rationale, this requirement is based on aircraft performance limits at the time TCAS was created. It is also based on minimum horizontal and vertical separation requirements. The safety analysis originally performed on TCAS is based on these assumptions. If aircraft performance limits change or if there are proposed changes in airspace management, as is now occurring in new Reduced Vertical Separation Minimums (RVSM), hazard analysis to determine the safety of such changes will require the design rationale and the tracing from safety constraints to specific system design features as recorded in intent specifications. Without such documentation, the cost of reanalysis could be enormous and in some cases even impractical. In addition, the links between design and operations and user manuals in level 6 will ease updating when design changes are made.

In a traditional System Safety program, much of this information is found in or can be derived from the hazard log, but it needs to be pulled out and provided in a form that makes it easy to locate and use in operations. Recording design rationale and assumptions in intent specifications allows using that information both as the criteria under which enforcement of the safety constraints is predicated and in the inevitable upgrades and changes that will need to be made during operations. Chapter 10 shows how to identify and record the necessary information.

The design of the operational safety controls are based on assumptions about the conditions during operations. Examples include assumptions about how the operators will operate the system and the environment (both social and physical) in which the system will operate. These conditions may change. Therefore, not only must the assumptions and design rationale be conveyed to those who will operate the system, but there also need to be safeguards against changes over time that violate those assumptions.

The changes may be in the behavior of the system itself:

- Physical changes: the equipment may degrade or not be maintained properly.
- Human changes: human behavior and priorities usually change over time.
- Organizational changes: change is a constant in most organizations, including changes in the safety control structure itself, or in the physical and social environment within which the system operates or with which it interacts.

Controls need to be established to reduce the risk associated with all these types of changes.

The safeguards may be in the design of the system itself or in the design of the operational safety control structure. Because operational safety depends on the accuracy of the assumptions and models underlying the design and hazard analysis processes, the operational system should be monitored to ensure that:

1. The system is constructed, operated, and maintained in the manner assumed by the designers.
2. The models and assumptions used during initial decision making and design are correct.
3. The models and assumptions are not violated by changes in the system, such as workarounds or unauthorized changes in procedures, or by changes in the environment.

Designing the operations safety control structure requires establishing controls and feedback loops to (1) identify and handle flaws in the original hazard analysis and system design and (2) to detect unsafe changes in the system during operations before the changes lead to losses. Changes may be intentional or they may be unintended and simply normal changes in system component behavior or the environment over time. Whether intended or unintended, system changes that violate the safety constraints must be controlled.

12.2 Detecting Development Process Flaws during Operations

Losses can occur due to flaws in the original assumptions and rationale underlying the system design. Errors may also have been made in the hazard analysis process used during system design. During operations, three goals and processes to achieve these goals need to be established:

1. Detect safety-related flaws in the system design and in the safety control structure, hopefully before major losses, and fix them.

2. Determine what was wrong in the development process that allowed the flaws to exist and improve that process to prevent the same thing from happening in the future.

3. Determine whether the identified flaws in the process might have led to other vulnerabilities in the operational system.

If losses are to be reduced over time and companies are not going to simply engage in constant firefighting, then mechanisms to implement learning and continual improvement are required. Identified flaws must not only be fixed (symptom removal), but the larger operational and development safety control structures must be improved, as well as the process that allowed the flaws to be introduced in the first place. The overall goal is to change the culture from *a fixing orientation*—identifying and eliminating deviations or symptoms of deeper problems—to a *learning orientation* where systemic causes are included in the search for the source of safety problems [33].

To accomplish these goals, a feedback control loop is needed to regularly track and assess the effectiveness of the development safety control structure and its controls. Were hazards overlooked or incorrectly assessed as unlikely or not serious? Were some potential failures or design errors not included in the hazard analysis? Were identified hazards inappropriately accepted rather than being fixed? Were the designed controls ineffective? If so, why?

When numerical risk assessment techniques are used, operational experience can provide insight into the accuracy of the models and probabilities used. In various studies of the DC-10 by McDonnell Douglas, the chance of engine power loss with resulting slat damage during takeoff was estimated to be less than one in a billion flights. However, this highly improbable event occurred four times in DC-10s in the first few years of operation without raising alarm bells before it led to an accident and changes were made. Even one event should have warned someone that the models used might be incorrect. Surprisingly little scientific evaluation of probabilistic risk assessment techniques has ever been conducted [115], yet these techniques are regularly taught to most engineering students and widely used in industry. Feedback loops to evaluate the assumptions underlying the models and the assessments produced are an obvious way to detect problems.

Most companies have an accident/incident analysis process that identifies the proximal failures that led to an incident, for example, a flawed design of the pressure relief valve in a tank. Typical follow-up would include replacement of that valve with an improved design. On top of fixing the immediate problem, companies should have procedures to evaluate and potentially replace all the uses of that pressure relief valve design in tanks throughout the plant or company. Even better would be to reevaluate pressure relief valve design for all uses in the plant, not just in tanks.

But for long-term improvement, a causal analysis—CAST or something similar—needs to be performed on the process that created the flawed design and that process improved. If the development process was flawed, perhaps in the hazard analysis or design and verification, then fixing that process can prevent a large number of incidents and accidents in the future.

Responsibility for this goal has to be assigned to an appropriate component in the safety control structure and feedback-control loops established. Feedback may come from accident and incident reports as well as detected and reported design and behavioral anomalies. To identify flaws before losses occur, which is clearly desirable, audits and performance assessments can be used to collect data for validating and informing the safety design and analysis process without waiting for a crisis. There must also be feedback channels to the development safety control structure so that appropriate information can be gathered and used to implement improvements. The design of these control loops is discussed in the rest of this chapter. Potential challenges in establishing such control loops are discussed in the next chapter on management.

12.3 Managing or Controlling Change

Systems are not static but instead are dynamic processes that are continually adapting to achieve their ends and to react to changes in themselves and their environment. In STAMP, adaptation or change is assumed to be an inherent part of any system, particularly those that include humans and organizational components: Humans and organizations optimize and change their behavior, adapting to the changes in the world and environment in which the system operates.

To avoid losses, not only must the original design enforce the safety constraints on system behavior, but the safety control structure must continue to enforce them as changes to the designed system, including the safety control structure itself, occur over time.

While engineers usually try to anticipate potential changes and to design for changeability, the bulk of the effort in dealing with change must necessarily occur during operations. Controls are needed both to prevent unsafe changes and to detect them if they occur.

In the friendly fire example in chapter 5, the AWACS controllers stopped handing off helicopters as they entered and left the no-fly zone. They also stopped using the Delta Point system to describe flight plans, although the helicopter pilots assumed the coded destination names were still being used and continued to provide them. Communication between the helicopters and the AWACS controllers was seriously degraded although nobody realized it. The basic safety constraint that all aircraft in the no-fly zone and their locations would be known to the AWACS controllers became over time untrue as the AWACS controllers optimized their procedures. This type of change is normal; it needs to be identified by checking that the assumptions upon which safety is predicated remain true over time.

The deviation from assumed behavior during operations was not, in the friendly fire example, detected until after an accident. Obviously, finding the deviations at this time is less desirable than using audits, and other types of feedback mechanisms to detect hazardous changes, that is, those that violate the safety constraints, before losses occur. Then something needs to be done to ensure that the safety constraints are enforced in the future.

Controls are required for both intentional (planned) and unintentional changes.

12.3.1   Planned Changes

Intentional system changes are a common factor in accidents, including physical, process, and safety control structure changes [115]. The Flixborough explosion provides an example of a temporary physical change resulting in a major loss: Without first performing a proper hazard analysis, a temporary pipe was used to replace a reactor that had been removed to repair a crack. The crack itself was the result of a previous process modification [54]. The Walkerton water contamination loss in appendix C provides an example of a control structure change when the government water testing lab was privatized without considering how that would affect feedback to the Ministry of the Environment.

Before any planned changes are made, including organizational and safety control structure changes, their impact on safety must be evaluated. Whether this process is expensive depends on how the original hazard analysis was performed and particularly how it was documented. Part of the rationale behind the design of intent specifications was to make it possible to retrieve the information needed.

While implementing change controls limits flexibility and adaptability, at least in terms of the time it takes to make changes, the high accident rate associated with intentional changes attests to the importance of controlling them and the high level of risk being assumed by not doing so. Decision makers need to understand these risks before they waive the change controls.

Most systems and industries do include such controls, usually called Management of Change (MOC) procedures. But the large number of accidents occurring after system changes without evaluating their safety implies widespread nonenforcement of these controls. Responsibility needs to be assigned for ensuring compliance with the MOC procedures so that change analyses are conducted and the results are not ignored. One way to do this is to reward people for safe behavior when they choose safety over other system goals and to hold them accountable when they choose to ignore the MOC procedures, even when no accident results. Achieving this goal, in turn, requires management commitment to safety (see chapter 13), as does just about every aspect of building and operating a safe system.

12.3.2 Unplanned Changes

While dealing with planned changes is relatively straightforward (even if difficult to enforce), unplanned changes that move systems toward states of higher risk are less straightforward. There need to be procedures established to prevent or detect changes that impact the ability of the operations safety control structure and the designed controls to enforce the safety constraints.

As noted earlier, people will tend to optimize their performance over time to meet a variety of goals. If an unsafe change is detected, it is important to respond quickly. People incorrectly reevaluate their perception of risk after a period of success. One way to interrupt this risk-reevaluation process is to intervene quickly to stop it before it leads to a further reduction in safety margins or a loss occurs. But that requires an alerting function to provide feedback to someone who is responsible for ensuring that the safety constraints are satisfied.

At the same time, change is a normal part of any system. Successful systems are continually changing and adapting to current conditions. Change should be allowed as long as it does not violate the basic constraints on safe behavior and therefore increase risk to unacceptable levels. While in the short term relaxing the safety constraints may allow other system goals to be achieved to a greater degree, in the longer term accidents and losses can cost a great deal more than the short-term gains.

The key is to allow flexibility in how safety goals are achieved, but not flexibility in violating them, and to provide the information that creates accurate risk perception by decision makers.

Detecting migration toward riskier behavior starts with identifying baseline requirements. The requirements follow from the hazard analysis. These requirements may be general ("Equipment will not be operated above the identified safety-critical limits" or "Safety-critical equipment must be operational when the system is operating") or specifically tied to the hazard analysis ("AWACS operators must always hand off aircraft when they enter and leave the no-fly zone" or "Pilots must always follow the TCAS alerts and continue to do so until they are canceled").

The next step is to assign responsibility to appropriate places in the safety control structure to ensure the baseline requirements are not violated, while allowing changes that do not raise risk. If the baseline requirements make it impossible for the system to achieve its goals, then instead of waiving them, the entire safety control structure should be reconsidered and redesigned. For example, consider the foam shedding problems on the Space Shuttle. Foam had been coming off the external tank for most of the operational life of the Shuttle. During development, a hazard had been identified and documented related to the foam damaging the thermal control surfaces of the spacecraft. Attempts had been made to eliminate foam shedding, but none of the proposed fixes worked. The response was to simply waive the requirement before each flight. In fact, at the time of the *Columbia* loss, more than three thousand potentially critical failure modes were regularly waived on the pretext that nothing could be done about them and the Shuttle had to fly [74]. More than a third of these waivers had not been reviewed in the ten years before the accident.

After the *Columbia* loss, controls and mitigation measures for foam shedding were identified and implemented, such as changing the fabrication procedures and adding cameras and inspection and repair capabilities and other contingency actions. The same measures could, theoretically, have been implemented before the loss of *Columbia*. Most of the other waived hazards were also resolved in the aftermath of the accident. While the operational controls to deal with foam shedding raise the risk associated with a Shuttle accident above actually fixing the problem, the risk is lower than simply ignoring and waiting for the hazards to occur. Understanding and explicitly accepting risk is better than simply denying and ignoring it.

The NASA safety program and safety control structure had seriously degraded before both the *Challenger* and *Columbia* losses [117]. Waiving requirements interminably represents an abdication of the responsibility to redesign the system, including the controls during operations, after the current design is determined to be unsafe.

Is such a hard line approach impractical? SUBSAFE, the U.S. nuclear submarine safety program established after the *Thresher* loss, described in chapter 14, has not allowed waiving the SUBSAFE safety requirements for more than forty-five years, with one exception. In 1967, four years after SUBSAFE was established, SUBSAFE requirements for one submarine were waived in order to satisfy pressing Navy performance goals. That submarine and its crew were lost less than a year later. The same mistake has not been made again.

If there is absolutely no way to redesign the system to be safe and at the same time to satisfy the system requirements that justify its existence, then the existence of the system itself should be rethought and a major replacement or new design considered. After the first accident, much more stringent and perhaps unacceptable controls will be forced on operations. While the decision to live with risk is usually accorded to management, those who will suffer the losses should have a right to participate in that decision. Luckily, the choice is usually not so stark if flexibility is allowed in the way the safety constraints are maintained and long-term rather than short-term thinking prevails.

Like any set of controls, unplanned change controls involve designing appropriate control loops. In general, the process involves identifying the responsibility of the controller(s); collecting data (feedback); turning the feedback into useful information (analysis) and updating the process models; generating any necessary control actions and appropriate communication to other controllers; and measuring how effective the whole process is (feedback again).

12.4 Feedback Channels

Feedback is a basic part of STAMP and of treating safety as a control problem. Information flow is key in maintaining safety.

There is often a belief—or perhaps hope—that a small number of "leading indicators" can identify increasing risk of accidents, or, in STAMP terms, migration toward states of increased risk. It is unlikely that general leading indicators applicable to large industry segments exist or will be useful. The identification of system safety constraints does, however, provide the possibility of identifying leading indicators applicable to a specific system.

The desire to predict the future often leads to collecting a large amount of information based on the hope that something useful will be obtained and noticed. The NASA Space Shuttle program was collecting six hundred metrics a month before the loss of *Columbia*. Companies often collect data on occupational safety, such as days without a lost time accident, and they assume that these data reflect on system safety [17], which of course it does not. Not only is this misuse of data potentially misleading, but collecting information that may not be indicative of real risk diverts limited resources and attention from more effective risk-reduction efforts.

Poorly defined feedback can lead to a decrease in safety. As an incentive to reduce the number of accidents in the California construction industry, for example, workers with the best safety records—as measured by fewest reported incidents—were rewarded [126]. The reward created an incentive to withhold information about small accidents and near misses, and they could not therefore be investigated and the causes eliminated. Under-reporting of incidents created the illusion that the system was becoming safer, when instead risk had merely been muted. The inaccurate risk perception by management led to not taking the necessary control actions to reduce risk. Instead, the reporting of accidents should have been rewarded.

Feedback requirements should be determined with respect to the design of the organization's safety control structure, the safety constraints (derived from the system hazards) that must be enforced on system operation, and the assumptions and rationale underlying the system design for safety. They will be similar for different organizations only to the extent that the hazards, safety constraints, and system design are similar.

The hazards and safety constraints, as well as the causal information derived by the use of STPA, form the foundation for determining what feedback is necessary to provide the controllers with the information they need to satisfy their safety responsibilities. In addition, there must be mechanisms to ensure that feedback channels are operating effectively.

The feedback is used to update the controller's process models and understanding of the risks in the processes they are controlling, to update their control algorithms, and to execute appropriate control actions.

Sometimes, cultural problems interfere with feedback about the state of the controlled process. If the culture does not encourage sharing information and if there is a perception that the information can be used in a way that is detrimental to those providing it, then cultural changes will be necessary. Such changes require leadership and freedom from blame (see "Just Culture" in chapter 13). Effective feedback collection requires that those making the reports are convinced that the information will be used for constructive improvements in safety and not as a basis for criticism or disciplinary action. Resistance to airing dirty laundry is understandable, but this quickly transitions into an organizational culture where only good news is passed on for fear of retribution. Everyone's past experience includes individual mistakes, and avoiding repeating the same mistakes requires a culture that encourages sharing.

Three general types of feedback are commonly used: audits and performance assessments; reporting systems; and anomaly, incident, and accident investigation.

12.4.1 Audits and Performance Assessments

Once again, audits and performance assessments should start from the safety constraints and design assumptions and rationale. The goal should be to determine whether the safety constraints are being enforced in the operation of the system and whether the assumptions underlying the safety design and rationale are still true. Audits and performance assessments provide a chance to detect whether the behavior of the system and the system components still satisfies the safety constraints and whether the way the controllers think the system is working—as reflected in their process models—is accurate.

The entire safety control structure must be audited, not just the lower-level processes. Auditing the upper levels of the organization will require buy-in and commitment from management and an independent group at a high enough level to control audits as well as explicit rules for conducting them.

Audits are often less effective than they might be. When auditing is performed through contracts with independent companies, there may be subtle pressures on the audit team to be unduly positive or less than thorough in order to maintain their customer base. In addition, behavior or conditions may be changed in anticipation of an audit and then revert back to their normal state immediately afterward.

Overcoming these limitations requires changes in organizational culture and in the use of the audit results. Safety controllers (managers) must feel personal responsibility for safety. One way to encourage this view is to trust them and expect them to be part of the solution and to care about safety. "Safety is everyone's responsibility" must be more than an empty slogan, and instead a part of the organizational culture.

A *participatory audit* philosophy can have an important impact on these cultural goals. Some features of such a philosophy are:

- Audits should not be punitive. Audits need to be viewed as a chance to improve safety and to evaluate the process rather than a way to evaluate employees.

- To increase buy-in and commitment, those controlling the processes being audited should participate in creating the rules and procedures and understand the reasons for the audit and how the results will be used. Everyone should have a chance to learn from the audit without it having negative consequences—it should be viewed as an opportunity to learn how to improve.

- People from the process being audited should participate on the audit team. In order to get an outside but educated view, using process experts from other parts of the organization not directly being audited is a better approach than using outside audit companies. Various stakeholders in safety may be included such as unions. The goal should be to inculcate the attitude that this is *our* audit and a chance to improve *our* practices. Audits should be treated as a learning experience for everyone involved—including the auditors.

- Immediate feedback should be provided and solutions discussed. Often audit results are not available until after the audit and are presented in a written report. Feedback and discussion with the audit team during the audit are discouraged. One of the best times to discuss problems found and how to design solutions, however, is when the team is together and on the spot. Doing this will also reinforce the understanding that the goal is to improve the process, not to punish or evaluate those involved.

- All levels of the safety control structure should be audited, along with the physical process and its immediate operators. Accepting being audited and implementing improvements as a result—that is, leading by example—is a powerful way for leaders to convey their commitment to safety and to its improvement.

- A part of the audit should be to determine the level of safety knowledge and training that actually exists, not what managers believe exists or what exists in the training programs and user manuals. These results can be fed back into the training materials and education programs. Under no circumstances, of course, should such assessments be used in a negative way or one that is viewed as punitive by those being assessed.

Because these rules for audits are so far from common practice, they may be viewed as unrealistic. But this type of audit is carried out today with great success. See chapter 14 for an example. The underlying philosophy behind these practices is that most people do not want to harm others and have innate belief in safety as a goal. The problems arise when other goals are rewarded or emphasized over safety. When safety is highly valued in an organizational culture, obtaining buy-in is usually not difficult. The critical step lies in conveying that commitment.

12.4.2 Anomaly, Incident, and Accident Investigation

Anomaly, incident, and accident investigations often focus on a single "root" cause and look for contributory causes near the events. The belief that there is a root cause, sometimes called *root cause seduction* [32], is powerful because it provides an illusion of control. If the root cause can simply be eliminated and if that cause is low in the safety control structure, then changes can easily be made that will eliminate accidents without implicating management or requiring changes that are costly or disruptive to the organization. The result is that physical design characteristics or low-level operators are usually identified as the root cause.

Causality is, however, much more complex than this simple but very entrenched belief, as has been argued throughout this book. To effect high-leverage policies and changes that are able to prevent large classes of future losses, the weaknesses in the entire safety control structure related to the loss need to be identified and the control structure redesigned to be more effective.

In general, effective learning from experience requires a change from a fixing orientation to a continual learning and improvement culture. To create such a culture requires high-level leadership by management, and sometimes organizational changes.

Chapter 11 describes a way to perform better analyses of anomalies, incidents, and accidents. But having a process is not enough; the process must be embedded in an organizational structure that allows the successful exploitation of that process. Two important organizational factors will impact the successful use of CAST: training and follow-up.

Applying systems thinking to accident analysis requires training and experience. Large organizations may be able to train a group of investigators or teams to perform CAST analyses. This group should be managerially and financially independent. Some managers prefer to have accident/incident analysis reports focus on the low-level system operators and physical processes and the reports never go beyond those factors. In other cases, those involved in accident analysis, while well-meaning, have too limited a view to provide the perspective required to perform an adequate causal analysis. Even when intentions are good and local skills and knowledge are available, budgets may be so tight and pressures to maintain performance schedules so high that it is difficult to find the time and resources to do a thorough causal analysis using local personnel. Trained teams with independent budgets can overcome some of these obstacles. But while the leaders of investigations and causal analysis can be independent, participation by those with local knowledge is also important.

A second requirement is *follow-up*. Often the process stops after recommendations are made and accepted. No follow-up is provided to ensure that the recommendations are implemented or that the implementations were effective. Deadlines and assignment of responsibility for making recommendations, as well as responsibility for ensuring that they are made, are required. The findings in the causal analysis should be an input to future audits and performance assessments. If the same or similar causes recur, then that itself requires an analysis of why the problem was not fixed when it first was detected. Was the fix unsuccessful? Did the system migrate back to the same high-risk state because the underlying causal factors were never successfully controlled? Were factors missed in the original causal analysis? Trend analysis is important to ensure that progress is being made in controlling safety.

12.4.3 Reporting Systems

Accident reports very often note that before a loss, someone detected an anomaly but never reported it using the official reporting system. The response in accident investigation reports is often to recommend that the requirement to use reporting systems be emphasized to personnel or to provide additional training in using them. This response may be effective for a short time, but eventually people revert back to their prior behavior. A basic assumption about human behavior in this book (and in systems approaches to human factors) is that human behavior can usually be explained by looking at the system in which the human is operating. The reason in the system design for the behavior must be determined and changed: Simply trying to force people to behave in ways that are unnatural for them will usually be unsuccessful.

So the first question to ask is why people do not use reporting systems and to fix those factors. One obvious reason is that they may be designed poorly. They may require extra, time-consuming steps, such as logging into a web-based system, that are not part of their normal operating procedures or environment. Once they get to the website, they may be faced with a poorly designed form that requires them to provide a lot of extraneous information or does not allow the flexibility necessary to enter the information they want to provide.

A second reason people do not report is that the information they provided in the past appeared to go into a black hole, with nobody responding to it. There is little incentive to continue to provide information under these conditions, particularly when the reporting system is time-consuming and awkward to use.

A final reason for lack of reporting is a fear that the information provided may be used against them or there are other negative repercussions such as a necessity to spend time filling out additional reports.

Once the reason for failing to use reporting systems is understood, the solutions usually become obvious. For example, the system may need to be redesigned so it is easy to use and integrated into normal work procedures. As an example, email is becoming a primary means of communication at work. The first natural response in finding a problem is to contact those who can fix it, not to report it to some database where there is no assurance it will be processed quickly or get to the right people. A successful solution to this problem used on one large air traffic control system was to require only that the reporter add an extra "cc:" on their emails in order to get it reported officially to safety engineering and those responsible for problem reports [94].

In addition, the receipt of a problem report should result in both an acknowledgment of receipt and a thank-you. Later, when a resolution is identified, information should be provided to the reporter of the problem about what was done about it. If there is no resolution within a reasonable amount of time, that too should be acknowledged. There is little incentive to use reporting systems if the reporters do not think the information will be acted upon.

Most important, an effective reporting system requires that those making the reports are convinced the information will be used for constructive improvements in safety and not as a basis for criticism or disciplinary action. If reporting is considered to have negative consequences for the reporter, then anonymity may be necessary and a written policy provided for the use of such reporting systems, including the rights of the reporters and how the reported information will be used. Much has been written about this aspect of reporting systems (e.g., see Dekker [51]). One warning is that trust is hard to gain and easy to lose. Once it is lost, regaining it is even harder than getting buy-in at the beginning.

When reporting involves an outside regulatory agency or industry group, protection of safety information and proprietary data from disclosure and use for purposes other than improving safety must be provided.

Designing effective reporting systems is very difficult. Examining two successful efforts, in nuclear power and in commercial aviation, along with the challenges they face is instructive.

Nuclear Power

Operators of nuclear power plants in the United States are required to file a Licensee Event Report (LER) with the Nuclear Regulatory Commission (NRC) whenever an irregular event occurs during plant operation. While the NRC collected an enormous amount of information on the operating experience of plants in this way, the data were not consistently analyzed until after the Three Mile Island (TMI) accident. The General Accounting Office (GAO) had earlier criticized the NRC for this failure, but no corrective action was taken until after the events at TMI [98].

The system also had a lack of closure: important safety issues were raised and studied to some degree, but were not carried through to resolution [115]. Many of the conditions involved in the TMI accident had occurred previously at other plants but nothing had been done about correcting them. Babcock and Wilcox, the engineering firm for TMI, had no formal procedures to analyze ongoing problems at plants they had built or to review the LERs on their plants filed with the NRC.

The TMI accident sequence started when a pilot-operated relief valve stuck open. In the nine years before the TMI incident, eleven of those valves had stuck open at other plants, and only a year before, a sequence of events similar to those at TMI had occurred at another U.S. plant.

The information needed to prevent TMI was available, including the prior incidents at other plants, recurrent problems with the same equipment at TMI, and engineers' critiques that operators had been taught to do the wrong thing in specific circumstances, yet nothing had been done to incorporate this information into operating practices.

In reflecting on TMI, the utility's president, Herman Dieckamp, said:

> To me that is probably one of the most significant learnings of the whole accident [TMI] the degree to which the inadequacies of that experience feedback loop . . . significantly contributed to making us and the plant vulnerable to this accident [98].

As a result of this wake-up call, the nuclear industry initiated better evaluation and follow-up procedures on LERs. It also created the Institute for Nuclear Power Operations (INPO) to promote safety and reliability through external reviews of performance and processes, training and accreditation programs, events analysis, sharing of operating information and best practices, and special assistance to member utilities. The IAEA (International Atomic Energy Agency) and World Association of Nuclear Operators (WANO) share these goals and serve similar functions worldwide.

The reporting system now provides a way for operators of each nuclear power plant to reflect on their own operating experience in order to identify problems, interpret the reasons for these problems, and select corrective actions to ameliorate the problems and their causes. Incident reviews serve as important vehicles for self-analysis, knowledge sharing across boundaries inside and outside specific plants, and development of problem-resolution efforts. Both INPO and the NRC issue various letters and reports to make the industry aware of incidents as part of operating experience feedback, as does IAEA's Incident Reporting System.

The nuclear engineering experience is not perfect, of course, but real strides have been made since the TMI wakeup call, which luckily occurred without major human losses. To their credit, an improvement and learning effort was initiated and has continued. High-profile incidents like TMI are rare, but smaller scale self-analyses and problem-solving efforts follow detection of small defects, near misses, and precursors and negative trends. Occasionally the NRC has stepped in and required changes. For example, in 1996 the NRC ordered the Millstone nuclear power plant in Connecticut to remain closed until management could demonstrate a "safety conscious work environment" after identified problems were allowed to continue without remedial action [34].

Commercial Aviation

The highly regarded ASRS (Aviation Safety Reporting System) has been copied by many individual airline information systems. Although much information is now collected, there still exist problems in evaluating and learning from it. The breadth and type of information acquired is much greater than the NRC reporting system described above. The sheer number of ASRS reports and the free form entry of the information make evaluation very difficult. There are few ways implemented to determine whether the report was accurate or evaluated the problem correctly. Subjective causal attribution and inconsistency in terminology and information included in the reports makes comparative analysis and categorization difficult and sometimes impossible.

Existing categorization schemes have also become inadequate as technology has changed, for example, with increased use of digital technology and computers in aircraft and ground operations. New categorizations are being implemented, but that creates problems when comparing data that used older categorization schemes.

Another problem arising from the goal to encourage use of the system is in the accuracy of the data. By filing an ASRS report, a limited form of indemnity against punishment is assured. Many of the reports are biased by personal protection considerations, as evidenced by the large percentage of the filings that report FAA regulation violations. For example, in a NASA Langley study of reported helicopter incidents in the ASRS over a nine-year period, nonadherence to FARs (Federal Aviation Regulations) was by far the largest category of reports. The predominance of FAR violations in the incident data may reflect the motivation of the ASRS reporters to obtain immunity from perceived or real violations of FARs and not necessarily the true percentages.

But with all these problems and limitations, most agree that the ASRS and similar industry reporting systems have been very successful and the information obtained extremely useful in enhancing safety. For example, reported unsafe airport conditions have been corrected quickly and improvements in air traffic control and other types of procedures made on the basis of ASRS reports.

The success of the ASRS has led to the creation of other reporting systems in this industry. The Aviation Safety Action Program (ASAP) in the United States, for example, encourages air carrier and repair station personnel to voluntarily report safety information to be used to develop corrective actions for identified safety concerns. An ASAP involves a partnership between the FAA and the certified organization (called the *certificate holder*) and may also include a third party, such as the employees' labor organization. It provides a vehicle for employees of the ASAP participants to identify and report safety issues to management and to the FAA without fear that the FAA will use the reports accepted under the program to take legal enforcement action against them or the company or that companies will use the information to take disciplinary action against the employee.

Certificate holders may develop ASAP programs and submit them to the FAA for review and acceptance. Ordinarily, programs are developed for specific employee groups, such as members of the flightcrew, flight attendants, mechanics, or dispatchers. The FAA may also suggest, but not require, that a certificate holder develop an ASAP to resolve an identified safety problem.

When ASAP reports are submitted, an event review committee (ERC) reviews and analyzes them. The ERC usually includes a management representative from the certificate holder, a representative from the employee labor association (if applicable), and a specially trained FAA inspector. The ERC considers each ASAP report for acceptance or denial, and if accepted, analyzes the report to determine the necessary controls to put in place to respond to the identified problem.

Single ASAP reports can generate corrective actions and, in addition, analysis of aggregate ASAP data can also reveal trends that require action. Under an ASAP, safety issues are resolved through corrective action rather than through punishment or discipline.

To prevent abuse of the immunity provided by ASAP programs, reports are accepted only for inadvertent regulatory violations that do not appear to involve an intentional disregard for safety and events that do not appear to involve criminal activity, substance abuse, or intentional falsification.

Additional reporting programs provide for sharing data that is collected by airlines for their internal use. FOQA (Flight Operational Quality Assurance) is an example. Air carriers often instrument their aircraft with extensive flight data recording systems or use pilot generated checklists and reports for gathering information internally to improve operations and safety. FOQA provides a voluntary means for the airlines to share this information with other airlines and with the FAA so that national trends can be monitored and the FAA can target its resources to address the most important operational risk issues.[1]

In contrast with the ASAP voluntary reporting of single events, FOQA programs allow the accumulation of accurate operational performance information covering all flights by multiple aircraft types such that single events or overall patterns of aircraft performance data can be identified and analyzed. Such aggregate data can determine trends specific to aircraft types, local flight path conditions, and overall flight performance trends for the commercial aircraft industry. FOQA data has been used to identify the need for changing air carrier operating procedures for specific aircraft fleets and for changing air traffic control practices at certain airports with unique traffic pattern limitations.

FOQA and other such voluntary reporting programs allow early identification of trends and changes in behavior (i.e., migration of systems toward states of increasing risk) before they lead to accidents. Follow-up is provided to ensure that unsafe conditions are effectively remediated by corrective actions.

A cornerstone of FOQA programs, once again, is the understanding that aggregate data provided to the FAA will be kept confidential and the identity of reporting personnel or airlines will remain anonymous. Data that could be used to identify flight crews are removed from the electronic record as part of the initial processing of the collected data. Air carrier FOQA programs, however, typically provide a gatekeeper who can securely retrieve identifying information for a limited amount of time, in order to enable follow-up requests for additional information from the specific flight crew associated with a FOQA event. The gatekeeper is typically a line captain designated by the air carrier's pilot association. FOQA programs usually involve agreements between pilot organizations and the carriers that define how the collected information can be used.

12.5   Using the Feedback

Once feedback is obtained, it needs to be used to update the controllers' process models and perhaps control algorithms. The feedback and its analysis may be passed to others in the control structure who need it.

Information must be provided in a form that people can learn from, apply to their daily jobs, and use throughout the system life cycle.

Various types of analysis may be performed by the controller on the feedback, such as trend analysis. If flaws in the system design or unsafe changes are detected, obviously actions are required to remedy the problems.

---

1. FOQA is voluntary in the United States but required in some countries.

In major accidents, precursors and warnings are almost always present but ignored or mishandled. While what appear to be warnings are sometimes simply a matter of hindsight, sometimes clear evidence does exist. In 1982, two years before the Bhopal accident, for example, an audit was performed that identified many of the deficiencies involved in the loss. The audit report noted such factors related to the later tragedy such as filter-cleaning operations without using slip blinds, leaking valves, and bad pressure gauges. The report recommended raising the capability of the water curtain and pointed out that the alarm at the flare tower was nonoperational and thus any leakage could go unnoticed for a long time. The report also noted that a number of hazardous conditions were known and allowed to persist for considerable amounts of time or inadequate precautions were taken against them. In addition, there was no follow-up to ensure that deficiencies were corrected. According to the Bhopal manager, all improvements called for in the report had been implemented, but obviously that was either untrue or the fixes were ineffective.

As with accidents and incidents, warning signs or anomalies also need to be analyzed using CAST. Because practice will naturally deviate from procedures, often for very good reasons, the gap between procedures and practice needs to be monitored and understood [50].

12.6 Education and Training

Everyone in the safety control structure, not just the lower-level controllers of the physical systems, must understand their roles and responsibilities with respect to safety and why the system—including the organizational aspects of the safety control structure—was designed the way it was.

People, both managers and operators, need to understand the risks they are taking in the decisions they make. Often bad decisions are made because the decision makers have an incorrect assessment of the risks being assumed, which has implications for training. Controllers must know exactly what to look for, not just be told to look for "weak signals," a common suggestion in the HRO literature. Before a bad outcome occurs, weak signals are simply noise; they take on the appearance of signals only in hindsight, when their relevance becomes obvious. Telling managers and operators to "be mindful of weak signals" simply creates a pretext for blame after a loss event occurs. Instead, the people involved need to be knowledgeable about the hazards associated with the operation of the system if we expect them to recognize the precursors to an accident. Knowledge turns unidentifiable weak signals into identifiable strong signals. People need to know what to look for.

Decision makers at all levels of the safety control structure also need to understand the risks they are taking in the decisions they make: Training should include not just *what* but *why*. For good decision making about operational safety, decision makers must understand the system hazards and their responsibilities with respect to avoiding them. Understanding the safety rationale, that is, the "why," behind the system design will also have an impact on combating complacency and unintended changes leading to hazardous states. This rationale includes understanding why previous accidents occurred. The Columbia Accident Investigation Board was surprised at the number of NASA engineers in the Space Shuttle program who had never read the official *Challenger* accident report [74]. In contrast, everyone in the U.S. nuclear Navy has training about the *Thresher* loss every year.

Training should not be a one-time event for employees but should be continual throughout their employment, if only as a reminder of their responsibilities and the system hazards. Learning about recent events and trends can be a focus of this training.

Finally, assessing for training effectiveness, perhaps during regular audits, can assist in establishing an effective improvement and learning process.

With highly automated systems, an assumption is often made that less training is required. In fact, training requirements go up (not down) in automated systems, and they change their nature. Training needs to be more extensive and deeper when using automation. One of the reasons for this requirement is that human operators of highly automated systems not only need a model of the current process state and how it can change state but also a model of the automation and its operation, as discussed in chapter 8.

To control complex and highly automated systems safely, operators (controllers) need to learn more than just the procedures to follow: If we expect them to control and monitor the automation, they must also have an in-depth understanding of the controlled physical process and the logic used in any automated controllers they may be supervising. System controllers—at all levels—need to know:

- The system hazards and the reason behind safety-critical procedures and operational rules.

- The potential result of removing or overriding controls, changing prescribed procedures, and inattention to safety-critical features and operations: Past accidents and their causes should be reviewed and understood.

- How to interpret feedback: Training needs to include different combinations of alerts and sequences of events, not just single events.

- How to think flexibly when solving problems: Controllers need to be provided with the opportunity to practice problem solving.

- General strategies rather than specific responses: Controllers need to develop skills for dealing with unanticipated events.

• How to test hypotheses in an appropriate way: To update mental models, human controllers often use hypothesis testing to understand the system state better and update their process models. Such hypothesis testing is common with computers and automated systems where documentation is usually so poor and hard to use that experimentation is often the only way to understand the automation behavior and design. Such testing can, however, lead to losses. Designers need to provide operators with the ability to test hypotheses safely and controllers must be educated on how to do so.

Finally, as with any system, emergency procedures must be overlearned and continually practiced. Controllers must be provided with operating limits and specific actions to take in case they are exceeded. Requiring operators to make decisions under stress and without full information is simply another way to ensure that they will be blamed for the inevitable loss event, usually based on hindsight bias. Critical limits must be established and provided to the operators, and emergency procedures must be stated explicitly.

12.7   Creating an Operations Safety Management Plan

The operations safety management plan is used to guide operational control of safety. The plan describes the objectives of the operations safety program and how they will be achieved. It provides a baseline to evaluate compliance and progress. Like every other part of safety program, the plan will need buy-in and oversight.

The organization should have a template and documented expectations for operations safety management plans, but this template may need to be tailored for particular project requirements.

The information need not all be contained in one document, but there should be a central reference with pointers to where the information can be found. As is true for every other part of the safety control structure, the plan should include review procedures for the plan itself as well as how the plan will be updated and improved through feedback from experience.

Some things that might be included in the plan:

• General Considerations
  – Scope and objectives
  – Applicable standards (company, industry)
  – Documentation and reports
  – Review of plan and progress reporting procedures
• Safety Organization (safety control structure)
  – Personnel qualifications and duties

- Staffing and manpower
- Communication channels
- Responsibility, authority, accountability (functional organization, organizational structure)
- Information requirements (feedback requirements, process model, updating requirements)
- Subcontractor responsibilities
- Coordination
- Working groups
- System safety interfaces with other groups, such as maintenance and test, occupational safety, quality assurance, and so on.
- Procedures
  - Problem reporting (processes, follow-up)
  - Incident and accident investigation
    · Procedures
    · Staffing (participants)
    · Follow-up (tracing to hazard and risk analyses, communication)
  - Testing and audit program
    · Procedures
    · Scheduling
    · Review and follow-up
    · Metrics and trend analysis
    · Operational assumptions from hazard and risk analyses
  - Emergency and contingency planning and procedures
  - Management of change procedures
  - Training
  - Decision making, conflict resolution
- Schedule
  - Critical checkpoints and milestones
  - Start and completion dates for tasks, reports, reviews
  - Review procedures and participants
- Safety Information System
  - Hazard and risk analyses, hazard logs (controls, review and feedback procedures)

    – Hazard tracking and reporting system

    – Lessons learned

    – Safety data library (documentation and files)

    – Records retention policies

  • Operations hazard analysis

    – Identified hazards

    – Mitigations for hazards

  • Evaluation and planned use of feedback to keep the plan up-to-date and improve it over time

12.8   Applying STAMP to Occupational Safety

Occupational safety has, traditionally, not taken a systems approach but instead has focused on individuals and changing their behavior. In applying systems theory to occupational safety, more emphasis would be placed on understanding the impact of system design on behavior and would focus on changing the system rather than people. For example, vehicles used in large plants could be equipped with speed regulators rather than depending on humans to follow speed limits and then punishing them when they do not. The same design for safety principles presented in chapter 9 for human controllers apply to designing for occupational safety.

  With the increasing complexity and automation of our plants, the line between occupational safety and engineering safety is blurring. By designing the system to be safe despite normal human error or judgment errors under competing work pressures, workers will be better protected against injury while fulfilling their job responsibilities.

13 Managing Safety and the Safety Culture

The key to effectively accomplishing any of the goals described in the previous chapters lies in management. Simply having better tools is not enough if they are not used. Studies have shown that management commitment to the safety goals is the most important factor distinguishing safe from unsafe systems and companies [101]. Poor management decision making can undermine any attempts to improve safety and ensure that accidents continue to occur.

This chapter outlines some of the most important management factors in reducing accidents. The first question is why managers should care about and invest in safety. The answer, in short, is that safety pays and investment in safety provides large returns over the long run.

If managers understand the importance of safety in achieving organizational goals and decide they want to improve safety in their organizations, then three basic organizational requirements are necessary to achieve that goal. The first is an effective safety control structure. Because of the importance of the safety culture in how effectively the safety control structure operates, the second requirement is to implement and sustain a strong safety culture. But even the best of intentions will not suffice without the appropriate information to carry them out, so the last critical factor is the safety information system.

The previous chapters in this book focus on what needs to be done during design and operations to control safety and enforce the safety constraints. This chapter describes the overarching role of management in this process.

13.1 Why Should Managers Care about and Invest in Safety?

Most managers do care about safety. The problems usually arise because of misunderstandings about what is required to achieve high safety levels and what the costs really are if safety is done right. Safety need not entail enormous financial or other costs.

A classic myth is that safety conflicts with achieving other goals and that tradeoffs are necessary to prevent losses. In fact, this belief is totally wrong. Safety is a prerequisite for achieving most organizational goals, including profits and continued existence.

History is replete with examples of major accidents leading to enormous financial losses and the demise of companies as a result. Even the largest global corporations may not be able to withstand the costs associated with such losses, including loss of reputation and customers. After all these examples, it is surprising that few seem to learn from them about their own vulnerabilities. Perhaps it is in the nature of mankind to be optimistic and to assume that disasters cannot happen to us, only to others. In addition, in the simpler societies of the past, holding governments and organizations responsible for safety was less common. But with loss of control over our own environment and its hazards, and with rising wealth and living standards, the public is increasingly expecting higher standards of behavior with respect to safety.

The "conflict" myth arises because of a misunderstanding about how safety is achieved and the long-term consequences of operating under conditions of high risk. Often, with the best of intentions, we simply do the wrong things in our attempts to improve safety. It's not a matter of lack of effort or resources applied, but how they are used that is the problem. Investments in safety need to be funneled to the most effective activities in achieving it.

Sometimes it appears that organizations are playing a sophisticated version of Whack-a-Mole, where symptoms are found and fixed but not the processes that allow these symptoms to occur. Enormous resources may be expended with little return on the investment. So many incidents occur that they cannot all be investigated in depth, so only superficial analysis of a few is attempted. If, instead, a few were investigated in depth and the systemic factors fixed, the number of incidents would decrease by orders of magnitude.

Such groups find themselves in continual firefighting mode and eventually conclude that accidents are inevitable and investments to prevent them are not cost-effective, thus, like Sisyphus, condemning themselves to traverse the same vicious circle in perpetuity. Often they convince themselves that their industry is just more hazardous than others and that accidents in their world are inevitable and are the price of productivity.

This belief that accidents are inevitable and occur because of random chance arises from our own inadequate efforts to prevent them. When accident causes are examined in depth, using the systems approach in this book, it becomes clear that there is nothing random about them. In fact, we seem to have the same accident over and over again, with only the symptoms differing, but the causes remaining fairly constant. Most of these causes could be eliminated, but they are not. The precipitating immediate factors, like a stuck valve, may have some randomness associated with them, such as which valve actually precipitates a loss. But there is nothing random about systemic factors that have not been corrected and exist over long periods of time, such as flawed valve design and analysis or inadequate maintenance practices.

As described in previous chapters, organizations tend to move inexorably toward states of higher risk under various types of performance pressures until an accident becomes inevitable. Under external or internal pressures, projects start to violate their own rules: "We'll do it just this once—it's critical that we get this procedure finished today." In the Deepwater Horizon oil platform explosion of 2010, cost pressures led to not following standard safety procedures and, in the end, to enormous financial losses [18]. Similar dynamics occurred, with slightly different pressures, in the *Columbia* Space Shuttle loss where the tensions among goals were created by forces largely external to NASA. What appear to be short-term conflicts of other organizational goals with safety goals, however, may not exist over the long term, as witnessed in both these cases.

When operating at elevated levels of risk, the only question is which of many potential events will trigger the loss. Before the *Columbia* accident, NASA manned space operations was experiencing a slew of problems in the orbiters. The head of the NASA Manned Space Program at the time misinterpreted the fact that they were finding and fixing problems and wrote a report that concluded risk had been reduced by more than a factor of five [74]. The same unrealistic perception of risk led to another report in 1995 recommending that NASA "restructure and reduce overall safety, reliability, and quality assurance elements" [105].

Figure 13.1 shows some of the dynamics at work.[1] The model demonstrates the major sources of the high risk in the Shuttle program at the time of the *Columbia* loss. In order to get the funding needed to build and operate the space shuttle, NASA had made unachievable performance promises. The need to justify expenditures and prove the value of manned space flight has been a major and consistent tension between NASA and other governmental entities: The more missions the Shuttle could fly, the better able the program was to generate funding. Adding to these pressures was a commitment to get the International Space Station construction complete by February 2004 (called "core complete"), which required deliveries of large items that could only be carried by the shuttle. The only way to meet the deadline was to have no launch delays, a level of performance that had never previously been achieved [117]. As just one indication of the pressure, computer screen savers were mailed to managers in NASA's human spaceflight program that depicted a clock counting down (in seconds) to the core complete deadline [74].

---

1. Appendix D explains how to read system dynamics models, for those unfamiliar with them.

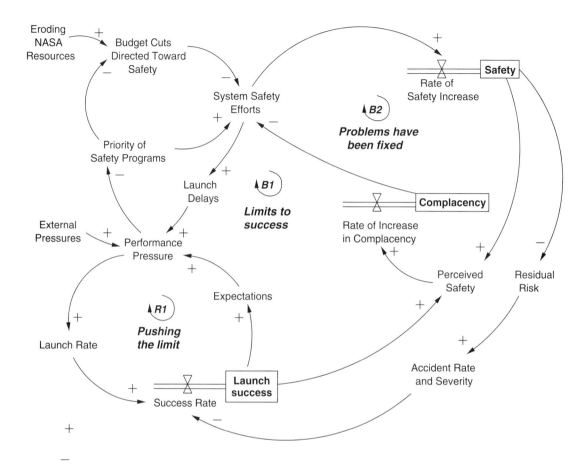

Figure 13.1
A simplified model of the dynamics of safety and performance pressures leading up to the *Columbia* loss. For a complete model, see [125].

The control loop in the lower left corner of figure 13.1, labeled R1 or *Pushing the Limit*, shows how as external pressures increased, performance pressure increased, which led to increased launch rates and thus success in meeting the launch rate expectations, which in turn led to increased expectations and increasing performance pressures. This reinforcing loop represents an unstable system and cannot be maintained indefinitely, but NASA is a "can-do" organization that believes anything can be accomplished with enough effort [136].

The upper left loop represents the Space Shuttle safety program, which when operating effectively is meant to balance the risks associated with loop R1. The external influences of budget cuts and increasing performance pressures, however, reduced the priority of safety procedures and led to a decrease in system safety efforts.

Adding to the problems is the fact that system safety efforts led to launch delays when problems were found, which created another reason for reducing the priority of the safety efforts in the face of increasing launch pressures.

While reduction in safety efforts and lower prioritization of safety concerns may lead to accidents, accidents usually do not occur for a while so false confidence is created that the reductions are having no impact on safety and therefore pressures increase to reduce the efforts and priority even further as the external and internal performance pressures mount.

The combination of the decrease in safety efforts along with loop B2 in which fixing the problems that were being found increased complacency, which also contributed to reduction of system safety efforts, eventually led to a situation of unrecognized high risk.

When working at such elevated levels of risk, the only question is which of many potential events will trigger the loss. The fact that it was the foam and not one of the other serious problems identified both before and after the loss was the only random part of the accident. At the time of the *Columbia* accident, NASA was regularly flying the Shuttle with many uncontrolled hazards; the foam was just one of them.

Often, ironically, our successful efforts to eliminate or reduce accidents contribute to the march toward higher risk. Perception of the risk associated with an activity often decreases over a period of time when no losses occur even though the real risk has not changed at all. This misperception leads to reducing the very factors that are preventing accidents because they are seen as no longer needed and available to trade off with other needs. The result is that risk increases until a major loss occurs. This vicious cycle needs to be broken to prevent accidents. In STAMP terms, the weakening of the safety control structure over time needs to be prevented or detected before the conditions occur that lead to a loss.

System migration toward states of higher risk is potentially controllable and detectable [167]. The migration results from weakening of the safety control structure. To achieve lasting results, strong operational safety efforts are needed that provide protection from and appropriate responses to the continuing environmental influences and pressures that tend to degrade safety over time and that change the safety control structure and the behavior of those in it.

The experience in the nuclear submarine community is a testament to the fact that such dynamics can be overcome. The SUBSAFE program (described in the next chapter) was established after the loss of the *Thresher* in 1963. Since that time, no submarine in the SUBSAFE program, that is, satisfying the SUBSAFE requirements, has been lost, although such losses were common before SUBSAFE was established.

The leaders in SUBSAFE describe other benefits beyond preventing the loss of critical assets. Because those operating the submarines have complete confidence in their ships, they can focus solely on the completion of their mission. The U.S. nuclear submarine program's experience over the past forty-five years belies the myth that increasing safety necessarily decreases system performance. Over a sustained period, a safer operation is generally more efficient. One reason is that stoppages and delays are eliminated.

Examples can also be found in private industry. As just one example, because of a number of serious accidents, OSHA tried to prohibit the use of power presses where employees had to place one or both hands beneath the ram during the production cycle [96]. After vehement protests that the expense would be too great in terms of reduced productivity, the requirement was dropped: Preliminary motion studies showed that reduced production would result if all loading and unloading were done with the die out from under the ram. Some time after OSHA gave up on the idea, one manufacturer who used power presses decided, purely as a safety and humanitarian measure, to accept the production penalty. Instead of reducing production, however, the effect was to increase production from 5 to 15 percent, even though the machine cycle was longer. Other examples of similar experiences can be found in *Safeware* [115].

The belief that safer systems cost more or that building safety in from the beginning necessarily requires unacceptable compromises with other goals is simply not justified. The costs, like anything else, depend on the methods used to achieve increased safety. In another ironic twist, in the attempt to avoid making tradeoffs with safety, systems are often designed to optimize mission goals and safety devices added grudgingly when the design is complete. This approach, however, is the most expensive and least effective that could be used. The costs are much less and in fact can be eliminated if safety is built into the system design from the beginning rather than added on or retrofitted later, usually in the form of redundancy or elaborate protection systems. Eliminating or reducing hazards early in design often results in a simpler design, which in itself may reduce both risk and costs. The reduced risk makes it more likely that the mission or system goals will be achieved.

Sometimes it takes a disaster to "get religion" but it should not have to. This chapter was written for those managers who are wise enough to know that investment in safety pays dividends, even before this fact is brought home (usually too late) by a tragedy.

13.2   General Requirements for Achieving Safety Goals

Escaping from the Whack-a-Mole trap requires identifying and eliminating the systemic factors behind accidents. Some common reasons why safety efforts are often not cost-effective were identified in chapter 6, including:

- Superficial, isolated, or misdirected safety engineering activities, such as spending most of the effort proving the system is safe rather than making it so
- Starting too late
- Using techniques inappropriate for today's complex systems and new technology
- Focusing only on the technical parts of the system, and
- Assuming systems are static throughout their lifetime and decreasing attention to safety during operations

Safety needs to be managed and appropriate controls established. The major ingredients of effective safety management include:

- Commitment and leadership
- A corporate safety policy
- Risk awareness and communication channels
- Controls on system migration toward higher risk
- A strong corporate safety culture
- A safety control structure with appropriate assignment of responsibility, authority, and accountability
- A safety information system
- Continual improvement and learning
- Education, training, and capability development

Each of these is described in what follows.

13.2.1 Management Commitment and Leadership

Top management concern about safety is the most important factor in discriminating between safe and unsafe companies matched on other variables [100]. This commitment must be genuine, not just a matter of sloganeering. Employees need to feel they will be supported if they show concern for safety. An Air Force study of system safety concluded:

> Air Force top management support of system safety has not gone unnoticed by contractors. They now seem more than willing to include system safety tasks, not as "window dressing" but as a meaningful activity [70, pp. 5–11].

The B1-B program is an example of how this result was achieved. In that development program, the program manager or deputy program manager chaired the meetings of the group where safety decisions were made. "An unmistaken image of the importance of system safety in the program was conveyed to the contractors" [70, p. 5].

A manager's open and sincere concern for safety in everyday dealings with employees and contractors can have a major impact on the reception given to safety-related activities [157]. Studies have shown that top management's support for and participation in safety efforts is the most effective way to control and reduce accidents [93]. Support for safety is shown by personal involvement, by assigning capable people and giving them appropriate objectives and resources, by establishing comprehensive organizational safety control structures, and by responding to initiatives by others.

13.2.2 Corporate Safety Policy

A policy is a written statement of the wisdom, intentions, philosophy, experience, and belief of an organization's senor managers that states the goals for the organization and guides their attainment [93]. The corporate safety policy provides employees with a clear, shared vision of the organization's safety goals and values and a strategy to achieve them. It documents and shows managerial priorities where safety is involved.

The author has found companies that justify not having a safety policy on the grounds that "everyone knows safety is important in our business." While safety may seem important for a particular business, management remaining mute on their policy conveys the impression that tradeoffs are acceptable when safety seems to conflict with other goals. The safety policy provides a way for management to clearly define the priority between conflicting goals they expect to be used in decision making. The safety policy should define the relationship of safety to other organizational goals and provide the scope for discretion, initiative, and judgment in deciding what should be done in specific situations.

Safety policy should be broken into two parts. The first is a short and concise statement of the safety values of the corporation and what is expected from employees with respect to safety. Details about how the policy will be implemented should be separated into other documents.

A complete safety policy contains such things as the goals of the safety program; a set of criteria for assessing the short- and long-term success of that program with respect to the goals; the values to be used in tradeoff decisions; and a clear statement of responsibilities, authority, accountability, and scope. The policy should be explicit and state in clear and understandable language what is expected, not a set of lofty goals that cannot be operationalized. An example sometimes found (as noted in the previous chapter) is a policy for employees to "be mindful of weak signals": This policy provides no useful guidance on what to do—both "mindful" and "weak signals" are undefined and undefinable. An alternative might be, "If you see something that you think is unsafe, you are responsible for reporting it immediately." In addition, employees need to be trained on the hazards in the processes they control and what to look for.

Simply having a safety policy is not enough. Employees need to believe the safety policy reflects true commitment by management. The only way this commitment can be effectively communicated is through actions by management that demonstrate that commitment. Employees need to feel that management will support them when they make reasonable decisions in favor of safety over alternative goals. Incentives and reward structures must encourage the proper handling of tradeoffs between safety and other goals. Not only the formal rewards and rules but also the informal rules (social processes) of the organizational culture must support the overall safety policy. A practical test is whether employees believe that company management will support them if they choose safety over the demands of production [128].

To encourage proper decision making, the flexibility to respond to safety problems needs to be built into the organizational procedures. Schedules, for example, should be adaptable to allow for uncertainties and possibilities of delay due to legitimate safety concerns, and production goals must be reasonable.

Finally, not only must a safety policy be defined, it must be disseminated and followed. Management needs to ensure that safety receives appropriate attention in decision making. Feedback channels must be established and progress in achieving the goals should be monitored and improvements identified, prioritized, and implemented.

See 431, 427

13.2.3   Communication and Risk Awareness

Awareness of the risk in the controlled process is a major component of safety-related decision making by controllers. The problem is that risk, when defined as the severity of a loss event combined with its likelihood, is not calculable or knowable. It can only be estimated from a set of variables, some of which may be unknown, or the information to evaluate likelihood of these variables may be lacking or incorrect. But decisions need to be made based on this unknowable property.

In the absence of accurate information about the state of the process, risk perception may be reevaluated downward as time passes without an accident. In fact, risk probably has not changed, only our perception of it. In this trap, risk is assumed to be reflected by a lack of accidents or incidents and not by the state of the safety control structure.

When STAMP is used as the foundation of the safety program, safety and risk are *a function of the effectiveness of the controls to enforce safe system behavior*, that is, the safety constraints and the control structure used to enforce those constraints.

Poor safety-related decision making on the part of management, for example, is commonly related to inadequate feedback and inaccurate process models. As such, risk is potentially knowable and not some amorphous property denoted by probability estimates. This new definition of risk can be used to create new risk assessment procedures.

While lack of accidents could reflect a strong safety control structure, it may also simply reflect delays between the relaxation of the controls and negative consequences. The delays encourage relaxation of more controls, which then leads to accidents. The basic problem is inaccurate risk perception and calculating risk using the wrong factors. This process is behind the frequently used but rarely defined label of "complacency." Complacency results from inaccurate process models and risk awareness.

Risk perception is directly related to *communication* and *feedback*. The more and better the information we have about the potential causes of accidents in our system and the state of the controls implemented to prevent them, the more accurate will be our perception of risk. Consider the loss of an aircraft when it took off from the wrong runway in Lexington, Kentucky, in August 2006. One of the factors in the accident was that construction was occurring and the pilots were confused about temporary changes in taxi patterns. Although similar instances of crew confusion had occurred in the week before the accident, there were no effective communication channels to get this information to the proper authorities. After the loss, a small group of aircraft maintenance workers told the investigators that they also had experienced confusion when taxiing to conduct engine tests—they were worried that an accident could happen, but did not know how to effectively notify people who could make a difference [142].

Another communication disconnect in this accident leading to a misperception of risk involved a misunderstanding by management about the staffing of the control tower at the airport. Terminal Services management had ordered the airport air traffic control management to both reduce control tower budgets and to ensure separate staffing of the tower and radar functions. It was impossible to comply with both directives. Because of an ineffective feedback mechanism, management did not know about the impossible and dangerous goal conflicts they had created or that the resolution of the conflict was to reduce the budget and ignore the extra staffing requirements.

Another example occurred in the Deepwater Horizon accident. Reports after the accident indicated that workers felt comfortable raising safety concerns and ideas for safety improvement to managers on the rig, but they felt that they could not raise concerns at the divisional or corporate level without reprisal. In a confidential survey of workers on Deepwater Horizon taken *before* the oil platform exploded, workers expressed concerns about safety:

"I'm petrified of dropping anything from heights not because I'm afraid of hurting anyone (the area is barriered off), but because I'm afraid of getting fired," one worker wrote. "The company is always using fear tactics," another worker said. "All these games and your mind gets tired." Investigators also said "nearly everyone among the workers they interviewed believed that Transocean's system for tracking health and safety issues on the rig was *counter productive*." Many workers entered fake data to try to circumvent the system, known as See, Think, Act, Reinforce, Track (or START). As a result, the company's perception of safety on the rig was distorted, the report concluded [27, p. A1]

Formal methods of operation and strict hierarchies can limit communication. When information is passed up hierarchies, it may be distorted, depending on the interests of managers and the way they interpret the information. Concerns about safety may even be completely silenced as it passes up the chain of command. Employees may not feel comfortable going around a superior who does not respond to their concerns. The result may be a misperception of risk, leading to inadequate control actions to enforce the safety constraints.

In other accidents, reporting and feedback systems are simply unused for a variety of reasons. In many losses, there was evidence that a problem occurred in time to prevent the loss, but there was either no communication channel established for getting the information to those who could understand it and to those making decisions or, alternatively, the problem-reporting channel was ineffective or simply unused.

Communication is critical in both providing information and executing control actions and in providing feedback to determine whether the control actions were successful and what further actions are required. Decision makers need accurate and timely information. Channels for information dissemination and feedback need to be established that include a means for comparing actual performance with desired performance and ensuring that required action is taken.

In summary, both the design of the communication channels and the communication dynamics must be considered as well as potential feedback delays. As an example of communication dynamics, reliance on face-to-face verbal reports during group meetings is a common method of assessing lower-level operations [189], but, particularly when subordinates are communicating with superiors, there is a tendency for adverse situations to be underemphasized [20].

13.2.4 Controls on System Migration toward Higher Risk

One of the key assumptions underlying the approach to safety described in this book is that systems adapt and change over time. Under various types of pressures, that adaptation often moves in the direction of higher risk. The good news is, as stated earlier, that adaptation is predictable and potentially controllable. The safety control structure must provide protection from and appropriate responses to the continuing influences and pressures that tend to degrade safety over time. More specifically, the potential reasons for and types of migration toward higher risk need to be identified and controls instituted to prevent it. In addition, audits and performance assessments based on the safety constraints identified during system development can be used to detect migration and the violation of the constraints as described in chapter 12.

One way to prevent such migration is to anchor safety efforts beyond short-term program management pressures. At one time, NASA had a strong agency-wide system safety program with common standards and requirements levied on everyone. Over time, agency-wide standards were eviscerated, and programs were allowed to set their own standards under the control of the program manager. While the manned space program started out with strong safety standards, under budget and performance pressures they were progressively weakened [117].

As one example, a basic requirement for an effective operational safety program is that all potentially hazardous incidents during operations are thoroughly investigated. Debris shedding had been identified as a potential hazard during Shuttle development, but the standard for performing hazard analyses in the Space Shuttle program was changed to specify that hazards would be revisited *only* when there was a new design or the Shuttle design was changed, not after an anomaly (such as foam shedding) occurred [117].

After the *Columbia* accident, safety standards in the Space Shuttle program (and the rest of NASA) were effectively *anchored* and protected from dilution over time by moving responsibility for them outside the projects.

13.2.5 Safety, Culture, and Blame

The high-level goal in managing safety is to create and maintain an effective safety control structure. Because of the importance of safety culture in how the control structure operates, achieving this goal requires implementing and sustaining a strong safety culture.

Proper function of the safety control structure relies on decision making by the controllers in the structure. Decision making always rests upon a set of industry or organizational values and assumptions. A *culture* is a set of shared values and norms, a way of looking at and interpreting the world and events around us and of taking action in a social context. Safety culture is that subset of culture that reflects the general attitude and approaches to safety and risk management.

Shein divides culture into three levels (figure 13.2) [188]. At the top are the surface-level cultural artifacts or routine aspects of everyday practice including hazard analyses and control algorithms and procedures. The second, middle level is the stated organizational rules, values, and practices that are used to create the top-level artifacts, such as safety policy, standards, and guidelines. At the lowest level is the often invisible but pervasive underlying deep cultural operating assumptions

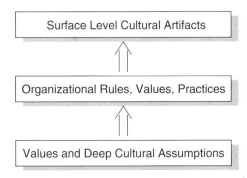

Figure 13.2
The three levels of an organizational culture.

upon which actions are taken and decisions are made and thus upon which the upper levels rest.

Trying to change safety outcomes by simply changing the organizational structures—including policies, goals, missions, job descriptions, and standard operating procedures—may lower risk over the short term, but superficial fixes that do not address the set of shared values and social norms are very likely to be undone over time. Changes are required in the organizational values that underlie people's behavior.

Safety culture is primarily set by the leaders of the organization as they establish the basic values under which decisions will be made. This fact explains why leadership and commitment by leaders is critical in achieving high levels of safety.

To *engineer* a safety culture requires identifying the desired organizational safety principles and values and then establishing a safety control structure to achieve those values and to sustain them over time. Sloganeering or jawboning is not enough: all aspects of the safety control structure must be engineered to be in alignment with the organizational safety principles, and the leaders must be committed to the stated policies and principles related to safety in the organization.

Along with leadership and commitment to safety as a basic value of the organization, achieving safety goals requires open communication. In an interview after the *Columbia* loss, the new center director at Kennedy Space Center suggested that the most important cultural issue the Shuttle program faced was establishing a feeling of openness and honesty with all employees, where everybody's voice was valued. Statements during the *Columbia* accident investigation and messages posted to the NASA Watch website describe a lack of trust of NASA employees to speak up. At the same time, a critical observation in the CAIB report focused on the engineers' claims that the managers did not hear the engineers' concerns [74]. The report concluded that this was in part due to the managers not asking or listening. Managers created barriers against dissenting opinions by stated preconceived conclusions based on subjective knowledge and experience rather than on solid data. Much of the time they listened to those who told them what they wanted to hear. One indication about the poor communication around safety and the atmosphere at the time were statements in the 1995 Kraft report [105] that dismissed concerns about Space Shuttle safety by accusing those who made them as being partners in an unneeded "safety shield conspiracy."

Unhealthy work atmospheres with respect to safety and communication are not limited to NASA. Carroll documents a similarly dysfunctional safety culture at the Millstone nuclear power plant [33]. An NRC review in 1996 concluded the safety culture at the plant was dangerously flawed: it did not tolerate dissenting views and stifled questioning attitudes among employees.

Changing such interaction patterns is not easy. Management style can be addressed through training, mentoring, and proper selection of people to fill management positions, but trust is hard to gain and easy to lose. Employees need to feel psychologically safe about reporting concerns and to believe that managers can be trusted to hear their concerns and to take appropriate action, while managers have to believe that employees are worth listening to and worthy of respect.

The difficulty is in getting people to change their view of reality. Gareth Morgan, a social anthropologist, defines culture as an ongoing, proactive process of reality construction. According to this view, organizations are socially constructed realities that rest as much in the heads and minds of their members as they do in concrete sets of rules and regulations. Morgan asserts that organizations are "sustained by belief systems that emphasize the importance of rationality" [139]. This myth of rationality "helps us to see certain patterns of action as legitimate, credible, and normal, and hence to avoid the wrangling and debate that would arise if we were to recognize the basic uncertainty and ambiguity underlying many of our values and actions" [139].

For both the *Challenger* and *Columbia* accidents, as well as most other major accidents where decision making was flawed, the decision makers saw their actions as rational. Understanding and preventing poor decision making under conditions of uncertainty requires providing environments and tools that help to stretch our belief systems and to see patterns that we do not necessarily want to see.

Some common types of dysfunctional safety cultures can be identified that are common to industries or organizations. Hopkins coined the term "culture of denial" after investigating accidents in the mining industry, but mining is not the only industry in which denial is pervasive. In such cultures, risk assessment is unrealistic and credible warnings are dismissed without appropriate action. Management only wants to hear good news and may ensure that is what they hear by punishing bad news, sometimes in a subtle way and other times not so subtly. Often arguments are made in these industries that the conditions are inherently more dangerous than others and therefore little can be done about improving safety or that accidents are the price of productivity and cannot be eliminated. Of course, this rationale is untrue but it is convenient.

A second type of dysfunctional safety culture might be termed a "paperwork culture." In these organizations, employees spend all their time proving the system is safe but little time actually doing the things necessary to make it so. After the Nimrod aircraft loss in Afghanistan in 2006, the accident report noted a "culture of paper safety" at the expense of real safety [78].

So what are the aspects of a good safety culture, that is, the core values and norms that allow us to make better decisions around safety?

- Safety commitment is valued.
- Safety information is surfaced without fear and incident analysis is conducted without blame.
- Incidents and accidents are valued as an important window into systems that are not functioning as they should—triggering in-depth and uncircumscribed causal analysis and improvement actions.
- There is a feeling of openness and honesty, where everyone's voice is respected. Employees feel that managers are listening.
  · There is trust among all parties.
  · Employees feel psychologically safe about reporting concerns.
  · Employees believe that managers can be trusted to hear their concerns and will take appropriate action.
  · Managers believe that employees are worth listening to and are worthy of respect.

Common ingredients of a safety culture based on these values include management commitment to safety and the safety values, management involvement in achieving the safety goals, employee empowerment, and appropriate and effective incentive structures and reporting systems.

When these ingredients form the basis of the safety culture, the organization has the following characteristics:

- Safety is integrated into the dominant culture; it is not a separate subculture.
- Safety is integrated into both development and operations. Safety activities employ a mixture of top-down engineering or reengineering and bottom-up process improvement.
- Individuals have required knowledge, skills, and ability.

- Early warning systems for migration toward states of high risk are established and effective.

- The organization has a clearly articulated safety vision, values and procedures, shared among the stakeholders.

- Tensions between safety priorities and other system priorities are addressed through a constructive, negotiated process.

- Key stakeholders (including all employees and groups such as unions) have full partnership roles and responsibilities regarding safety.

- Passionate, effective leadership exists at all levels of the organization (particularly the top), and all parts of the safety control structure are committed to safety as a high priority for the organization.

- Effective communication channels exist for disseminating safety information.

- High levels of visibility of the state of safety (i.e., risk awareness) exist at all levels of the safety control structure through appropriate and effective feedback.

- The results of operating experience, process hazard analyses, audits, near misses, or accident investigations are used to improve operations and the safety control structure.

- Deficiencies found during assessments, audits, inspections, and incident investigation are addressed promptly and tracked to completion.

The Just Culture Movement

The Just Culture movement is an attempt to avoid the type of unsafe cultural values and professional interactions that have been implicated in so many accidents. Its origins are in aviation although some in the medical community, particularly hospitals, have also taken steps down this road. Much has been written on Just Culture—only a summary is provided here. The reader is directed in particular to Dekker's book *Just Culture* [51], which is the source of much of what follows in this section.

A foundational principle of Just Culture is that the difference between a safe and unsafe organization is how it deals with reported incidents. This principle stems from the belief that an organization can benefit more by learning from mistakes than by punishing people who make them.

In an organization that promotes such a Just Culture [51]:

- Reporting errors and suggesting changes is normal, expected, and without jeopardy for anyone involved.

- A mistake or incident is not seen as a failure but as a free lesson, an opportunity to focus attention and to learn.

- Rather than making people afraid, the system makes people participants in change and improvement.
- Information provided in good faith is not used against those who report it.

Most people have a genuine concern for the safety and quality of their work. If through reporting problems they contribute to visible improvements, few other motivations or exhortations to report are necessary. In general, empowering people to affect their work conditions and making the reporters of safety problems part of the change process promotes their willingness to shoulder their responsibilities and to share information about safety problems.

Beyond the obvious safety implications, a Just Culture may improve morale, commitment to the organization, job satisfaction, and willingness to do extra, to step outside their role. It encourages people to participate in improvement efforts and gets them actively involved in creating a safer system and workplace.

There are several reasons why people may not report safety problems, which were covered in chapter 12. To summarize, the reporting channels may be difficult or time consuming to use, they may feel there is no point in reporting because the organization will not do anything anyway or they may fear negative consequences in reporting. Each of these reasons must be and can be mitigated through better system design. Reporting should be easy and not require excessive time or effort that takes away from direct job responsibilities. There must be responses made both to the initial report that indicates it was received and read and later information should be provided about the resolution of the reported problem.

Promoting a Just Culture requires getting away from blame and punishment as a solution to safety problems. One of the new assumptions in chapter 2 for an accident model and underlying STAMP was:

> Blame is the enemy of safety. Focus should instead be on understanding how the entire system behavior led to the loss and not on who or what to blame.

Blame and punishment discourage reporting problems and mistakes so improvements can be made to the system. As has been argued throughout this book, changing the system is the best way to achieve safety, not trying to change people.

When blame is a primary component of the safety culture, people stop reporting incidents. This basic understanding underlies the Aviation Safety Reporting System (ASRS) where pilots and others are given protection from punishment if they report mistakes (see chapter 12). A decision was made in establishing the ASRS and other aviation reporting systems that organizational and industry learning from mistakes was more important than punishing people for them. If most errors stem from the design of the system or can be prevented by changing the design of the system, then blaming the person who made the mistake is misplaced anyway.

A culture of blame creates a climate of fear that makes people reluctant to share information. It also hampers the potential to learn from incidents; people may even tamper with safety recording devices, turning them off, for example. A culture of blame interferes with regulatory work and the investigation of accidents because people and organizations are less willing to cooperate. The role of lawyers can impede safety efforts and actually make accidents more likely: Organizations may focus on creating paper trails instead of utilizing good safety engineering practices. Some companies avoid standard safety practices under the advice of their lawyers that this will protect them in legal proceedings, thus almost guaranteeing that accidents and legal proceedings will occur.

Blame and the overuse of punishment as a way to change behavior can directly lead to accidents that might not have otherwise occurred. As an example, a train accident in Japan—the 2005 Fukuchiyama line derailment— occurred when a train driver was on the phone trying to ensure that he would not be reported for a minor infraction. Because of this distraction, he did not slow down for a curve, resulting in the deaths of 106 passengers and the train driver along with injury of 562 passengers [150]. Blame and punishment for mistakes causes stress and isolation and makes people perform less well.

The alternative is to see mistakes as an indication of an organizational, operational, educational, or political problem. The question then becomes what should be done about the problem and who should bear responsibility for implementing the changes. The mistake and any harm from it should be acknowledged, but the response should be to lay out the opportunities for reducing such mistakes by everyone (not just this particular person), and the responsibilities for making changes so that the probability of it happening again is reduced. This approach allows people and organizations to move forward to prevent mistakes in the future and not just focus on punishing past behavior [51]. Punishment is usually not a long-term deterrent for mistakes if the system in which the person operates has not changed the reason for the mistake. Just Culture principles allow us to learn from minor incidents instead of waiting until tragedies occur.

A common misunderstanding is that a Just Culture means a lack of accountability. But, in reality, it is just the opposite. Accountability is increased in a Just Culture by not simply assigning responsibility and accountability to the person at the bottom of the safety control structure who made the direct action involved in the mistake. All components of the safety control structure involved are held accountable including (1) those in operations who contribute to mistakes by creating operational pressures and providing inadequate oversight to ensure safe procedures are being followed, and (2) those in development who create a system design that contributes to mistakes.

The difference in a Just Culture is not in the accountability for safety problems but how accountability is implemented. Punishment is an appropriate response to gross negligence and disregard for other people's safety, which, of course, applies to everyone in the safety control structure, including higher-level management and developers as well as the lower level controllers. But if mistakes were made or inadequate controls over safety provided because of flaws in the design of the controlled system or the safety control structure, then punishment is not the appropriate response—fixing the system or the safety control structure is. Dekker has suggested that accountability be defined in terms of responsibility for finding solutions to the system design problems from which the mistakes arose [51].

Overcoming our cultural bias to punish people for their mistakes and the common belief that punishment is the only way to change behavior can be very difficult. But the payoff is enormous if we want to significantly reduce accident rates. Trust is a critical requirement for encouraging people to share their mistakes and safety problems with others so something can be done before major losses occur.

13.2.6   Creating an Effective Safety Control Structure

In some industries, the safety control structure is called the safety management system (SMS). In civil aviation, ICAO (International Civil Aviation Authority) has created standards and recommended practices for safety management systems and individual countries have strongly recommended or required certified air carriers to establish such systems in order to control organizational factors that contribute to accidents.

There is no right or wrong design of a safety control structure or SMS. Most of the principles for design of safe control loops in chapter 9 also apply here. The culture of the industry and the organization will play a role in what is practical and effective. There are some general rules of thumb, however, that have been found to be important in practice.

General Safety Control Structure Design Principles

Making everyone responsible for safety is a well-meaning misunderstanding of what is required. While, of course, everyone should try to behave safely and to achieve safety goals, someone has to be assigned responsibility for ensuring that the goals are being achieved. This lesson was learned long ago in the U.S. Intercontinental Ballistic Missile System (ICBM). Because safety was such an important consideration in building the early 1950s missile systems, safety was not assigned as a specific responsibility, but was instead considered to be everyone's responsibility. The large number of resulting incidents, particularly those involving the interfaces between subsystems, led to the understanding that safety requires leadership and focus.

There needs to be assignment of responsibility for ensuring that hazardous behaviors are eliminated or, if not possible, mitigated in design and operations. Almost all attention during development is focused on what the system and its components are supposed to do. System safety engineering is responsible for ensuring that adequate attention is also paid to what the system is *not* supposed to do and verifying that hazardous behavior will not occur. It is this unique focus that has made the difference in systems where safety engineering successfully identified problems that were not found by the other engineering processes.

At the other extreme, safety efforts may be assigned to a separate group that is isolated from critical decision making. During system development, responsibility for safety may be concentrated in a separate quality assurance group rather than in the system engineering organization. During operations, safety may be the responsibility of a staff position with little real power or impact on line operations.

The danger inherent in this isolation of the safety efforts is argued repeatedly throughout this book. To be effective, the safety efforts must have impact, and they must be integrated into mainstream system engineering and operations.

Putting safety into the quality assurance organization is the worst place for it. For one thing, it sets up the expectation that safety is an after-the-fact or auditing activity only: safety must be intimately integrated into design and decision-making activities. Safety permeates every part of development and operations. While there may be staff positions performing safety functions that affect everyone at their level of the organization and below, safety must be integrated into all of engineering development and line operations. Important safety functions will be performed by most everyone, but someone needs the responsibility to ensure that they are being carried out effectively.

At the same time, independence is also important. The CAIB report addresses this issue:

> Organizations that successfully operate high-risk technologies have a major characteristic in common: they place a premium on safety and reliability by structuring their programs so that technical and safety engineering organizations own the process of determining, maintaining, and waiving technical requirements with a voice that is equal to yet independent of Program Managers, who are governed by cost, schedule, and mission-accomplishment goals [74, p. 184].

Besides associating safety with after-the-fact assurance and isolating it from system engineering, placing it in an assurance group can have a negative impact on its stature, and thus its influence. Assurance groups often do not have the prestige necessary to have the influence on decision making that safety requires. A case can be made that the centralization of system safety in quality assurance at NASA, matrixed to other parts of the organization, was a major factor in the decline of the safety culture preceding the *Columbia* loss. Safety was neither fully independent nor sufficiently influential to prevent the loss events [117].

Safety responsibilities should be assigned at every level of the organization, although they will differ from level to level. At the corporate level, system safety responsibilities may include defining and enforcing corporate safety policy, and establishing and monitoring the safety control structure. In some organizations that build extremely hazardous systems, a group at the corporate or headquarters level certify these systems as safe for use. For example, the U.S. Navy has a Weapons Systems Explosives Safety Review Board that assures the incorporation of explosive safety criteria in all weapon systems by reviews conducted throughout all the system's life cycle phases. For some companies, it may be reasonable to have such a review process at more than just the highest level.

Communication is important because safety motivated changes in one subsystem may affect other subsystems and the system as a whole. In military procurement groups, oversight and communication is enhanced through the use of *safety working groups*. In establishing any oversight process, two extremes must be avoided: "getting into bed" with the project and losing objectivity or backing off too far and losing insight. Working groups are an effective way of avoiding these extremes. They assure comprehensive and unified planning and action while allowing for independent review and reporting channels.

Working groups usually operate at different levels of the organization. As an example, the Navy Aegis[2] system development, a very large and complex system, included a System Safety Working Group at the top level chaired by the Navy Principal for Safety, with the permanent members being the prime contractor's system safety lead and representatives from various Navy offices. Contractor representatives attended meetings as required. Members of the group were responsible for coordinating safety efforts within their respective organizations, for reporting the status of outstanding safety issues to the group, and for providing information to the Navy Weapons Systems Explosives Safety Review Board. Working groups also functioned at lower levels, providing the necessary coordination and communication for that level and to the levels above and below.

A surprisingly large percentage of the reports on recent aerospace accidents have implicated improper transition from an oversight to an insight process (for example, see [193, 215, 153]). This transition implies the use of different levels of feedback control and a change from prescriptive management control to management by objectives, where the objectives are interpreted and satisfied according to the local context. For these accidents, the change in management role from oversight to insight seems to have been implemented simply as a reduction in personnel and budgets without assuring that anyone was responsible for specific critical tasks.

---

2. The Aegis Combat System is an advanced command and control and weapon control system that uses powerful computers and radars to track and guide weapons to destroy enemy targets.

Assigning Responsibilities

An important question is what responsibilities should be assigned to the control structure components. The list below is derived from the author's experience on a large number and variety of projects. Many also appear in accident report recommendations, particularly those generated using CAST.

The list is meant only to be a starting point for those establishing a comprehensive safety control structure and a checklist for those who already have sophisticated safety management systems. It should be supplemented using other sources and experiences.

The list does not imply that each responsibility will be assigned to a single person or group. The responsibilities will probably need to be separated into multiple individual responsibilities and assigned throughout the safety control structure, with one group actually implementing the responsibilities and others above them supervising, leading (directing), or overseeing the activity. Of course, each responsibility assumes the need for associated authority and accountability plus the controls, feedback, and communication channels necessary to implement the responsibility. The list may also be useful in accident and incident analysis to identify inadequate controls and control structures.

Management and General Responsibilities

- Provide leadership, oversight, and management of safety at all levels of the organization.

- Create a corporate or organizational safety policy. Establish criteria for evaluating safety-critical decisions and implementing safety controls. Establish distribution channels for the policy. Establish feedback channels to determine whether employees understand it, are following it, and whether it is effective. Update the policy as needed.

- Establish corporate or organizational safety standards and then implement, update, and enforce them. Set minimum requirements for safety engineering in development and operations and oversee the implementation of those requirements. Set minimum physical and operational standards for hazardous operations.

- Establish incident and accident investigation standards and ensure recommendations are implemented and effective. Use feedback to improve the standards.

- Establish management of change requirements for evaluating all changes for their impact on safety, including changes in the safety control structure. Audit the safety control structure for unplanned changes and migration toward states of higher risk.

- Create and monitor the organizational safety control structure. Assign responsibility, authority, and accountability for safety.

- Establish working groups.

- Establish robust and reliable communication channels to ensure accurate management risk awareness of the development system design and the state of the operating process.

- Provide physical and personnel resources for safety-related activities. Ensure that those performing safety-critical activities have the appropriate skills, knowledge, and physical resources.

- Create an easy-to-use problem reporting system and then monitor it for needed changes and improvements.

- Establish safety education and training for all employees and establish feedback channels to determine whether it is effective along with processes for continual improvement. The education should include reminders of past accidents and causes and input from lessons learned and trouble reports. Assessment of effectiveness may include information obtained from knowledge assessments during audits.

- Establish organizational and management structures to ensure that safety-related technical decision making is independent from programmatic considerations, including cost and schedule.

- Establish defined, transparent, and explicit resolution procedures for conflicts between safety-related technical decisions and programmatic considerations. Ensure that the conflict resolution procedures are being used and are effective.

- Ensure that those who are making safety-related decisions are fully informed and skilled. Establish mechanisms to allow and encourage all employees and contractors to contribute to safety-related decision making.

- Establish an assessment and improvement process for safety-related decision making.

- Create and update the organizational safety information system.

- Create and update safety management plans.

- Establish communication channels, resolution processes, and adjudication procedures for employees and contractors to surface complaints and concerns about the safety of the system or parts of the safety control structure that are not functioning appropriately. Evaluate the need for anonymity in reporting concerns.

Development

- Implement special training for developers and development managers in safety-guided design and other necessary skills. Update this training as events occur and more is learned from experience. Create feedback, assessment, and improvement processes for the training.

- Create and maintain the hazard log.

- Establish working groups.

- Design safety into the system using system hazards and safety constraints. Iterate and refine the design and the safety constraints as the design process proceeds. Ensure the system design includes consideration of how to reduce human error.

- Document operational assumptions, safety constraints, safety-related design features, operating assumptions, safety-related operational limitations, training and operating instructions, audits and performance assessment requirements, operational procedures, and safety verification and analysis results. Document both what and why, including tracing between safety constraints and the design features to enforce them.

- Perform high-quality and comprehensive hazard analyses to be available and usable when safety-related decisions need to be made, starting with early decision making and continuing through the system's life. Ensure that the hazard analysis results are communicated in a timely manner to those who need them. Establish a communication structure that allows communication downward, upward, and sideways (i.e., among those building subsystems). Ensure that hazard analyses are updated as the design evolves and test experience is acquired.

- Train engineers and managers to use the results of hazard analyses in their decision making.

- Maintain and use hazard logs and hazard analyses as experience with the system is acquired. Ensure communication of safety-related requirements and constraints to everyone involved in development.

- Gather lessons learned in operations (including accident and incident reports) and use them to improve the development processes. Use operating experience to identify flaws in the development safety controls and implement improvements.

Operations

- Develop special training for operators and operations management to create needed skills and update this training as events occur and more is learned from experience. Create feedback, assessment, and improvement processes for this training. Train employees to perform their jobs safely, understand proper use of safety equipment, and respond appropriately in an emergency.

- Establish working groups.

- Maintain and use hazard logs and hazard analyses during operations as experience is acquired.

- Ensure all emergency equipment and safety devices are operable at all times during hazardous operations. Before safety-critical, nonroutine, potentially hazardous operations are started, inspect all safety equipment to ensure it is operational, including the testing of alarms.

- Perform an in-depth investigation of any operational anomalies, including hazardous conditions (such as water in a tank that will contain chemicals that react to water) or events. Determine why they occurred before any potentially dangerous operations are started or restarted. Provide the training necessary to do this type of investigation and proper feedback channels to management.

- Create management of change procedures and ensure they are being followed. These procedures should include hazard analyses on all proposed changes and approval of all changes related to safety-critical operations. Create and enforce policies about disabling safety-critical equipment.

- Perform safety audits, performance assessments, and inspections using the hazard analysis results as the preconditions for operations and maintenance. Collect data to ensure safety policies and procedures are being followed and that education and training about safety is effective. Establish feedback channels for leading indicators of increasing risk.

- Use the hazard analysis and documentation created during development and passed to operations to identify leading indicators of migration toward states of higher risk. Establish feedback channels to detect the leading indicators and respond appropriately.

- Establish communication channels from operations to development to pass back information about operational experience.

- Perform in-depth incident and accident investigations, including all systemic factors. Assign responsibility for implementing all recommendations. Follow up to determine whether recommendations were fully implemented and effective.

- Perform independent checks of safety-critical activities to ensure they have been done properly.

- Prioritize maintenance for identified safety-critical items. Enforce maintenance schedules.

- Create and enforce policies about disabling safety-critical equipment and making changes to the physical system.

- Create and execute special procedures for the startup of operations in a previously shutdown unit or after maintenance activities.

- Investigate and reduce the frequency of spurious alarms.

- Clearly mark malfunctioning alarms and gauges. In general, establish procedures for communicating information about all current malfunctioning equipment to operators and ensure the procedures are being followed. Eliminate all barriers to reporting malfunctioning equipment.

- Define and communicate safe operating limits for all safety-critical equipment and alarm procedures. Ensure that operators are aware of these limits. Assure that operators are rewarded for following the limits and emergency procedures, even when it turns out no emergency existed. Provide for tuning the operating limits and alarm procedures over time as required.

- Ensure that spare safety-critical items are in stock or can be acquired quickly.

- Establish communication channels to plant management about all events and activities that are safety-related. Ensure management has the information and risk awareness they need to make safe decisions about operations.

- Ensure emergency equipment and response is available and operable to treat injured workers.

- Establish communication channels to the community to provide information about hazards and necessary contingency actions and emergency response requirements.

13.2.7 The Safety Information System

The safety information system is a critical component in managing safety. It acts as a source of information about the state of safety in the controlled system so that controllers' process models can be kept accurate and coordinated, resulting in better decision making. Because it in essence acts as a shared process model or a source for updating individual process models, accurate and timely feedback and data are important. After studying organizations and accidents, Kjellan concluded that an effective safety information system ranked second only to top management concern about safety in discriminating between safe and unsafe companies matched on other variables [101].

Setting up a long-term information system can be costly and time consuming, but the savings in terms of losses prevented will more than make up for the effort. As an example, a Lessons Learned Information System was created at Boeing for commercial jet transport structural design and analysis. The time constants are large in this industry, but they finally were able to validate the system after using it in the design of the 757 and 767 [87]. A tenfold reduction in maintenance costs due to corrosion and fatigue were attributed to the use of recorded lessons learned from past designs. All the problems experienced in the introduction of new carbon-fiber aircraft structures like the B787 show how valuable such learning from the past can be and the problems that result when it does not exist.

Lessons learned information systems in general are often inadequate to meet the requirements for improving safety: collected data may be improperly filtered and thus inaccurate, methods may be lacking for the analysis and summarization of causal data, information may not be available to decision makers in a form that is meaningful to them, and such long-term information system efforts may fail to survive after the original champions and initiators move on to different projects and management does not provide the resources and leadership to continue the efforts. Often, lots of information is collected about occupational safety because it is required for government reports but less for engineering safety.

Setting up a safety information system for a single project or product may be easier. The effort starts in the development process and then is passed on for use in operations. The information accumulated during the safety-driven design process provides the baseline for operations, as described in chapter 12. For example, the identification of critical items in the hazard analysis can be used as input to the maintenance process for prioritization. Another example is the use of the assumptions underlying the hazard analysis to guide the audit and performance assessment process. But first the information needs to be recorded and easily located and used by operations personnel.

In general, the safety information system includes

- A safety management plan (for both development and operations)
- The status of all safety-related activities
- The safety constraints and assumptions underlying the design, including operational limitations
- The results of the hazard analyses (hazard logs) and performance audits and assessments
- Tracking and status information on all known hazards
- Incident and accident investigation reports and corrective actions taken
- Lessons learned and historical information
- Trend analysis

One of the first components of the safety information system for a particular project or product is a safety program plan. This plan describes the objectives of the program and how they will be achieved. In addition to other things, the plan provides a baseline to evaluate compliance and progress. While the organization may have a general format and documented expectations for safety management plans, this template may need to be tailored for specific project requirements. The plan should include review procedures for the plan itself as well as how the plan will be updated and improved through feedback from experience.

All of the information in the safety information system will probably not be in one document, but there should be a central location containing pointers to where all the information can be found. Chapter 12 contains a list of what should be in an operations safety management plan. The overall safety management plan will contain similar information with some additions for development.

When safety information is being shared among companies or with regulatory agencies, there needs to be protection from disclosure and use of proprietary data for purposes other than safety improvement.

13.2.8   Continual Improvement and Learning

Processes and structures need to be established to allow continual improvement and learning. Experimentation is an important part of the learning process, and trying new ideas and approaches to improving safety needs to be allowed and even encouraged.

In addition, accidents and incidents should be treated as opportunities for learning and investigated thoroughly, as described in chapter 11. Learning will be inhibited if a thorough understanding of the systemic factors involved is not sought.

Simply identifying the causal factors is not enough: recommendations to eliminate or control these factors must be created along with concrete plans for implementing the recommendations. Feedback loops are necessary to ensure that the recommendations are implemented in a timely manner and that controls are established to detect and react to reappearance of those same causal factors in the future.

13.2.9   Education, Training, and Capability Development

If employees understand the intent of the safety program and commit to it, they are more likely to comply with that intention rather than simply follow rules when it is convenient to do so.

Some properties of effective training programs are presented in chapter 12. Everyone involved in controlling a potentially dangerous process needs to have safety training, not just the low-level controllers or operators. The training must include not only information about the hazards and safety constraints to be implemented in the control structure and the safety controls, but also about priorities and how decisions about safety are to be made.

One interesting option is to have managers serve as teachers [46]. In this education program design, training experts help manage group dynamics and curriculum development, but the training itself is delivered by the project leaders. Ford Motor Company used this approach as part of what they term their Business Leadership Initiative and have since extended it as part of the Safety Leadership Initiative. They found that employees pay more attention to a message delivered by their boss than by a trainer or safety official. By learning to teach the materials, supervisors and managers are also more likely to absorb and practice the key principles [46].

13.3 Final Thoughts

Management is key to safety. Top-level management sets the culture, creates the safety policy, and establishes the safety control structure. Middle management enforces safe behavior through the designed controls.

Most people want to run safe organizations, but they may misunderstand the tradeoffs required and how to accomplish the goals. This chapter and the book as a whole have tried to correct misperceptions and provide advice on how to create safer products and organizations. The next chapter provides a real-life example of a successful systems approach to safety.

14 SUBSAFE: An Example of a Successful Safety Program

This book is filled with examples of accidents and of what not to do. One possible conclusion might be that despite our best efforts accidents are inevitable in complex systems. That conclusion would be wrong. Many industries and companies are able to avoid accidents: the nuclear Navy SUBSAFE program is a shining example. By any measure, SUBSAFE has been remarkably successful: In nearly fifty years since the beginning of SUBSAFE, no submarine in the program has been lost.

Looking at a successful safety program and trying to understand why it has been successful can be very instructive.[1] This chapter looks at the history of the program and what it is, and proposes some explanations for its great success. SUBSAFE also provides a good example of most of the principles expounded in this book.

Although SUBSAFE exists in a government and military environment, most of the important components could be translated into the commercial, profit-making world. Also note that the success is not related to small size—there are 40,000 people involved in the U.S. submarine safety program, a large percentage of whom are private contractors and not government employees. Both private and public shipyards are involved. SUBSAFE is distributed over large parts of the United States, although mostly on the coasts (for obvious reasons). Five submarine classes are included, as well as worldwide naval operations.

14.1 History

The SUBSAFE program was created after the loss of the nuclear submarine *Thresher*. The USS *Thresher* was the first ship of her class and the leading edge of U.S. submarine technology, combining nuclear power with modern hull design and newly designed equipment and components. On April 10, 1963, while performing a

---

1. I am particularly grateful to Rear Admiral Walt Cantrell, Al Ford, and Commander Jim Hassett for their insights on and information about the SUBSAFE program.

deep test dive approximately two hundred miles off the northeastern coast of the United States, the USS *Thresher* was lost at sea with all persons aboard: 112 naval personnel and 17 civilians died.

The head of the U.S. nuclear Navy, Admiral Hyman Rickover, gathered his staff after the *Thresher* loss and ordered them to design a program that would ensure such a loss never happened again. The program was to be completed by June and operational by that December. To date, that goal has been achieved. Between 1915 and 1963, the U.S. had lost fifteen submarines to noncombat causes, an average of one loss every three years, with a total of 454 casualties. *Thresher* was the first nuclear submarine lost, the worst submarine disaster in history in terms of lives lost (figure 14.1).

SUBSAFE was established just fifty-four days after the loss of *Thresher*. It was created on June 3, 1963, and the program requirements were issued on December 20 of that same year. Since that date, no SUBSAFE-certified submarine has ever been lost.

One loss did occur in 1968—the USS *Scorpion*—but it was not SUBSAFE certified. In a rush to get *Scorpion* ready for service after it was scheduled for a major overhaul in 1967, the Chief of Naval Operations allowed a reduced overhaul process and deferred the required SUBSAFE inspections. The design changes deemed necessary after the loss of *Thresher* were not made, such as newly designed central valve control and emergency blow systems, which had not operated properly on *Thresher*. Cold War pressures prompted the Navy to search for ways to reduce the duration of overhauls. By not following SUBSAFE requirements, the Navy reduced the time *Scorpion* was out of commission.

In addition, the high quality of the submarine components required by SUBSAFE, along with intensified structural inspections, had reduced the availability of critical parts such as seawater piping [8]. A year later, in May 1968, *Scorpion* was lost at sea. Although some have attributed its loss to a Soviet attack, a later investigation of the debris field revealed the most likely cause of the loss was one of its own torpedoes exploding inside the torpedo room [8]. After the *Scorpion* loss, the need for SUBSAFE was reaffirmed and accepted.

The rest of this chapter outlines the SUBSAFE program and provides some hypotheses to explain its remarkable success. The reader will notice that much of the program rests on the same systems thinking fundamentals advocated in this book.

Details of the *Thresher* Loss

The accident was thoroughly investigated including, to the Navy's credit, the systemic factors as well as the technical failures and deficiencies. Deep sea photography, recovered artifacts, and an evaluation of the *Thresher*'s design and operational

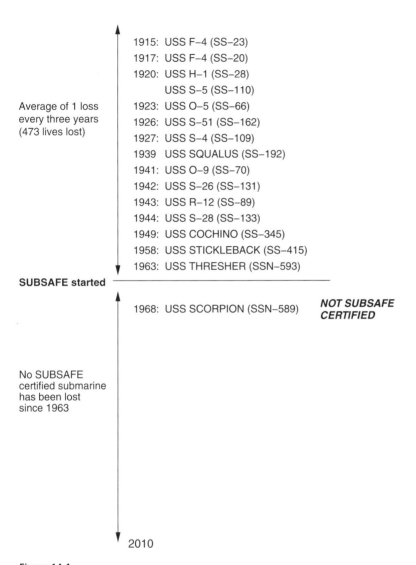

Figure 14.1
The history of noncombat U.S. submarine losses.

history led a court of inquiry to conclude that the failure of a deficient silver-braze joint in a salt water piping system, which relied on silver brazing instead of welding, led to flooding in the engine room. The crew was unable to access vital equipment to stop the flooding. As a result of the flooding, saltwater spray on the electrical components caused short circuits, shutdown of the nuclear reactor, and loss of propulsion. When the crew attempted to blow the main ballast tanks in order to surface, excessive moisture in the air system froze, causing a loss of airflow and inability to surface.

The accident report included recommendations to fix the design problems, for example, to add high-pressure air compressors to permit the emergency blow system to operate property. The finding that there were no centrally located isolation valves for the main and auxiliary seawater systems led to the use of flood-control levers that allowed isolation valves to be closed remotely from a central panel.

Most accident analyses stop at this point, particularly in that era. To their credit, however, the investigation continued and looked at why the technical deficiencies existed, that is, the management and systemic factors involved in the loss. They found deficient specifications, deficient shipbuilding practices, deficient maintenance practices, inadequate documentation of construction and maintenance actions, and deficient operational procedures. With respect to documentation, there appeared to be incomplete or no records of the work that had been done on the submarine and the critical materials and processes used.

As one example, *Thresher* had about three thousand silver-brazed pipe joints exposed to full pressure when the submarine was submerged. During her last shipyard maintenance, 145 of these joints were inspected on a "not-to-delay" vessel basis using what was then the new technique called ultrasonic testing. Fourteen percent of the 145 joints showed substandard joint integrity. Extrapolating these results to the entire complement of three thousand joints suggests that more than four hundred joints could have been substandard. The ship was allowed to go to sea in this condition. The *Thresher* loss investigators looked at whether the full scope of the joint problem had been determined and what rationale could have been used to allow the ship to sail without fixing the joints.

One of the conclusions of the accident investigation is that Navy risk management practices had not advanced as fast as submarine capability.

14.2   SUBSAFE Goals and Requirements

A decision was made in 1963 to concentrate the SUBSAFE program on the essentials, and a program was designed to provide maximum reasonable assurance of two things:

- Watertight integrity of the submarine's hull.
- Operability and integrity of critical systems to control and recover from a flooding hazard.

By being focused, the SUBSAFE program does not spread or dilute its focus beyond this stated purpose. For example, mission assurance is not a focus of SUBSAFE, although it benefits from it. Similarly, fire safety, weapons safety, occupational health and safety, and nuclear reactor systems safety are *not* in SUBSAFE. These additional concerns are handled by regular System Safety programs and mission assurance activities focused on the additional hazards. In this way, the extra rigor required by SUBSAFE is limited to those activities that ensure U.S. submarines can surface and return to port safely in an emergency, making the program more acceptable and practical than it might otherwise be.

SUBSAFE requirements, as documented in the SUBSAFE manual, permeate the entire submarine community. These requirements are invoked in design, construction, operations, and maintenance and cover the following aspects of submarine development and operations:

- Administrative
- Organizational
- Technical
- Unique design
- Material control
- Fabrication
- Testing
- Work control
- Audits
- Certification

These requirements are invoked in design contracts, construction contracts, overhaul contracts, the fleet maintenance manual and spare parts procurement specifications, and so on.

Notice that the requirements encompass not only the technical aspects of the program but the administrative and organizational aspects as well. The program requirements are reviewed periodically and renewed when deemed necessary. The Submarine Safety Working Group, consisting of the SUBSAFE Program Directors from all SUBSAFE facilities around the country, convenes twice a year to discuss program issues of mutual concern. This meeting often leads to changes and improvements to the program.

14.3 SUBSAFE Risk Management Fundamentals

SUBSAFE is founded on a basic set of risk management principles, both technical and cultural. These fundamentals are:

- Work discipline: Knowledge of and compliance with requirements
- Material control: The correct material installed correctly
- Documentation: (1) Design products (specifications, drawings, maintenance standards, system diagrams, etc.), and (2) objective quality evidence (defined later)
- Compliance verification: Inspections, surveillance, technical reviews, and audits
- Learning from inspections, audits, and nonconformances

These fundamentals, coupled with a questioning attitude and what those in SUBSAFE term a *chronic uneasiness*, are credited for SUBSAFE success. The fundamentals are taught and embraced throughout the submarine community. The members of this community believe that it is absolutely critical that they do not allow themselves to drift away from the fundamentals.

The Navy, in particular, expends a lot of effort in assuring compliance verification with the SUBSAFE requirements. A common saying in this community is, "Trust everybody, but check up." Whenever a significant issue arises involving compliance with SUBSAFE requirements, including material defects, system malfunctions, deficient processes, equipment damage, and so on, the Navy requires that an initial report be provided to Naval Sea Systems Command (NAVSEA) headquarters within twenty-four hours. The report must describe what happened and must contain preliminary information concerning apparent root cause(s) and immediate corrective actions taken. Beyond providing the information to prevent recurrence, this requirement also demonstrates top management commitment to safety and the SUBSAFE program.

In addition to the technical and managerial risk management fundamentals listed earlier, SUBSAFE also has cultural principles built into the program:

- A questioning attitude
- Critical self-evaluation
- Lessons learned and continual improvement
- Continual training
- Separation of powers (a management structure that provides checks and balances and assures appropriate attention to safety)

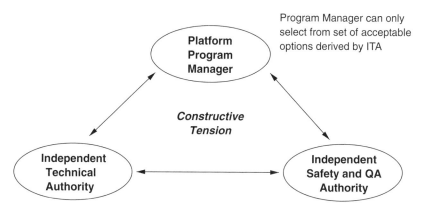

Figure 14.2
SUBSAFE separation of powers ("three-legged stool").

As is the case with most risk management programs, the foundation of SUBSAFE is the personal integrity and responsibility of those individuals who are involved in the program. The cement bonding this foundation is the selection, training, and cultural mentoring of those individuals who perform SUBSAFE work. Ultimately, these people attest to their adherence to technical requirements by documenting critical data, parameters, statements and their personal signature verifying that work has been properly completed.

14.4   Separation of Powers

SUBSAFE has created a unique management structure they call *separation of powers* or, less formally, the *three-legged stool* (figure 14.2). This structure is the cornerstone of the SUBSAFE program. Responsibility is divided among three distinct entities providing a system of checks and balances.

The new construction and in-service Platform Program Managers are responsible for the cost, schedule, and quality of the ships under their control. To ensure that safety is not traded off under cost and schedule pressures, the Program Managers can only select from a set of acceptable design options. The Independent Technical Authority has the responsibility to approve those acceptable options.

The third leg of the stool is the Independent Safety and Quality Assurance Authority. This group is responsible for administering the SUBSAFE program and for enforcing compliance. It is staffed by engineers with the authority to question and challenge the Independent Technical Authority and the Program Managers on their compliance with SUBSAFE requirements.

The Independent Technical Authority (ITA) is responsible for establishing and assuring adherence to technical standards and policy. More specifically, they:

- Set and enforce technical standards.
- Maintain technical subject matter expertise.
- Assure safe and reliable operations.
- Ensure effective and efficient systems engineering.
- Make unbiased, independent technical decisions.
- Provide stewardship of technical and engineering capabilities.

Accountability is important in SUBSAFE and the ITA is held accountable for exercising these responsibilities.

This management structure only works because of support from top management. When Program Managers complain that satisfying the SUBSAFE requirements will make them unable to satisfy their program goals and deliver new submarines, SUBSAFE requirements prevail.

14.5 Certification

In 1963, a SUBSAFE certification *boundary* was defined. Certification focuses on the structures, systems, and components that are critical to the watertight integrity and recovery capability of the submarine.

Certification is also strictly based on what the SUBSAFE program defines as *Objective Quality Evidence* (OQE). OQE is defined as any statement of fact, either quantitative or qualitative, pertaining to the quality of a product or service, based on observations, measurements, or tests *that can be verified.* Probabilistic risk assessment, which usually cannot be verified, is not used.

OQE is evidence that deliberate steps were taken to comply with requirements. It does not matter who did the work or how well they did it, if there is no OQE then there is no basis for certification.

The goal of certification is to provide maximum reasonable assurance through the initial SUBSAFE certification and by maintaining certification throughout the submarine's life. SUBSAFE inculcates the basic STAMP assumption that systems change throughout their existence. SUBSAFE certification is not a one-time activity but has to be maintained over time: SUBSAFE certification is a process, not just a final step. This rigorous process structures the construction program through a specified sequence of events leading to formal authorization for sea trials and delivery to the Navy. Certification then applies to the maintenance and operations programs and must be maintained throughout the life of the ship.

Figure 14.3
The four components of SUBSAFE certification.

14.5.1   Initial Certification

Initial certification is separated into four elements (figure 14.3):

1. *Design certification:*   Design certification consists of design product approval and design review approval, both of which are based on OQE. For design product approval, the OQE is reviewed to confirm that the appropriate technical authority has approved the design products, such as the technical drawings. Most drawings are produced by the submarine design yard. Approval may be given by the Navy's Supervisor of Shipbuilding, which administers and oversees the contract at each of the private shipyards, or, in some cases, the NAVSEA may act as the review and approval technical authority. Design approval is considered complete only after the proper technical authority has reviewed the OQE and at that point the design is certified.

2. *Material certification:*   After the design is certified, the material procured to build the submarine must meet the requirements of that design. Technical specifications must be embodied in the purchase documents. Once the material is received, it goes through a rigorous receipt inspection process to confirm and certify that it meets the technical specifications. This process usually involves examining the vendor-supplied chemical and physical OQE for the material. Records of chemical assay results, heat treatment applied to the material, and nondestructive testing conducted on the material constitute OQE.

3. *Fabrication certification:*   Once the certified material is obtained, the next step is fabrication where industrial processes such as machining, welding, and assembly are used to construct components, systems, and ships. OQE is used to document the industrial processes. Separately, and prior to actual fabrication of the final product, the facility performing the work is certified in the industrial processes necessary to perform the work. An example is a specific

high-strength steel welding procedure. In addition to the weld procedure, the individual welder using this particular process in the actual fabrication receives documented training and successfully completes a formal qualification in the specific weld procedure to be used. Other industrial processes have similar certification and qualification requirements. In addition, steps are taken to ensure that the measurement devices, such as temperature sensors, pressure gauges, torque wrenches, micrometers, and so on, are included in a robust calibration program at the facility.

4. *Testing certification:* Finally, a series of tests is used to prove that the assembly, system, or ship meets design parameters. Testing occurs throughout the fabrication of a submarine, starting at the component level and continuing through system assembly, final assembly, and sea trials. The material and components may receive any of the typical nondestructive tests, such as radiography, magnetic particle, and representative tests. Systems are also subjected to strength testing and operational testing. For certain components, destructive tests are performed on representative samples.

Each of these certification elements is defined by detailed, documented SUBSAFE requirements.

At some point near the end of the new construction period, usually lasting five or so years, every submarine obtains its initial SUBSAFE certification. This process is very formal and preceded by scrutiny and audit conducted by the shipbuilder, the supervising authority, and finally, by a NAVSEA Certification Audit Team assembled and led by the Office of Safety and Quality Assurance at NAVSEA. The initial certification is in the end granted at the flag officer level.

14.5.2 Maintaining Certification

After the submarine enters the fleet, SUBSAFE certification must be maintained through the life of the slip. Three tools are used: the Reentry Control (REC) Process, the Unrestricted Operations Maintenance Requirements Card (URO MRC) program, and the audit program.

The Reentry Control (REC) process carefully controls work and testing within the SUBSAFE boundary, that is, the structures, systems, and components that are critical to the watertight integrity and recovery capability of the submarine. The purpose of REC is to provide maximum reasonable assurance that the areas disturbed have been restored to their fully certified condition. The procedures used provide an identifiable, accountable, and auditable record of the work performed.

REC control procedures have three goals: (1) to maintain work discipline by identifying the work to be performed and the standards to be met, (2) to establish personal accountability by having the responsible personnel sign their names on the reentry control document, and (3) to collect the OQE needed for maintaining certification.

The second process, the Unrestricted Operations Maintenance Requirements Card (URO MRC) program, involves periodic inspections and tests of critical items to ensure they have not degraded to an unacceptable level due to use, age, or environment. In fact, URO MRC did not originate with SUBSAFE, but was developed to extend the operating cycle of USS *Queenfish* by one year in 1969. It now provides the technical basis for continued unrestricted operation of submarines to test depth.

The third aspect of maintaining certification is the audit program. Because the audit process is used for more general purposes than simply maintaining certification, it is considered in a separate section.

14.6 Audit Procedures and Approach

Compliance verification in SUBSAFE is treated as a process, not just one step in a process or program. The Navy demands that each Navy facility participate fully in the process, including the use of inspection, surveillance, and audits to confirm their own compliance. Audits are used to verify that this process is working. They are conducted either at fixed intervals or when a specific condition is found to exist that needs attention.

Audits are multi-layered: they exist at the contractor and shipyard level, at the local government level, and at Navy headquarters. Using the terminology adopted in this book, responsibilities are assigned to all the components of the safety control structure as shown in figure 14.4. Contractors and shipyard responsibilities include implementing specified SUBSAFE requirements, establishing processes for controlling work, establishing processes to verify compliance and certify its own work, and presenting the certification OQE to the local government oversight authority. The processes established to verify compliance and certify their work include a quality management system, surveillance, inspections, witnessing critical contractor work (contractor quality assurance), and internal audits.

Local government oversight responsibilities include surveillance, inspections, assuring quality, and witnessing critical contractor work, audits of the contractor, and certifying the work of the contractor to Navy headquarters.

The responsibilities of Navy headquarters include establishing and specifying SUBSAFE requirements, verifying compliance with the requirements, and providing SUBSAFE certification for each submarine. Compliance is verified through two types of audits: (1) ship-specific and (2) functional or facility audits.

A ship-specific audit looks at the OQE associated with an individual ship to ensure that the material condition of that submarine is satisfactory for sea trial and

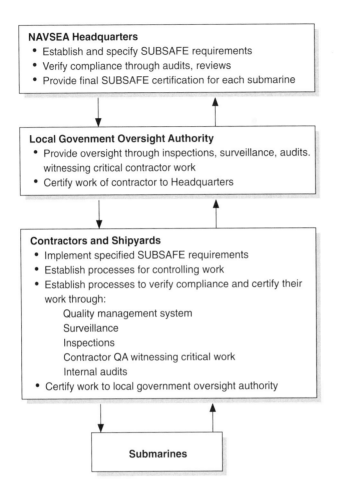

Figure 14.4
Responsibility assignments in the SUBSAFE compliance control structure.

unrestricted operations. This audit represents a significant part of the certification process that a submarine's condition meets SUBSAFE requirements and is safe to go to sea.

Functional or facility audits (such as contractors or shipyards) include reviews of policies, procedures, and practices to confirm compliance with the SUBSAFE program requirements, the health of processes, and the capability of producing certifiable hardware or design products.

Both types of audits are carried out with structured audit plans and qualified auditors.

The audit philosophy is part of the reason for SUBSAFE success. Audits are treated as a *constructive, learning* experience. Audits start from the assumption that policies, procedures, and practices are in compliance with requirements. The goal of the audit is to confirm that compliance. Audit findings must be based on a clear violation of requirements or must be identified as an "operational improvement."

The objective of audits is "to make our submarines safer" not to evaluate individual performance or to assign blame. Note the use of the word "our": the SUBSAFE program emphasizes common safety goals and group effort to achieve them. Everyone owns the safety goals and is assumed to be committed to them and working to the same purpose. SUBSAFE literature and training talks about those involved as being part of a "very special family of people who design, build, maintain, and operate our nation's submarines."

To this end, audits are a peer review. A typical audit team consists of twenty to thirty people with approximately 80 percent of the team coming from various SUBSAFE facilities around the country and the remaining 20 percent coming from NAVSEA headquarters. An audit is considered a team effort—the facility being audited is expected to help the audit team make the audit report as accurate and meaningful as possible.

Audits are conducted under rules of continuous communication—when a problem is found, the emphasis is on full understanding of the identified problem as well as identification of potential solutions. Deficiencies are documented and adjudicated. Contentious issues sometimes arise, but an attempt is made to resolve them during the audit process.

A significant byproduct of a SUBSAFE audit is the learning experience it provides to the auditors as well as those being audited. Expected results include cross-pollination of successful procedures and process improvements. The rationale behind having SUBSAFE participants on the audit team is not only their understanding of the SUBSAFE program and requirements, but also their ability to learn from the audits and apply that learning to their own SUBSAFE groups.

The current audit philosophy is a product of experience and learning. Before 1986, only ship-specific audits were conducted, not facility or headquarters audits. In 1986, there was a determination that they had gotten complacent and were assuming that once an audit was completed, there would be no findings if a follow-up audit was performed. They also decided that the ship-specific audits were not rigorous or complete enough. In STAMP terms, only the lowest level of the safety control structure was being audited and not the other components. After that time, biennial audits were conducted at all levels of the safety control structure, even the highest levels of management. A biennial NAVSEA internal audit gives the field activities a chance to evaluate operations at headquarters. Headquarters personnel must be willing to accept and resolve audit findings just like any other member of the nuclear submarine community.

One lesson learned has been that developing a robust compliance verification program is difficult. Along the way they learned that (1) clear ground rules for audits must be established, communicated, and adhered to; (2) it is not possible to "audit in" requirements; and (3) the compliance verification organization must be equal with the program managers and the technical authority. In addition, they determined that not just anyone can do SUBSAFE work. The number of activities authorized to perform SUBSAFE activities is strictly controlled.

14.7   Problem Reporting and Critiques

SUBSAFE believes that lessons learned are integral to submarine safety and puts emphasis on problem reporting and critiques. Significant problems are defined as those that affect ship safety, cause significant damage to the ship or its equipment, delay ship deployment or incur substantial cost increase, or involve severe personnel injury. Trouble reports are prepared for all significant problems encountered in the construction, repair, and maintenance of naval ships. Systemic problems and issues that constitute significant lessons learned for other activities can also be identified by trouble reports. Critiques are similar to trouble reports and are utilized by the fleet.

Trouble reports are distributed to all SUBSAFE responsible activities and are used to report significant problems to NAVSEA. NAVSEA evaluates the reports to identify SUBSAFE program improvements.

14.8   Challenges

The leaders of SUBSAFE consider their biggest challenges to be:

- *Ignorance:*   The state of not knowing;
- *Arrogance:*   Behavior based on pride, self-importance, conceit, or the assumption of intellectual superiority and the presumption of knowledge that is not supported by facts; and
- *Complacency:*   Satisfaction with one's accomplishments accompanied by a lack of awareness of actual dangers or deficiencies.

Combating these challenges is a "constant struggle every day" [69]. Many features of the program are designed to control these challenges, particularly training and education.

14.9   Continual Training and Education

Continual training and education are a hallmark of SUBSAFE. The goals are to:

- Serve as a reminder of the consequences of complacency in one's job.
- Emphasize the need to proactively correct and prevent problems.
- Stress the need to adhere to program fundamentals.
- Convey management support for the program.

Continual improvement and feedback to the SUBSAFE training programs comes not only from trouble reports and incidents but also from the level of knowledge assessments performed during the audits of organizations that perform SUBSAFE work.

Annual training is required for all headquarters SUBSAFE workers, from the apprentice craftsman to the admirals. A periodic refresher is also held at each of the contractor's facilities. At the meetings, a video about the loss of *Thresher* is shown and an overview of the SUBSAFE program and their responsibilities is provided as well as recent lessons learned and deficiency trends encountered over the previous years. The need to avoid complacency and to proactively correct and prevent problems is reinforced.

Time is also taken at the annual meetings to remind everyone involved about the history of the program. By guaranteeing that no one forgets what happened to USS *Thresher*, the SUBSAFE program has helped to create a culture that is conducive to strict adherence to policies and procedures. Everyone is recommitted each year to ensure that a tragedy like the one that occurred in 1963 never happens again. SUBSAFE is described by those in the program as "a requirement, an attitude, and a responsibility."

14.10   Execution and Compliance over the Life of a Submarine

The design, construction, and initial certification are only a small percentage of the life of the certified ship. The success of the program during the vast majority of the certified ship's life depends on the knowledge, compliance, and audit by those operating and maintaining the submarines. Without the rigor of compliance and sustaining knowledge from the petty officers, ship's officers, and fleet staff, all of the great virtues of SUBSAFE would "come to naught" [30]. The following anecdote by Admiral Walt Cantrell provides an indication of how SUBSAFE principles permeate the entire nuclear Navy:

> I remember vividly when I escorted the first group of NASA skeptics to a submarine and
> they figured they would demonstrate that I had exaggerated the integrity of the program

by picking a member of ship's force at random and asked him about SUBSAFE. The NASA folks were blown away. A second class machinist's mate gave a cogent, complete, correct description of the elements of the program and how important it was that all levels in the Submarine Force comply. That part of the program is essential to its success—just as much, if not more so, than all the other support staff effort [30].

14.11 Lessons to Be Learned from SUBSAFE

Those involved in SUBSAFE are very proud of their achievements and the fact that even after nearly fifty years of no accidents, the program is still strong and vibrant. On January 8, 2005, USS *San Francisco*, a twenty-six-year-old ship, crashed head-on into an underwater mountain. While several crew members were injured and one died, this incident is considered by SUBSAFE to be a success story: In spite of the massive damage to her forward structure, there was no flooding, and the ship surfaced and returned to port under her own power. There was no breach of the pressure hull, the nuclear reactor remained on line, the emergency main ballast tank blow system functioned as intended, and the control surfaces functioned properly. Those in the SUBSAFE program attribute this success to the work discipline, material control, documentation, and compliance verification exercised during the design, construction, and maintenance of USS *San Francisco*.

Can the SUBSAFE principles be transferred from the military to commercial companies and industries? The answer lies in why the program has been so effective and whether these factors can be maintained in other implementations of the principles more appropriate to non-military venues. Remember, of course, that private contractors form the bulk of the companies and workers in the nuclear Navy, and they seem to be able to satisfy the SUBSAFE program requirements. The primary difference is in the basic goals of the organization itself.

Some factors that can be identified as contributing to the success of SUBSAFE, most of which could be translated into a safety program in private industry are:

- Leadership support and commitment to the program.
- Management (NAVSEA) is not afraid to say "no" when faced with pressures to compromise the SUBSAFE principles and requirements. Top management also agrees to be audited for adherence to the principles of SUBSAFE and to correct any deficiencies that are found.
- Establishment of clear and written safety requirements.
- Education, not just training, with yearly reminders of the past, continual improvement, and input from lessons learned, trouble reports, and assessments during audits.
- Updating the SUBSAFE program requirements and the commitment to it periodically.

- Separation of powers and assignment of responsibility.
- Emphasis on rigor, technical compliance, and work discipline.
- Documentation capturing what they do and why they do it.
- The participatory audit philosophy and the requirement for objective quality evidence.
- A program based on written procedures, not personality-driven.
- Continual feedback and improvement. When something does not conform to SUBSAFE specifications, it must be reported to NAVSEA headquarters along with the causal analysis (including the systemic factors) of why it happened. Everyone at every level of the organization is willing to examine his or her role in the incident.
- Continual certification throughout the life of the ship; it is not a one-time event.
- Accountability accompanying responsibility. Personal integrity and personal responsibility is stressed. The program is designed to foster everyone's pride in his or her work.
- A culture of shared responsibility for safety and the SUBSAFE requirements.
- Special efforts to be vigilant against complacency and to fight it when it is detected.

Epilogue

In the simpler world of the past, classic safety engineering techniques that focus on preventing failures and chains of failure events were adequate. They no longer suffice for the types of systems we want to build, which are stretching the limits of complexity human minds and our current tools can handle. Society is also expecting more protection from those responsible for potentially dangerous systems.

Systems theory provides the foundation necessary to build the tools required to stretch our human limits on dealing with complexity. STAMP translates basic system theory ideas into the realm of safety and thus provides a foundation for our future.

As demonstrated in the previous chapter, some industries have been very successful in preventing accidents. The U.S. nuclear submarine program is not the only one. Others seem to believe that accidents are the price of progress or of profits, and they have been less successful. What seems to distinguish those experiencing success is that they:

- Take a systems approach to safety in both development and operations
- Have instituted a learning culture where they have effective learning from events
- Have established safety as a priority and understand that their long-term success depends on it

This book suggests a new approach to engineering for safety that changes the focus from "prevent failures" to "enforce behavioral safety constraints," from reliability to control. The approach is constructed on an extended model of accident causation that includes more than the traditional models, adding those factors that are increasingly causing accidents today. It allows us to deal with much more complex systems. What is surprising is that the techniques and tools described in part III that are built on STAMP and have been applied in practice on extremely complex systems have been easier to use and much more effective than the old ones.

Others will improve these first tools and techniques. What is critical is the overall philosophy of safety as a function of *control*. This philosophy is not new: It stems from the prescient engineers who created System Safety after World War II in the military aviation and ballistic missile defense systems. What they lacked, and what we have been hindered in our progress by not having, is a more powerful accident causality model that matches today's new technology and social drivers. STAMP provides that. Upon this foundation and using systems theory, new more powerful hazard analysis, design, specification, system engineering, accident/incident analysis, operations, and management techniques can be developed to engineer a safer world.

Mueller in 1968 described System Safety as "organized common sense" [109]. I hope that you have found that to be an accurate description of the contents of this book. In closing I remind you of the admonition by Bertrand Russell: "A life without adventure is likely to be unsatisfying, but a life in which adventure is allowed to take any form it will is sure to be short" [179, p. 21].

APPENDIXES

A Definitions

People have been arguing about them for decades, so it is unlikely that everyone will agree with all (or perhaps even any) of the following definitions. They reflect, however, the use of these terms in this book.

Accident An undesired and unplanned event that results in a loss (including loss of human life or injury, property damage, environmental pollution, and so on).

Hazard A system state or set of conditions that, together with a particular set of worst-case environment conditions, will lead to an accident (loss).

Hazard Analysis The process of identifying hazards and their potential causal factors.

Hazard Assessment The process involved in determining the hazard level.

Hazard Level A function of the hazard *severity* (worst case damage that could result from the hazard given the environment in its most unfavorable state) and the *likelihood* (qualitative or quantitative) of its occurrence (figure A.1).

Risk Analysis The process of identifying risk factors and their potential causal factors.

Risk Assessment The process of determining the risk level (quantifying risk).

Risk Factors Factors leading to an accident, including both hazards and the conditions or states of the environment associated with that hazard leading to an accident.

Risk Level A function of the hazard level combined with (1) the likelihood of the hazard leading to an accident and (2) hazard exposure or duration.

Safety Freedom from accidents (loss events).

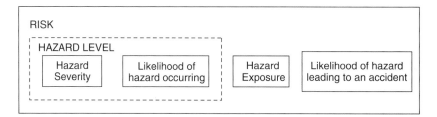

Figure A.1
The components of risk.

System Safety Engineering   The system engineering processes used to prevent accidents by identifying and eliminating or controlling hazards. Note that hazards are not the same as failures; dealing with failures is usually the province of reliability engineering.

B   The Loss of a Satellite

On April 30, 1999, at 12:30 EDT, a Titan IV B-32 booster equipped with a Centaur TC-14 upper stage was launched from Cape Canaveral, Florida. The mission was to place a Milstar-3 satellite into geosynchronous orbit. Milstar is a joint services satellite communications system that provides secure, jam-resistant, worldwide communications to meet wartime requirements. It was the most advanced military communications satellite system to that date. The first Milstar satellite was launched February 7, 1994, and the second was launched November 5, 1995. This mission was to be the third launch.

As a result of some anomalous events, the Milstar satellite was placed in an incorrect and unusable low elliptical final orbit, as opposed to the intended geosynchronous orbit. Media interest was high because this mishap was the third straight Titan IV failure and because there had been recent failures of other commercial space launches. In addition, this accident is believed to be one of the most costly unmanned losses in the history of Cape Canaveral Launch Operations. The Milstar satellite cost about $800 million, and the launcher an additional $433 million.

To its credit, the accident investigation board went beyond the usual chain-of-events model and instead interpreted the accident in terms of a complex and flawed process:

> Failure of the Titan IV B-32 mission is due to a failed software development, testing, and quality assurance process for the Centaur upper stage. That failed process did not detect and correct a human error in the manual entry of the I1(25) roll rate filter constant entered in the Inertial Measurement System flight software file. The value should have been entered as −1.992476, but was entered as −0.1992476. Evidence of the incorrect I1(25) constant appeared during launch processing and the launch countdown, but its impact was not sufficiently recognized or understood and, consequently, not corrected before launch. The incorrect roll rate filter constant zeroed any roll rate data, resulting in the loss of roll axis control, which then caused loss of yaw and pitch control. The loss of attitude control caused excessive firings of the Reaction Control system and subsequent hydrazine depletion. Erratic vehicle flight during the Centaur main engine burns caused the Centaur to achieve an orbit apogee and perigee much lower than desired, which resulted in the Milstar separating in a useless low final orbit. [153]

Fully understanding this accident requires understanding why the error in the roll rate filter constant was introduced in the load tape, why it was not found during the load tape production process and internal review processes, why it was not found during the extensive independent verification and validation effort applied to this software, and why it was not detected during operations at the launch site—in other words, why the safety control structure was ineffective in each of these instances.

Figure B.1 shows the hierarchical control model of the accident, or at least those parts that can be gleaned from the official accident report.[1] Lockheed Martin Astronautics (LMA) was the prime contractor for the mission. The Air Force Space and Missile Systems Center Launch Directorate (SMC) was responsible for insight and administration of the LMA contract. Besides LMA and SMC, the Defense Contract Management Command (DCMC) played an oversight role, but the report is not clear about what exactly this role was beyond a general statement about responsibility for contract management, software surveillance, and overseeing the development process.

LMA designed and developed the flight control software, while Honeywell was responsible for the Inertial Measurement System (IMS) software. This separation of control, combined with poor coordination, accounts for some of the problems that occurred. Analex was the independent verification and validation (IV&V) contractor, while Aerospace Corporation provided independent monitoring and evaluation. Ground launch operations at Cape Canaveral Air Station (CCAS) were managed by the Third Space Launch Squadron (3SLS).

Once again, starting from the physical process and working up the levels of control, a STAMP analysis examines each level for the flaws in the process at that level that provided inadequate control of safety in the process level below. The process flaws at each level are then examined and explained in terms of a potential mismatch in models between the controller's model of the process and the real process, incorrect design of the control algorithm, lack of coordination among the control activities, deficiencies in the reference channel, and deficiencies in the feedback or monitoring channel. When human decision making is involved, the analysis results must also include information about the context in which the decisions were made and the information available (and necessary information *not* available) to the decision makers.

One general thing to note in this accident is that there were a large number of redundancies in each part of the process to prevent the loss, but they were not effective. Sometimes (as in this case), built-in redundancy itself causes complacency and overconfidence and is a critical factor in the accident process. The use of redundancy

---

1. Some details of the control structure may be incorrect because they were not detailed in the report, but the structure is close enough for the purpose of this chapter.

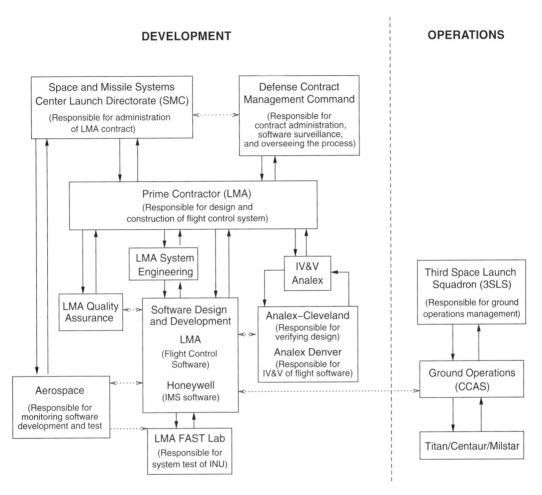

Figure B.1
Hierarchical control structure.

to provide protection against losses must include a detailed analysis of coverage and any potential gaps in the safety control provided by the redundancy.

B.1 The Physical Process

Components of the Physical Process: The Lockheed Martin Astronautics (LMA) Titan IV B is a heavy-lift space launch vehicle used to carry government payloads such as Defense Support Program, Milstar, and National Reconnaissance Office satellites into space. It can carry up to 47,800 pounds into low-earth orbit and up to

INU (Inertial Navigation Unit)

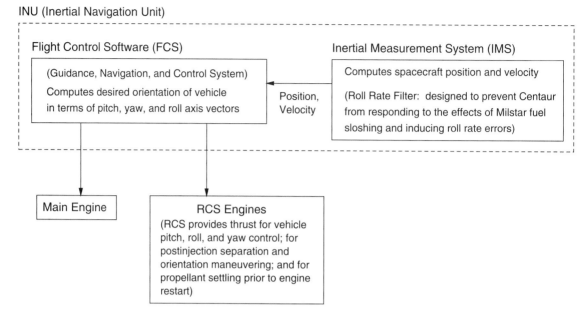

Figure B.2
Technical process control structure for INU.

12,700 pounds into a geosynchronous orbit. The vehicle can be launched with no upper stage or with one of two optional upper stages, providing greater and varied capability.

The LMA Centaur is a cryogenic, high-energy upper stage. It carries its own guidance, navigation, and control system, which measures the Centaur's position and velocity on a continuing basis throughout flight. It also determines the desired orientation of the vehicle in terms of pitch, yaw, and roll axis vectors. It then issues commands to the required control components to orient the vehicle in the proper attitude and position, using the main engine or the Reaction Control System (RCS) engines (figure B.2). The main engines are used to control thrust and velocity. The RCS provides thrust for vehicle pitch, yaw, and roll control, for post-injection separation and orientation maneuvers, and for propellant settling prior to engine restart.

System Hazards: (1) The satellite does not reach a useful geosynchronous orbit; (2) the satellite is damaged during orbit insertion maneuvers and cannot provide its intended function.

Description of the Process Controller (the INU):   The Inertial Navigation Unit (INU) has two parts (figure B.2): (1) the Guidance, Navigation, and Control System (the Flight Control Software or FCS) and (2) an Inertial Measurement System

(IMS). The FCS computes the desired orientation of the vehicle in terms of the pitch, yaw, and roll axis vectors and issues commands to the main engines and the reaction control system to control vehicle orientation and thrust. To accomplish this goal, the FCS uses position and velocity information provided by the IMS. The component of the IMS involved in the loss is a roll rate filter, which is designed to prevent the Centaur from responding to the effects of Milstar fuel sloshing and thus inducing roll rate errors.

Safety Constraint on FCS that was Violated: The FCS must provide the attitude control, separation, and orientation maneuvering commands to the main engines and the RCS system necessary to attain geosynchronous orbit.

Safety Constraints on IMS that were Violated: The position and velocity values provided to the FCS must not be capable of leading to a hazardous control action. The roll rate filter must prevent the Centaur from responding to the effects of fuel sloshing and inducing roll rate errors.

B.2   Description of the Proximal Events Leading to the Loss

There were three planned burns during the Centaur flight. The first burn was intended to put the Centaur into a parking orbit. The second would move the Centaur into an elliptical transfer orbit that was to carry the Centaur and the satellite to geosynchronous orbit. The third and final burn would circularize the Centaur in its intended geosynchronous orbit. A coast phase was planned between each burn. During the coast phase, the Centaur was to progress under its own momentum to the proper point in the orbit for the next burn. The Centaur would also exercise a roll sequence and an attitude control maneuver during the coast periods to provide passive thermal control and to settle the main engine propellants in the bottom of the tanks.

First Burn: The first burn was intended to put the Centaur into a parking orbit. The IMS transmitted a zero or near zero roll rate to the Flight Control software, however, due to the use of an incorrect roll rate filter constant. With no roll rate feedback, the FCS provided inappropriate control commands that caused the Centaur to become unstable about the roll axis and not to roll to the desired first burn orientation. The Centaur began to roll back and forth, eventually creating sloshing of the vehicle liquid fuel in the tanks, which created unpredictable forces on the vehicle and adversely affected flow of fuel to the engines. By the end of the first burn (approximately 11 minutes and 35 seconds after liftoff), the roll oscillation began to affect the pitch and yaw rates of the vehicle as well. The FCS predicted an incorrect time for main engine shutdown due to the effect on the acceleration of the vehicle's tumbling and fuel sloshing. The incorrect shutdown in turn resulted in the Centaur not achieving its intended velocity during the first burn, and the vehicle was placed in an unintended park orbit.

First Coast Phase: During the coast phases, the Centaur was to progress under its own momentum to the proper point in the orbit for the next burn. During this coasting period, the FCS was supposed to command a roll sequence and an attitude control maneuver to provide passive thermal control and to settle the main engine propellants in the bottom of the tanks. Because of the roll instability and transients created by the engine shutdown, the Centaur entered this first coast phase tumbling. The FCS directed the RCS to stabilize the vehicle. Late in the park orbit, the Centaur was finally stabilized about the pitch and yaw axes, although it continued to oscillate about the roll axis. In stabilizing the vehicle, however, the RCS expended almost 85 percent of the RCS system propellant (hydrazine).

Second Burn: The FCS successfully commanded the vehicle into the proper attitude for the second burn, which was to put the Centaur and the satellite into an elliptical transfer orbit that would carry them to geosynchronous orbit. The FCS ignited the main engines at approximately one hour, six minutes, and twenty-eight seconds after liftoff. Soon after entering the second burn phase, however, inadequate FCS control commands caused the vehicle to again become unstable about the roll axis and to begin a diverging roll oscillation. Because the second burn is longer than the first, the excess roll commands from the FCS eventually saturated the pitch and yaw channels. At approximately two minutes into the second burn, pitch and yaw control was lost (as well as roll), causing the vehicle to tumble for the remainder of the burn. Due to its uncontrolled tumbling during the burn, the vehicle did not achieve the planned acceleration for transfer orbit.

Second Coast Phase (Transfer Orbit):   The RCS attempted to stabilize the vehicle, but it continued to tumble. The RCS depleted its remaining propellant approximately twelve minutes after the FCS shut down the second burn.

Third Burn: The goal of the third burn was to circularize the Centaur in its intended geosynchronous orbit. The FCS started the third burn at two hours, thirty-four minutes, and fifteen seconds after liftoff. It was started earlier and was shorter than had been planned. The vehicle tumbled throughout the third burn, but without the RCS there was no way to control it. Space vehicle separation was commanded at approximately two hours after the third burn began, resulting in the Milstar being placed in a useless low elliptical orbit, as opposed to the desired geosynchronous orbit (figure B.3).

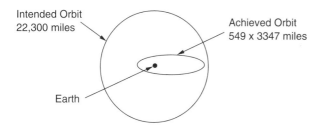

Intended Orbit
22,300 miles

Achieved Orbit
549 x 3347 miles

Earth

Figure B.3
Achieved orbit vs. intended orbit.

Post Separation: The mission director ordered early turn-on of the satellite in an attempt to save it, but the ground controllers were unable to contact the satellite for approximately three hours. Six hours and fourteen minutes after liftoff, control was acquired and various survival and emergency actions were taken. The satellite had been damaged from the uncontrolled vehicle pitch, yaw, and roll movements, however, and there were no possible actions the ground controllers could have taken in response to the anomalous events that would have saved the mission.

The mission was officially declared a failure on May 4, 1999, but personnel from LMA and the Air Force controlled the satellite for six additional days in order to place the satellite in a non-interfering orbit with minimum risk to operational satellites. It appears the satellite performed as designed, despite the anomalous conditions. It was shut down by ground control on May 10, 1999.

B.3 Physical Process and Automated Controller Failures and Dysfunctional Interactions

Figure B.4 shows the automated controller flaws leading to the accident. The Inertial Measurement System (IMS) process model was incorrect; specifically, there was an incorrect roll rate filter constant in the IMS software file (figure B.4) that led to a dysfunctional interaction with the flight control software.[2]

The Flight Control Software also operated correctly, that is, according to its requirements. However, it received incorrect input from the IMS, leading to an incorrect internal FCS model of the state of the spacecraft—the roll rate was

---

2. The load tape for a computer control program on a spacecraft contains the values specific to the mission being performed, that is, the model of the controlled process for that specific mission. The IMS software algorithm did not itself "fail," it operated as designed but it provided incorrect information to the flight control software because of the incorrect process model used to generate the information.

INU (Inertial Navigation Unit)

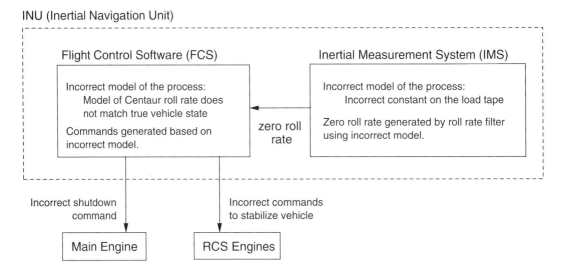

Figure B.4
Control flaws at the physical process and software controller levels.

thought to be zero or near zero when it was not. There was a mismatch between the FCS internal model of the process state and the real process state. This mismatch led to the RCS issuing incorrect control commands to the main engine (to shut down early) and to the RCS engines. Using STAMP terminology, the loss resulted from a dysfunctional interaction between the FCS and the IMS. Neither failed. They both operated correctly with respect to the instructions (including constants) and data provided.

The accident report does not explore whether the FCS software could have included sanity checks on the roll rate or vehicle behavior to detect that incorrect roll rates were being provided by the IMS. Even if the FCS did detect it was getting anomalous roll rates, there may not have been any recovery or fail-safe behavior that could have been designed into the system. Without more information about the Centaur control requirements and design, it is not possible to speculate about whether the Inertial Navigation Unit software (the IMS and FCS) might have been designed to be fault tolerant with respect to filter constant errors.

Process Models: Both the FCS and the IMS had process models that did not match the real process state, leading to the hazardous outputs from the software. The FCS model of the vehicle orientation did not match the actual orientation due to incorrect input about the state of a controlled variable (the roll rate). The IMS provided the bad input because of an incorrect model of the process, namely, the I1(25) constant used in the roll rate filter. That is, the feedback or monitoring channel of the FCS provided incorrect feedback about the roll rate.

This level of explanation of the flaws in the process (the vehicle and its flight behavior) as well as its immediate controller provides a description of the "symptom," but does not provide enough information about the factors involved to prevent reoccurrences. Simply fixing that particular flight tape does not solve any problems. We need to look at the higher levels of the control structure to understand the accident process well enough to prevent a recurrence. Some specific questions needing to be answered are: Why was the roll rate error not detected during launch operations? Why was an erroneous load tape created in the first place? Why was the error not detected in the regular verification and validation process or during the extensive independent verification and validation process? How did the error get past the quality assurance process? What role did program management play in the accident? This accident report does a much better job in answering these types of questions than the Ariane 5 report.

Figures B.5 and B.6 summarize the STAMP-based accident analysis.

B.4   Launch Site Operations

The function of launch site operations (figure B.6) is to monitor launchpad behavior and tests and to detect any critical anomalies prior to flight. Why was the roll-rate error not detected during launch operations?

Process Being Controlled: Preparations for launch at the launch site as well as the launch itself.

Safety-Constraint Violated: Critical variables (including those in software) must be monitored and errors detected before launch. Potentially hazardous anomalies detected at the launch site must be formally logged and thoroughly investigated and handled.

Context: Management had greatly reduced the number of engineers working launch operations, and those remaining were provided with few guidelines as to how they should perform their job. The accident report says that their tasks were not defined by their management so they used their best engineering judgment to determine which tasks they should perform, which variables they should monitor, and how closely to analyze the data associated with each of their monitoring tasks.

Controls: The controls are not described well in the report. From what is included, it does not appear that controls were implemented to monitor or detect software

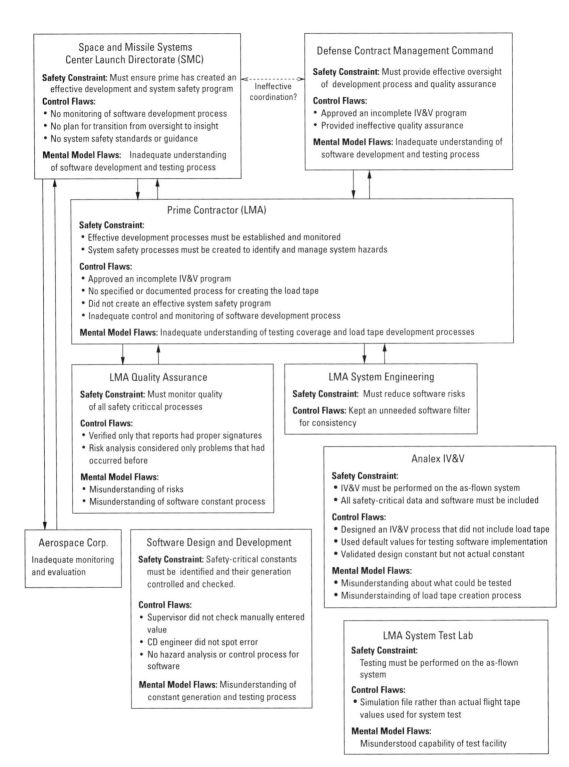

Space and Missile Systems Center Launch Directorate (SMC)

Safety Constraint: Must ensure prime has created an effective development and system safety program

Control Flaws:
- No monitoring of software development process
- No plan for transition from oversight to insight
- No system safety standards or guidance

Mental Model Flaws: Inadequate understanding of software development and testing process

Ineffective coordination?

Defense Contract Management Command

Safety Constraint: Must provide effective oversight of development process and quality assurance

Control Flaws:
- Approved an incomplete IV&V program
- Provided ineffective quality assurance

Mental Model Flaws: Inadequate understanding of software development and testing process

Prime Contractor (LMA)

Safety Constraint:
- Effective development processes must be established and monitored
- System safety processes must be created to identify and manage system hazards

Control Flaws:
- Approved an incomplete IV&V program
- No specified or documented process for creating the load tape
- Did not create an effective system safety program
- Inadequate control and monitoring of software development process

Mental Model Flaws: Inadequate understanding of testing coverage and load tape development processes

LMA Quality Assurance

Safety Constraint: Must monitor quality of all safety criticcal processes

Control Flaws:
- Verified only that reports had proper signatures
- Risk analysis considered only problems that had occurred before

Mental Model Flaws:
- Misunderstanding of risks
- Misunderstanding of software constant process

LMA System Engineering

Safety Constraint: Must reduce software risks

Control Flaws: Kept an unneeded software filter for consistency

Analex IV&V

Safety Constraint:
- IV&V must be performed on the as-flown system
- All safety-critical data and software must be included

Control Flaws:
- Designed an IV&V process that did not include load tape
- Used default values for testing software implementation
- Validated design constant but not actual constant

Mental Model Flaws:
- Misunderstanding about what could be tested
- Misunderstainding of load tape creation process

Aerospace Corp.

Inadequate monitoring and evaluation

Software Design and Development

Safety Constraint: Safety-critical constants must be identified and their generation controlled and checked.

Control Flaws:
- Supervisor did not check manually entered value
- CD engineer did not spot error
- No hazard analysis or control process for software

Mental Model Flaws: Misunderstanding of constant generation and testing process

LMA System Test Lab

Safety Constraint:
Testing must be performed on the as-flown system

Control Flaws:
- Simulation file rather than actual flight tape values used for system test

Mental Model Flaws:
Misunderstood capability of test facility

Figure B.5
STAMP model of development process.

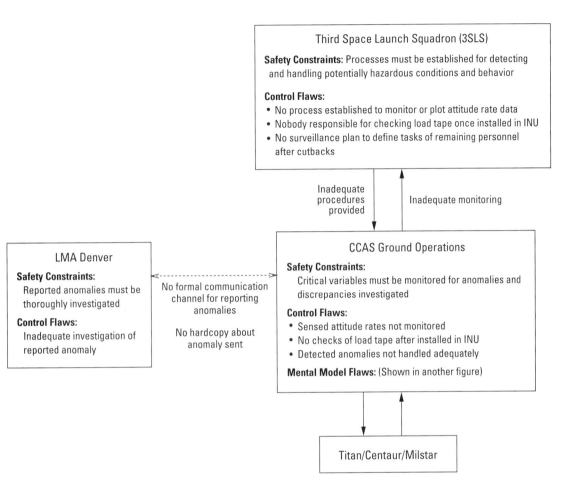

Figure B.6
STAMP model of launch operations process.

errors at the launch site although a large number of vehicle variables were monitored.

Roles and Responsibilities: The report is also not explicit about the roles and responsibilities of those involved. LMA had launch personnel at CCAS, including Product Integrity Engineers (PIEs). 3SLS had launch personnel to control the launch process as well as software to check process variables and to assist the operators in evaluating observed data.

Failures, Dysfunctional Interactions, Flawed Decisions, and Inadequate Control Actions: Despite clear indications of a problem with the roll rate information being produced by the IMS, it was not detected by some launch personnel who should have and detected but mishandled by others. Specifically:

1. One week before launch, LMA personnel at CCAS observed much lower roll rate filter values than they expected. When they could not explain the differences at their level, they raised their concerns to Denver LMA Guidance Product Integrity Engineers (PIEs), who were now at CCAS. The on-site PIEs could not explain the differences either, so they directed the CCAS personnel to call the control dynamics (CD) design engineers in Denver. On Friday, April 23, the LMA guidance engineer telephoned the LMA CD lead. The CD lead was not in his office so the guidance engineer left a voice mail stating she noticed a significant change in roll rate when the latest filter rate coefficients were entered. She requested a return call to her or to her supervisor. The guidance engineer also left an email for her supervisor at CCAS explaining the situation. Her supervisor was on vacation and was due back at the office on Monday morning, April 26, when the guidance engineer was scheduled to work the second shift. The CD lead and the CD engineer who originally specified the filter values listened to the voice mail from the guidance engineer. They called her supervisor at CCAS who had just returned from vacation. He was initially unable to find the email during their conversation. He said he would call back, so the CD engineer left the CD lead's office. The CD lead subsequently talked to the guidance engineer's supervisor after he found and read the email. The CD lead told the supervisor at CCAS that the filter values had changed in the flight tape originally loaded on April 14, 1999, and the roll rate output should also be expected to change. Both parties believed the difference in roll rates observed were attributable to expected changes with the delivery of the flight tape.

2. On the day of the launch, a 3SLS INU Product Integrity Engineer (PIE) at CCAS noticed the low roll rates and performed a rate check to see if the gyros were operating properly. Unfortunately, the programmed rate check used a default set of I1 constants to filter the measured rate and consequently reported that the gyros were sensing the earth rate correctly. If the sensed attitude rates had been monitored at that time or if they had been summed and plotted to ensure they were properly sensing the earth's gravitational rate, the roll rate problem could have been identified.

3. A 3SLS engineer also saw the roll rate data at the time of tower rollback, but was not able to identify the problem with the low roll rate. He had no documented requirement or procedures to review the data and no reference to compare to the roll rate actually being produced.

The communication channel between LMA Denver and the LMA engineers at CCAS was clearly flawed. The accident report provides no information about any established reporting channel from the LMA CCAS or LMA Denver engineers to a safety organization or up the management chain. No "alarm" system adequate to detect the problem or that it was not being adequately handled seems to have existed. The report says there was confusion and uncertainty from the time the roll rate anomaly was first raised by the CCAS LMA engineer in email and voice mail until it was "resolved" as to how it should be reported, analyzed, documented, and tracked because it was a "concern" and not a "deviation." There is no explanation of these terms nor any description of a formal problem reporting and handling system in the accident report.

Inadequate Control Algorithm: The accident report says that at this point in the prelaunch process, there was no process to monitor or plot attitude rate data, that is, to perform a check to see if the attitude filters were properly sensing the earth's rotation rate. Nobody was responsible for checking the load tape constants once the tape was installed in the INU at the launch site. Therefore, nobody was able to question the anomalous rate data recorded or correlate it to the low roll rates observed about a week prior to launch and on the day of launch. In addition, the LMA engineers at Denver never asked to see a hard copy of the actual data observed at CCAS, nor did they talk to the guidance engineer or Data Station Monitor at CCAS who questioned the low filter rates. They simply explained it away as attributable to expected changes associated with the delivery of the flight tape.

Process Model Flaws: Five models are involved here (see figure B.7):

1. Ground rate check software: The software used to do a rate check on the day of launch used default constants instead of the actual load tape. Thus there was a mismatch between the model used in the ground rate checking software and the model used by the actual IMS software.

2. Ground crew models of the development process: Although the report does not delve into this factor, it is very possible that complacency may have been involved and that the model of the thoroughness of the internal quality assurance and external IV&V development process in the minds of the ground operations personnel as well as the LMA guidance engineers who were informed of the observed anomalies right before launch did not match the real development process. There seemed to be no checking of the correctness of the software after the standard testing during development. Hardware failures are usually checked up to launch time, but often testing is assumed to have removed all software errors, and therefore further checks are not needed.

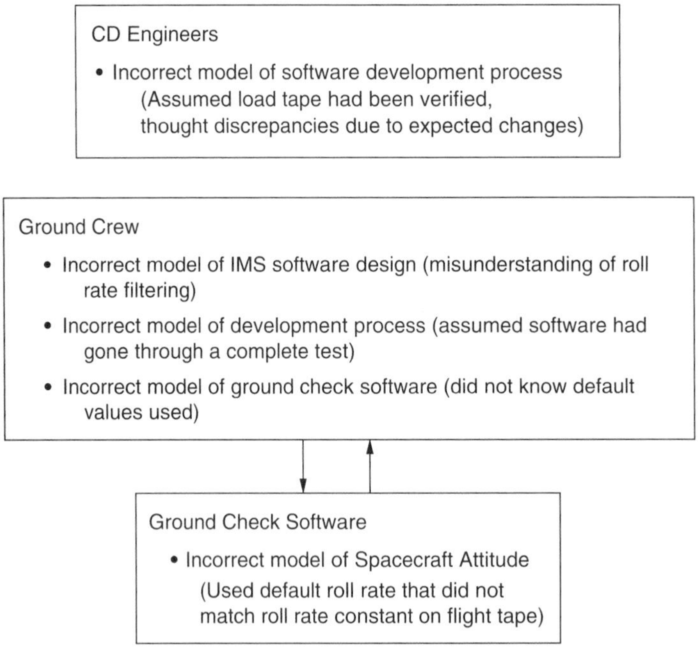

Figure B.7
The flawed process models used by the ground personnel and software.

3. Ground crew models of the IMS software design: The ground launch crew had an inadequate understanding of how the roll rate filters worked. No one other than the control dynamics engineers who designed the I1 roll rate constants understood their use or the impact of filtering the roll rate to zero. So when discrepancies were found before launch, nobody at the launch site understood the I1 roll rate filter design well enough to detect the error.

4. Ground crew models of the rate check software: Apparently, the ground crew was unaware that the checking software used default values for the filter constants.

5. CD engineers' model of the flight tape change: The control dynamics lead engineer at the launch site and her supervisor at LMA Denver thought that the roll rate anomalies were due to known changes in the flight tape. Neither went back to the engineers themselves to check this conclusion with those most expert in the details of the Centaur control dynamics.

Coordination: Despite several different groups being active at the launch site, nobody had been assigned responsibility for monitoring the software behavior after it was loaded into the INU. The accident report does not mention coordination problems, although it does say that there was a lack of understanding of each other's responsibilities between the LMA launch personnel (at CCAS) and the development personnel at LMA Denver and that this led to the concerns of the LMA personnel at CCAS not being adequately addressed.

A more general question that might have been investigated was whether the failure to act properly after detecting the roll rate problem involved a lack of coordination and communication problems between LMA engineers at CCAS and 3SLS personnel. Why did several people notice the problem with the roll rate but do nothing, and why were the anomalies they noticed not effectively communicated to those who could do something about it? Several types of coordination problems might have existed. For example, there might have been an overlap problem, with each person who saw the problem assuming that someone else was handling it or the problem might have occurred at the boundary between several people's responsibilities.

Feedback: There was a missing or inadequate feedback channel from the launch personnel to the development organization.

Tests right before launch detected the zero roll rate, but there was no formal communication channel established for getting that information to those who could understand it. Instead, voice mail and email were used. The report is not clear, but either there was no formal anomaly reporting and tracking system or it was not known or used by the process participants. Several recent aerospace accidents have involved the bypassing of formal anomaly reporting channels and the substitution of informal email and other communication—with similar results.

The LMA (Denver) engineers requested no hardcopy information about the reported anomaly and did not speak directly with the guidance engineer or data station monitor at CCAS.

B.5   Air Force Launch Operations Management

Air Force launch operations were managed by the Third Space Launch Squadron (3SLS).

Process Being Controlled: Activities of the CCAS personnel at the launch site (ground launch operations).

Safety Constraint: Processes must be established for detecting and handling potentially hazardous conditions and behavior detected during launch preparations.

Context: 3SLS management was transitioning from an *oversight* role to an *insight* one without a clear definition of what such a transition might mean or require.

Control Algorithm Flaws: After the ground launch personnel cutbacks, 3SLS management did not create a master surveillance plan to define the tasks of the remaining personnel (the formal insight plan was still in draft). In particular, there were no formal processes established to check the validity of the I1 filter constants or to monitor attitude rates once the flight tape was loaded into the INU at Cape Canaveral Air Station (CCAS) prior to launch. 3SLS launch personnel were provided with no documented requirement or procedures to review the data and no references with which to compare the observed data in order to detect anomalies.

Process Model: It is possible that misunderstandings (an incorrect model) about the thoroughness of the development process led to a failure to provide requirements and processes for performing software checks at the launch site. Complacency may also have been involved: A common assumption is that software does not fail and that software testing is exhaustive, and therefore additional software checking was not needed. However, this is speculation, as the report does not explain why management did not provide documented requirements and procedures to review the launch data nor ensure the availability of references for comparison so that discrepancies could be discovered.

Coordination: The lack of oversight led to a process that did not assign anyone the responsibility for some specific launch site tasks.

Feedback or Monitoring Channel: Apparently, launch operations management had no "insight" plan in place to monitor the performance of the launch operations process. There is no information included in the accident report about the process to monitor the performance of the launch operations process or what type of feedback was used (if any) to provide insight into the process.

B.6 Software/System Development of the Centaur Flight Control System

Too often, accident investigators stop at this point after identifying operational errors that, if they had not occurred, might have prevented the loss. Occasionally, operations management is faulted. To their credit, the accident investigation board in this case kept digging. To understand why an erroneous flight tape was created in the first place (and to learn how to prevent a similar occurrence in the future), the software and system development process associated with generating the tape needs to be examined.

Process Description: The INU consists of two major software components developed by different companies: LMA developed the Flight Control System software and was responsible for overall INU testing while Honeywell developed the IMS and was partially responsible for its software development and testing. The I1 constants are processed by the Honeywell IMS, but were designed and tested by LMA.

Safety Constraint Violated: Safety-critical constants must be identified and their generation controlled and checked.

Dysfunctional Interactions, Flawed Decisions, and Inadequate Control Actions: A software Constants and Code Words memo was generated by the LMA Control Dynamics (CD) group and sent to the LMA Centaur Flight Software (FS) group on December 23, 1997. It provided the intended and correct values for the first I1 constants in hardcopy form. The memo also allocated space for ten additional constants to be provided by the LMA Avionics group at a later time and specified a path and file name for an electronic version of the first thirty constants. The memo did not specify or direct the use of either the hardcopy or the electronic version for creating the constants database.

In early February 1999, the LMA Centaur FS group responsible for accumulating all the software and constants for the flight load tape was given discretion in choosing a baseline data file. The flight software engineer who created the database dealt with more than seven hundred flight constants generated by multiple sources, in differing formats, and at varying time (some with multiple iterations) all of which had to be merged into a single database. Some constant values came from electronic files that could be merged into the database, while others came from paper memos manually input into the database.

When the FS engineer tried to access the electronic file specified in the software Constants and Code Words Memo, he found the file no longer existed at the specified location on the electronic file folder because it was now over a year after the file had been originally generated. The FS engineer selected a different file as a baseline that only required him to change five I1 values for the digital roll rate filter (an algorithm with five constants). The filter was designed to prevent the Centaur from responding to the effects of Milstar fuel sloshing and inducing roll rate errors at 4 radians/second. During manual entry of those five I1 roll-rate filter values, the LMA FS engineer incorrectly entered or missed the exponent for the I1(25) constant. The correct value of the I1(25) filter constant was −1.992476. The exponent should have been a one, but instead was entered as a zero, making the entered constant one-tenth of the intended value, or −0.1992476. The flight software engineer's immediate supervisor did not check the manually entered values.

The only person who checked the manually input I1 filter rate values, besides the flight software engineer who actually input the data, was an LMA Control Dynamics engineer. The FS engineer who developed the Flight Load tape notified the CD engineer responsible for design of the first thirty I1 constants that the tape was completed and the printout of the constants was ready for inspection. The CD engineer went to the FS offices and looked at the hardcopy listing to perform the check and sign off the I1 constants. The manual and visual check consisted of comparing a list of I1 constants from appendix C of the Software Constants and Code Words

Memo to the paper printout from the Flight Load tape. The formats of the floating-point numbers (the decimal and exponent formats) were different on each of these paper documents for the three values crosschecked for each I1 constant. The CD engineer did not spot the exponent error for I1(25) and signed off that the I1 constants on the flight load tape were correct. He did not know that the design values had been inserted manually into the database used to build the flight tapes (remember, the values had been stored electronically but the original database no longer existed) and that they were never formally tested in any simulation prior to launch.

The CD engineer's immediate supervisor, the lead for the CD section, did not review the signoff report or catch the error. Once the incorrect filter constant went undetected in the signoff report, there were no other formal checks in the process to ensure the I1 filter rate values used in flight matched the designed filter.

Control Algorithm Flaws:

- A process input was missing (the electronic file specified in the Software Constants and Code Words memo), so an engineer regenerated it, making a mistake in doing so.

- Inadequate control was exercised over the constants process. No specified or documented software process existed for electronically merging all the inputs into a single file. There was also no formal, documented process to check or verify the work of the flight software engineer in creating the file. Procedures for creating and updating the database were left up to the flight software engineer's discretion.

- Once the incorrect filter constant went undetected in the signoff report, there were no other formal checks in the process to ensure the I1 filter rate values used in flight matched the designed filter.

- The hazard analysis process was inadequate, and no control was exercised over the potential hazard of manually entering incorrect constants, a very common human error. If system safety engineers had identified the constants as critical, then a process would have existed for monitoring the generation of these critical variables. In fact, neither the existence of a system safety program or any form of hazard analysis is mentioned in the accident report. If such a program had existed, one would think it would be mentioned.

  The report does say that quality assurance (QA) engineers performed a risk analysis, but they considered only those problems that had happened before:

  Their risk analysis was not based on determining steps critical to mission success, but on how often problems previously surfaced in particular areas on past launches. They determined software constant generation was low risk because there had not been previous problems in that area. They only verified that the signoff report containing the constants had all the proper signatures [153].

Considering only the causes of past accidents is not going to be effective for software problems or when new technology is introduced into a system. Computers are introduced, in fact, in order to make previously infeasible changes in functionality and design, which reduces the effectiveness of a "fly-fix-fly" approach to safety engineering. Proper hazard analyses examining all the ways the system components can contribute to an accident need to be performed.

Process Model Flaws: The accident report suggests that many of the various partners were confused about what the other groups were doing. The LMA software personnel who were responsible for creating the database (from which the flight tapes are generated) were not aware that IV&V testing did not use the as-flown (manually input) I1 filter constants in their verification and validation process. The LMA Control Dynamics engineer who designed the I1 rate filter also did not know that the design values were manually input into the database used to build the flight tapes and that the values were never formally tested in any simulation prior to launch.

While the failure of the LMA CD engineer who designed the I1 rate filter to find the error during his visual check was clearly related to the difficulty of checking long lists of differently formatted numbers, it also may have been partly due to less care being taken in the process due to an incorrect mental model: (1) he did not know the values were manually entered into the database (and were not from the electronic file he had created), (2) he did not know the load tape was never formally tested in any simulation prior to launch, and (3) he was unaware the load tape constants were not used in the IV&V process.

Coordination: The fragmentation and stovepiping in the flight software development process, coupled with the lack of comprehensive and defined system and safety engineering processes, resulted in poor and inadequate communication and coordination among the many partners and subprocesses.

Because the IMS software was developed by Honeywell, almost everyone outside of Honeywell (LMA control dynamics engineers, flight software engineers, product integrity engineers, SQA, IV&V, and DCMC personnel) focused on the FCS and had little knowledge of the IMS software.

B.7   Quality Assurance (QA)

Process Being Controlled: The quality of the guidance, navigation, and control system design and development.

Safety Constraint: QA must monitor the quality of all safety-critical processes.

Process Flaw: The internal LMA quality assurance processes did not detect the error in the role rate filter constant software file.

Control Algorithm Flaws: QA verified only that the signoff report containing the load tape constants had all the proper signatures, an obviously inadequate process. This accident is indicative of the problems with QA as generally practiced and why it is often ineffective. The LMA Quality Assurance Plan used was a top-level document that focused on verification of process completion, not on how the processes were executed or implemented. It was based on the original General Dynamics Quality Assurance Plan with recent updates to ensure compliance with ISO 9001. According to this plan, the LMA Software Quality Assurance staff was required only to verify that the signoff report containing the constants had all the proper signatures; they left the I1 constant generation and validation process to the flight software and control dynamics engineers. Software Quality Assurance involvement was limited to verification of software checksums and placing quality assurance stamps on the software products that were produced.

B.8   Developer Testing Process

Once the error was introduced into the load tape, it could potentially have been detected during verification and validation. Why did the very comprehensive and thorough developer and independent verification and validation process miss this error?

Safety Constraint Violated: Testing must be performed on the as-flown software (including load tape constants).

Flaws in the Testing Process: The INU (FCS and IMS) was never tested using the actual constants on the load tape:

- Honeywell wrote and tested the IMS software, but they did not have the actual load tape.
- The LMA Flight Analogous Simulation Test (FAST) lab was responsible for system test; it tested the compatibility and functionality of the flight control software and the Honeywell IMS. But the FAST lab testing used a 300 Hz filter simulation data file for IMS filters and not the flight tape values. The simulation data file was built from the original, correctly specified values of the designed constants (specified by the LMA CS engineer), not those entered by the software personnel in the generation of the flight load tape. Thus the mix of actual flight software and simulated filters used in the FAST testing did not contain the I1(25) error, and the error could not be detected by the internal LMA testing.

Process Model Mismatch: The testing capability that the current personnel thought the lab had did not match the real capability. The LMA FAST facility was used predominantly to test flight control software developed by LMA. The lab had been originally constructed with the capability to exercise the actual flight values for the I1 roll rate filter constants, but that capability was not widely known by the current FAST software engineers until after this accident; knowledge of the capability had been lost in the corporate consolidation and evolution process, so the current software engineers used a set of default roll rate filter constants. Later it was determined that had they used the actual flight values in their simulations prior to launch, they would have caught the error.

B.9   Independent Verification and Validation (IV&V)

Safety Constraint Violated: IV&V must be performed on the as-flown software and constants. All safety-critical data and software must be included in the IV&V process.

Dysfunctional Interactions: Each component of the IV&V process performed its function correctly, but the overall design of the process was flawed. In fact, it was designed in such a way that it was not capable of detecting the error in the role rate filter constant.

Analex was responsible for the overall IV&V effort of the flight software. In addition to designing the IV&V process, Analex-Denver performed the IV&V of the flight software to ensure the autopilot design was properly implemented in the software, while Analex-Cleveland verified the design of the autopilot but not its implementation. The "truth baseline" provided by LMA, per agreement between LMA and Analex, was generated from the constants verified in the signoff report.

In testing the flight software implementation, Analex-Denver used IMS default values instead of the actual I1 constants contained on the flight tape. Generic or default I1 constants were used because they believed the actual I1 constants could not be adequately validated in their rigid body simulations, that is, the rigid body simulation of the vehicle would not exercise the filters sufficiently. They found out after the mission failure that if they had used the actual I1 constants in their simulation, they would have found the order of magnitude error.

Analex-Denver also performed a range check of the program constants and the Class I flight constants and verified that format conversions were done correctly. However, the process did not require Analex-Denver to check the accuracy of the numbers in the truth baseline, only to do a range check and a bit-to-bit comparison against the firing tables, which contained the wrong constant. Thus the format conversions they performed simply compared the incorrect I1(25) value in the firing tables to the incorrect I1(25) value after the conversion, and they matched. They did not verify that the designed I1 filter constants were the ones actually used on the flight tape.

Analex-Cleveland had responsibility for verifying the functionality of the design constant but not the actual constant loaded into the Centaur for flight. That is, it was validating the design only, and not the "implementation" of the design. Analex-Cleveland received the Flight Dynamics and Control Analysis Report (FDACAR) containing the correct value for the roll filter constant. Their function was to validate the autopilot design values provided in the FDACAR. That does not include IV&V of the I1 constants in the flight format. The original design work was correctly represented by the constants in the FDACAR. In other words, the filter constant in question was listed in the FDACAR with its correct value of -1.992476, and not the value on the flight tape (-0.1992476).

Control Algorithm Flaws: Analex developed (with LMA and government approval) an IV&V program that did not verify or validate the I1 filter rate constants actually used in flight. The I1 constants file was not sent to Analex-Cleveland for autopilot validation because Analex-Cleveland performed only design validation. Analex-Denver used default values for testing and never validated the actual I1 constants used in flight.

Process Model Mismatches: The decision to use default values for testing (both by LMA FAST lab and by Analex-Denver) was based on a misunderstanding about the development and test environment and what was capable of being tested. Both the LMA FAST lab and Analex-Denver could have used the real load tape values, but did not think they could.

In addition, Analex-Denver, in designing the IV&V process, did not understand the generation or internal verification process for all the constants in the "truth baseline" provided to them by LMA. The Analex-Denver engineers were not aware that the I1 filter rate values provided originated from a manual input and might not be the same as those subjected to independent V&V by Analex-Cleveland.

None of the participants was aware that nobody was testing the software with the actual load tape values nor that the default values they used did not match the real values.

Coordination: This was a classic case of coordination problems. Responsibility was diffused among the various partners, without complete coverage. In the end, nobody tested the load tape, and everyone thought someone else was doing it.

B.10   Systems Engineering

System engineering at LMA was responsible for the identification and allocation of the functionality to be included in the system. In fact, the software filter involved in the loss was not needed and should have been left out instead of being retained, yet another example of asynchronous evolution. Why was that decision made? The filter was designed to prevent the Centaur from responding to the effects of Milstar fuel sloshing and inducing roll rate errors at 4 radians/second. Early in the design phase of the first Milstar satellite, the manufacturer asked to filter that frequency. The satellite manufacturer subsequently determined filtering was not required at that frequency and informed LMA. However, LMA decided to leave the filter in place for the first and subsequent Milstar flights for "consistency." No further explanation is included in the report.

B.11   Prime Contractor Project Management

Process Being Controlled: The activities involved in the development and assurance of the system and its components.

Safety Constraint: Effective software development processes must be established and monitored. System safety processes must be created to identify and manage system hazards.

Context: The Centaur software process was developed early in the Titan/Centaur program: Many of the individuals who designed the original process were no longer involved in it due to corporate mergers and restructuring (e.g., Lockheed, Martin Marietta, General Dynamics) and the maturation and completion of the Titan IV design and development. Much of the system and process history and design rationale was lost with their departure.

Control Algorithm Flaws:
- A flawed software development process was designed. For example, no process was provided for creating and validating the flight constants.
- LMA, as prime contractor, did not exert adequate control over the development process. The Accident Investigation Board could not identify a single process owner responsible for understanding, designing, documenting, or controlling configuration and ensuring proper execution of the process.
- An effective system safety program was not created.
- An inadequate IV&V program (designed by Analex-Denver) was approved and instituted that did not verify or validate the I1 filter rate constants used in flight.

Process Model Flaws: Nobody seemed to understand the overall software development process, and apparently all had a misunderstanding about the coverage of the testing process.

B.12 Defense Contract Management Command (DCMC)

Process Being Controlled: The report is vague, but apparently DCMC was responsible for contract administration, software surveillance, and overseeing the development process.

Control Inadequacies: The report says that DCMC approved an IV&V process with incomplete coverage and that there was a software quality assurance function operating at DCMC, but it operated without a detailed understanding of the overall process or program and therefore was ineffective.

Coordination: No information was provided in the accident report although coordination problems between SMC and DCMA may have been involved. Was each assuming the other was monitoring the overall process? What role did Aerospace Corporation play? Were there gaps in the responsibilities assigned to each of the many groups providing oversight here? How did the overlapping responsibilities fit together? What kind of feedback did DCMC use to perform its process monitoring?

B.13   Air Force Program Office

The Air Force Space and Missile Systems Center Launch Directorate (SMC) controlled development and launch.

Process Being Controlled: Management of the Titan/Centaur/Milstar development and launch control structures. SMC was responsible for "insight" and administration of the LMA contract.

Safety Constraint: SMC must ensure that the prime contractor creates an effective development and safety assurance program.

Context: Like 3SLS, the Air Force Space and Missile System Center Launch Directorate was transitioning from a task oversight to a process insight role and had, at the same time, undergone personnel reductions.

Control Algorithm Flaws:

- The SMC Launch Programs Directorate essentially had no personnel assigned to monitor or provide insight into the generation and verification of the software development process. The Program Office did have support from Aerospace Corporation to monitor the software development and test process, but that support had been cut by over 50 percent since 1994. The Titan Program Office had no permanently assigned civil service or military personnel and no full-time support to work the Titan/Centaur software. They decided that because the Titan/Centaur software was "mature, stable, and had not experienced problems in the past," they could best use their resources to address hardware issues.

- The transition from oversight to insight was not managed by a detailed plan. AF responsibilities under the insight concept had not been well defined, and requirements to perform those responsibilities had not been communicated to the workforce. In addition, implementation of the transition from an oversight role to an insight role was negatively affected by the lack of documentation and understanding of the software development and testing process. Similar flawed transitions to an "insight" role are a common factor in many recent aerospace accidents.

- The Titan Program Office did not impose any standards (e.g., Mil-Std-882) or process for safety. While one could argue about what particular safety standards and program could or should be imposed, it is clear from the complete lack of such a program that no guidance was provided. Effective control of safety requires that responsibility for safety be assigned at each level of the control structure. Eliminating this control leads to accidents. The report does not say whether responsibility for controlling safety was retained at the program office or whether it had been delegated to the prime contractor. But even if it had been delegated to LMA, the program office must provide overall leadership and monitoring of the effectiveness of the efforts. Clearly there was an inadequate safety program in this development and deployment project. Responsibility for detecting this omission lies with the program office.

In summary, understanding why this accident occurred and making the changes necessary to prevent future accidents requires more than simply identifying the proximate cause—a human error in transcribing long strings of digits. This type of error is well known and there should have been controls established throughout the process to detect and fix it. Either these controls were missing in the development and operations processes, or they were inadequately designed and executed.

While the accident report was more thorough than most, information that would have been helpful in understanding the entire accident process and generating more complete recommendations, was omitted. A STAMP-based accident analysis process provides assistance in determining what questions should be asked during the incident or accident investigation.

C A Bacterial Contamination of a Public Water Supply

In May 2000, in the small town of Walkerton, Ontario, Canada, some contaminants, largely *Escherichia coli* O157:H7 (the common abbreviation for which is *E. coli*) and *Campylobacter jejuni* entered the Walkerton water system through a municipal well. About half the people in the town of 4,800 became ill, and seven died [147]. The proximate events are presented first and then the STAMP analysis of the accident.

C.1 Proximate Events at Walkerton

The Walkerton Public Utilities Commission (WPUC) operated the Walkerton water system. Stan Koebel was the WPUC's general manager and his brother Frank its foreman. In May 2000, the water system was supplied by three groundwater sources: Wells 5, 6, and 7. The water pumped from each well was treated with chlorine before entering the distribution system.

The source of the contamination was manure that had been spread on a farm near Well 5. Unusually heavy rains from May 8 to May 12 carried the bacteria to the well. Between May 13 and May 15, Frank Koebel checked Well 5 but did not take measurements of chlorine residuals, although daily checks were supposed to be made.[1] Well 5 was turned off on May 15.

On the morning of May 15, Stan Koebel returned to work after having been away from Walkerton for more than a week. He turned on Well 7, but shortly after doing so, he learned a new chlorinator for Well 7 had not been installed, and the well was therefore pumping unchlorinated water directly into the distribution system. He did not turn off the well but instead allowed it to operate without chlorination until noon on Friday May 19, when the new chlorinator was installed.

---

1. Low chlorine residuals are a sign that contamination is overwhelming the disinfectant capacity of the chlorination process.

On May 15, samples from the Walkerton water distribution system were sent to A&L Labs for testing according to the normal procedure. On May 17, A&L Labs advised Mr. Koebel that samples from May 15 tested positive for *E. coli* and total coliforms. The next day, May 18, the first symptoms of widespread illness appeared in the community. Public inquiries about the water prompted assurances by Stan Koebel that the water was safe. By May 19 the scope of the outbreak had grown, and a pediatrician contacted the local health unit with a suspicion that she was seeing patients with symptoms of *E. coli*.

The Bruce–Grey–Owen Sound (BGOS) Health Unit, which is the government unit responsible for public health in the area, began an investigation. In two separate calls placed to Stan Koebel, the health officials were told that the water was "okay." At that time, Stan Koebel did not disclose the lab results from May 15, but he did start to flush and superchlorinate the system to try to destroy any contaminants in the water. The chlorine residuals began to recover. Apparently, Mr. Koebel did not disclose the lab results for a combination of two reasons: he did not want to reveal the unsafe practices he had engaged in from May 15 to May 17 (i.e., running Well 7 without chlorination), and he did not understand the serious and potentially fatal consequences of the presence of *E. coli* in the water system. He continued to flush and superchlorinate the water through the following weekend, successfully increasing the chlorine residuals. Ironically, it was not the operation of Well 7 without a chlorinator that caused the contamination; the contamination instead entered the system through Well 5 from May 12 until it was shut down on May 15.

On May 20, the first positive test for *E. coli* infection was reported, and the BGOS Health Unit called Stan Koebel twice to determine whether the infection might be linked to the water system. Both times, Stan Koebel reported acceptable chlorine residuals and failed to disclose the adverse test results. The Health Unit assured the public that the water was safe based on the assurances of Mr. Koebel.

That same day, a WPUC employee placed an anonymous call to the Ministry of the Environment (MOE) Spills Action Center, which acts as an emergency call center, reporting the adverse test results from May 15. On contacting Mr. Koebel, the MOE was given an evasive answer and Mr. Koebel still did not reveal that contaminated samples had been found in the water distribution system. The health unit contacted the Local Medical Officer, and he took over the investigation. The health unit took its own water samples and delivered them to the Ministry of Health laboratory in London (Ontario) for microbiological testing.

When asked by the MOE for documentation, Stan Koebel finally produced the adverse test results from A&L Laboratory and the daily operating sheets for Wells 5 and 6, but said he could not produce the sheet for Well 7 until the next day. Later, he instructed his brother Frank to revise the Well 7 sheet with the intention of concealing the fact that Well 7 had operated without a chlorinator. On Tuesday, May

23, Stan Koebel provided the altered daily operating sheet to the MOE. That same day, the health unit learned that two of the water samples it had collected on May 21 had tested positive for *E. coli*.

Without waiting for its own samples to be returned, on May 21 the BGOS health unit issued a boil-water advisory on local radio. About half of Walkerton's residents became aware of the advisory on May 21, with some members of the public still drinking the Walkerton town water as late as May 23. The first person died on May 22, a second on May 23, and two more on May 24. During this time, many children became seriously in and some victims will probably experience lasting damage to their kidneys as well as other long-term health effects. In all, seven people died, and more than 2,300 became ill.

Looking only at these proximate events and connecting them by some type of causal chain, it appears that this is a simple case of incompetence, negligence, and dishonesty by WPUC employees. In fact, the government representatives argued at the accident inquiry that Stan Koebel and the WPUC were solely responsible for the outbreak and that they were the only ones who could have prevented it. In May 2003, exactly three years after the accident, Stan and Frank Koebel were arrested for their part in the loss. But a systems-theoretic analysis using STAMP provides a much more informative and useful understanding of the accident besides simply blaming it only on the actions of the Koebel brothers.

C.2   System Hazards, System Safety Constraints, and Control Structure

As in the previous examples, the first step in creating a STAMP analysis is to identify the system hazards, the system safety constraints, and the hierarchical control structure in place to enforce the constraints.

The system hazard related to the Walkerton accident is public exposure to E. coli or other health-related contaminants through drinking water. This hazard leads to the following system safety constraint:

The safety control structure must prevent exposure of the public to contaminated water.

1. *Water quality must not be compromised.*

2. *Public health measures must reduce risk of exposure if water quality is compromised (e.g., boil-water advisories).*

Each component of the sociotechnical public water system safety control structure (figure C.1) plays a role in enforcing this general system safety constraint and will, in turn, have its own safety constraints to enforce that are related to its function in the overall system. For example, the Canadian federal government is responsible for establishing a nationwide public health system and ensuring it is operating effectively. Federal guidelines are provided to the provinces, but responsibility for water quality is primarily delegated to each individual province.

The provincial governments are responsible for regulating and overseeing the safety of the drinking water. They do this by providing budgets to the ministries involved—in Ontario these are the Ministry of the Environment (MOE), the Ministry of Health (MOH), and the Ministry of Agriculture, Food, and Rural Affairs— and by passing laws and adopting government policies affecting water safety.

According to the report on the official Inquiry into the Walkerton accident [147], the Ministry of Agriculture, Food, and Rural Affairs in Ontario is responsible for regulating agricultural activities with potential impact on drinking water sources. In fact, there was no watershed protection plan to protect the water system from agricultural runoff. Instead, the MOE was responsible for ensuring that the water systems could not be affected by such runoff.

The MOE has primary responsibility for regulating and for enforcing legislation, regulations, and policies that apply to the construction and operation of municipal water systems. Guidelines and objectives are set by the MOE, based on federal guidelines. They are enforceable through certificates of approval issued to public water utilities operators under the Ontario Water Resources Act. The MOE also has legislative responsibility for building and maintaining water treatment plants and has responsibility for public water system inspections and drinking water surveillance, for setting standards for certification of water systems, and for continuing education requirements for operators to maintain competence as knowledge about water safety increases.

The MOH supervises local health units, in this case, the Bruce–Grey–Owen Sound (BGOS) Department of Health, run by local officers of health. BGOS receives inputs from various sources, including hospitals, the local medical community, the MOH, and the WPUC, and in turn is responsible for issuing advisories and alerts if required to protect public health. Upon receiving adverse water quality reports from the government testing labs or the MOE, the local public health inspector in Walkerton would normally contact the WPUC to ensure that follow-up samples were taken and chlorine residuals maintained.

The public water system in Walkerton is run by the WPUC, which operates the wells and is responsible for chlorination and for measurement of chlorine residuals. Oversight is provided by elected commissioners. The commissioners are responsible for establishing and controlling the policies under which the WPUC operates, while the general manager (Stan Koebel) and staff are responsible for administering these policies in operating the water facility. Although theoretically also responsible for the public water system, the municipality left the operation of the water system to the WPUC.

Together, the safety constraints enforced by all of these system control components must be adequate to enforce the overall system safety constraints. Figure C.1 shows the overall theoretical water safety control structure in Ontario and the safety-related requirements and constraints for each system component.

Each component of the sociotechnical public water safety system plays a role in enforcing the system safety constraints. Understanding the accident requires again understanding the role in the accident scenario played by each level of the system's hierarchical control structure in the accident by not adequately enforcing its part of the safety constraint. For each component, the contribution to the accident is described in terms of the four conditions required for adequate control: the goal, the actions, the process or mental models, and feedback. At each level of control, the context in which the behaviors took place is also considered. It is not possible to understand human behavior without knowing the context in which it occurs and the behavior-shaping factors in the environment.

This first level of analysis provides a view of the limitations of the static control structure at the time of the accident. But systems are not static—they adapt and change over time. In STAMP, systems are treated as a dynamic process that is continually adapting to achieve its ends and to react to changes in itself and its environment. The original system design must not only enforce the system safety constraints, but the system must continue to enforce the constraints as changes occur. The analysis of accidents, therefore, requires understanding not only the flaws in the static control structure that allowed the safety constraints to be violated but also the changes to the safety control structure over time (the *structural dynamics*) and the dynamic processes behind these changes (the *behavioral dynamics*). Section C.8 analyzes the structural dynamics of the Walkerton accident.

C.3   Physical Process View of the Accident

As in other component interaction accidents, there were no physical failures involved. If, as in figure C.2, we draw the boundary of the physical system around the wells, the public water system, and public health, then one can describe the "cause" of the accident at the physical system level as the inability of the physical design to enforce the physical safety constraint in the face of an environmental disturbance, in this case the unusually heavy rains that resulted in the transport of contaminants from the fields to the water supply. The safety constraint being enforced at this level is that water must be free from unacceptable levels of contaminants.

Well 5 was a very shallow well: all of its water was drawn from an area between 5m and 8m below the surface. More significantly, the water was drawn from an area of bedrock, and the shallowness of the soil overburden above the bedrock along

System Hazard: Public is exposed to *E. coli* or other health related contaminants through drinking water.

System Safety Constraints: The safety control structure must prevent exposure of the public to contaminated water.

(1) Water quality must not be compromised.

(2) Public health measures must reduce risk of exposure if water quality is compromised
   (e.g., notification and procedures to follow)

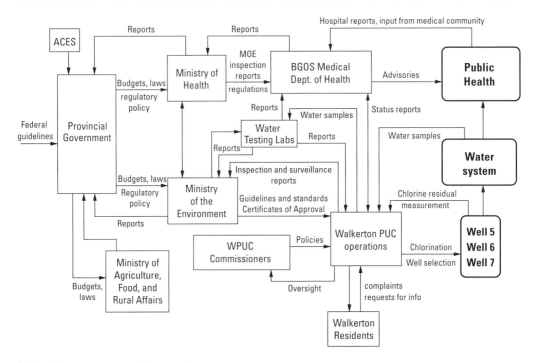

Safety Requirements and Constraints:

Federal Government
- Establish a nationwide public health system and ensure it is operating effectively.

Provincial Government
- Establish regulatory bodies and codes of responsibilities, authority, and accountability
- Provide adequate resources to regulatory bodies to carry out their responsibilities.
- Provide oversight and feedback loops to ensure that provincial regulatory bodies are doing their job adequately.
- Ensure adequate risk assessment is conducted and effective risk management plans are in place.

Ministry of the Environment
- Ensure that those in charge of water supplies are competent to carry out their responsibilities.
- Perform inspections and surveillance. Enforce compliance if problems found.
- Perform hazard analyses to identify vulnerabilities and monitor them.
- Perform continual risk evaluation for existing facilities and establish new controls if necessary.
- Establish criteria for determining whether a well is at risk.
- Establish feedback channels for adverse test results. Provide multiple paths.
- Enforce legislation, regulations and policies applying to construction and operation of municipal water systems.
- Establish certification and training requirements for water system operators.

ACES
- Provide stakeholder and public review and input on ministry standards

Ministry of Health
- Ensure adequate procedures exist for notification and risk abatement if water quality is compromised.

Water Testing Labs
- Provide timely reports on testing results to MOE, PUC, and and Medical Dept. of Health

WPUC Commissioners
- Oversee operations to ensure water quality is not compromised.

WPUC Operations Management
- Monitor operations to ensure that sample taking and reporting is accurate and adequate chlorination is being performed.

WPUC Operations
- Measure chlorine residuals.
- Apply adequate doses of cholorine to kill bacteria.

BGOS Medical Department of Health
- Provide oversight of drinking water quality.
- Follow up on adverse drinking water quality reports.
- Issue boil water advisories when necessary.

Figure C.1
The basic water safety control structure. Lines going into the left of a box are control lines. Lines from or to the top or bottom of a box represent information, feedback, or a physical flow. Rectangles with sharp corners are controllers, while rectangles with rounded corners represent plants.

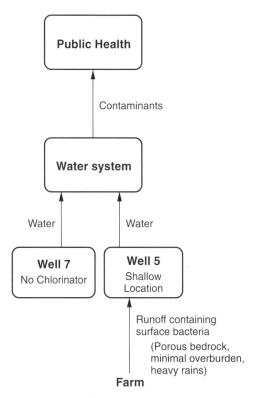

Figure C.2
The physical components of the water safety control structure.

with the fractured and porous nature of the bedrock itself made it possible for surface bacteria to make its way to Well 5.

C.4   First-Level Operations

Besides the physical system analysis, most hazard analysis techniques and accident investigations consider the immediate operators of the system. Figure C.3 shows the results of a STAMP analysis of the flaws by the lower operations levels at Walkerton that were involved in the accident.

The safety requirements and constraints on the operators of the local water system were that they must apply adequate doses of chlorine to kill bacteria and must measure chlorine residuals. Stan Koebel, the WPUC manager, and Frank Koebel, its foreman, were not qualified to hold their positions within the WPUC. Before 1993, there were no mandatory certification requirements, and after 1993 they were certified through a grandfathering process based solely on experience.

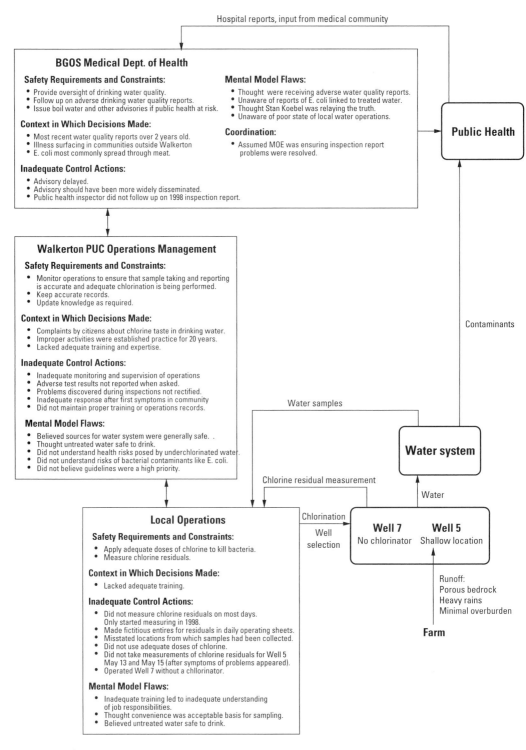

Figure C.3
The physical and operational components of the water safety control structure.

Stan Koebel knew how to operate the water system mechanically, but he lacked knowledge about the health risks associated with a failure to properly operate the system and of the importance of following the requirements for treatment and monitoring of the water quality. The inquiry report stated that many improper operating practices had been going on for years before Stan Koebel became manager: He simply left them in place. These practices, some of which went back twenty years, included misstating the locations at which samples for microbial testing were taken, operating wells without chlorination, making false entries in daily operating sheets, not measuring chlorine residuals daily, not adequately chlorinating the water, and submitting false annual reports to the MOE.

The operators of the Walkerton water system did not intentionally put the public at risk. Stan Koebel and the other WPUC employees believed the untreated water was safe and often drank it themselves at the well sites. Local residents also pressed the WPUC to decrease the amount of chlorine used because they objected to the taste of chlorinated water.

A second first-level control component was the local health units, in this case, the BGOS Department of Health. Local health units are supervised by the MOH and run by local Officers of Health to execute their role in protecting public health. The BGOS Medical Department of Health receives inputs (feedback) from various sources, including hospitals, the local medical community, the MOH, and the WPUC, and in turn is responsible for issuing advisories and alerts if required to protect public health. While the local health unit did issue a boil-water advisory on local radio, when it finally decided that the water system might be involved, this means of notifying the public was not very effective. Other more effective means could have been employed. One reason for the delay was simply that evidence was not strong that the water system was the source of the contamination. *E. coli* is most often spread by meat, which is why it is commonly called the "hamburger disease." In addition, some reported cases of illness came from people who did not live in the Walkerton water district. Finally, the local health inspector had no reason to believe that there were problems with the way the Walkerton water system was operated.

An important event related to the accident occurred in 1996, when the government water testing laboratories were privatized. Previously, water samples were sent to government laboratories for testing. These labs then shared the results with the appropriate government agencies as well as the local operators. Upon receiving adverse water quality reports from the government testing labs or the MOE, the local public health inspector in Walkerton would contact the WPUC to ensure that follow-up samples were taken and chlorine residuals maintained.

After water testing laboratory services for municipalities were assumed by the private sector in 1996, the MOH health unit for the Walkerton area sought assurances from the MOE's local office that it would continue to be notified of all adverse water quality results relating to community water systems. It received that assurance, both in correspondence and at a meeting, but it did not receive adverse water test reports. Without feedback about any problems in the water system, the local public health authorities assumed everything was fine.

In fact, there *were* warnings of problems. Between January and April of 2000 (the months just prior to the May *E. coli* outbreak), the lab that tested Walkerton's water repeatedly detected coliform bacteria—an indication that surface water was getting into the water supply. The lab notified the MOE on five separate occasions. The MOE in turn phoned the WPUC, was assured the problems were being fixed, and let it go at that. The MOE did not inform the local Walkerton Medical Office of Health, however, as by law it was required to do.

The WPUC changed water-testing laboratories in May 2000. The new laboratory, A&L Canada Laboratories East, was unaware of any notification guidelines. In fact, they considered test results to be confidential and thus improper to send to anyone but the client (in this case, the WPUC manager Stan Koebel).

In 1998, the BGOS health unit did receive a report on an MOE inspection of the Walkerton water system that showed some serious problems did exist. When the local Walkerton public health inspector read the report, he filed it, assuming that the MOE would ensure that the problems identified were properly addressed. Note the coordination problems here in an area of overlapping control. Both the MOE and the local public health inspector should have followed up on the 1998 inspection report, but there was no written protocol instructing the public health inspector on how to respond to adverse water quality or water system inspection reports. The MOE also lacked such protocols. Once again, the local public health authorities received no feedback that indicated water system operations were problematic.

Looking only at the physical system and local operations, it appears that the accident was simply the result of incompetent water system operators, who initially lied to protect their jobs (but who were unaware of the potentially fatal consequences of their lies) made worse by an inadequate response by the local health unit. If the goal is to find someone to blame, this conclusion is reasonable. If, however, the goal is to understand why the accident occurred in order to make effective changes (beyond simply firing the Koebel brothers) in order to prevent repetitions in the future or to learn how to prevent accidents in other situations, then a more complete study of the larger water safety control structure within which the local operations is embedded is necessary.

C.5   Municipal Government

Figure C.4 summarizes the flaws in the municipal water system control structure that allowed the dysfunctional interactions and thus the accident to occur.

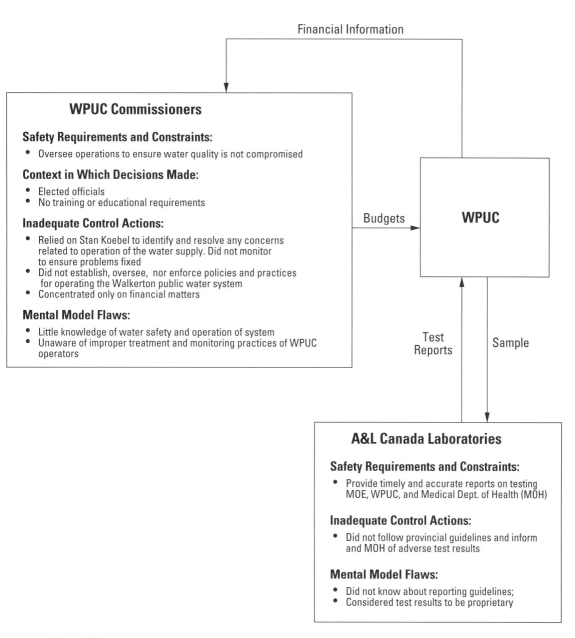

Figure C.4
The municipal control structure and its contribution to the accident.

Operating conditions on the public water system should theoretically have been imposed by the municipality, the WPUC commissioners, and the WPUC manager. The municipality left the operation of the water system to the WPUC. The commissioners, who were elected, over the years became more focused on the finances of the PUC than the operations. They had little or no training or knowledge of water system operations or even water quality itself. Without such knowledge and with their focus on financial issues, they gave all responsibility for operations to the manager of the WPUC (Stan Koebel) and provided no other operational oversight.

The WPUC commissioners received a copy of the 1998 inspection report but did nothing beyond asking for an explanation from Stan Koebel and accepting his word that he would correct the deficient practices. They never followed up to make sure he did. The mayor of Walkerton and the municipality also received the report but they assumed the WPUC would take care of the problems.

C.6   Provincial Regulatory Agencies (Ministries)

The MOE has primary responsibility for regulating and for enforcing legislation, regulations, and policies that apply to the construction and operation of municipal water systems. Guidelines and objectives are set by the MOE, based on federal guidelines that are enforceable through certificates of approval issued to public water utility operators.

Walkerton Well 5 was built in 1978 and issued a certificate of approval by the MOE in 1979. Despite potential problems—the groundwater supplying the well was recognized as being vulnerable to surface contamination—no explicit operating conditions were imposed at the time.

Although the original certificate of approval for Well 5 did not include any special operating conditions, over time MOE practices changed. By 1992, the MOE had developed a set of model operating conditions for water treatment and monitoring that were routinely attached to new certificates of approval for municipal water systems. There was no effort, however, to determine whether such conditions should be attached to existing certificates, such as the one for Well 5.

The provincial water quality guidelines were amended in 1994 to require the continuous monitoring of chlorine residuals and turbidity for wells supplied by a groundwater source that was under the direct influence of surface water (as was Walkerton's Well 5). Automatic monitoring and shutoff valves would have mitigated the operational problems at Walkerton and prevented the deaths and illness associated with the *E. coli* contamination in May 2000 if the requirement had been enforced in existing wells. However, at the time, there was no program or policy to review existing wells to determine whether they met the requirements for continuous monitoring. In addition, MOE inspectors were not directed to notify well

Ministry of the Environment

Safety Requirements and Constraints:

- Ensure those in charge of water supplies are competent to carry out their responsibilities.
- Perform inspections and enforce compliance if problems found.
- Perform hazard analyses to provide information about where vulnerabilities are and monitor them.
- Perform continual risk evaluation of existing facilities and establish new controls if necessary.
- Establish criteria for determining whether a well is at risk.
- Establish feedback channels for adverse test results. Provide multiple paths so that dysfunctional paths cannot prevent reporting.
- Enforce legislation, regulations, and policies applying to construction and operation of municipal water systems.
- Establish certification and training requirements for water system operators.

Context in Which Decisions Made:

- Critical information about history of known vulnerable water sources not easily accessible.
- Budget cuts and staff reductions.

Inadequate Control Actions:

- No legally enforceable measures taken to ensure that concerns identified in inspections are addressed. Weak response to repeated violations uncovered in periodic inspections.
- Relied on voluntary compliance with regulations and guidelines.
- No systematic review of existing certificates of approval to determine if conditions should be added for continuous monitoring.
- Did not retroactively apply new approvals program to older facilities when procedures changed in 1992.
- Did not require continuous monitoring of existing facilities when ODWO amended in 1994.
- MOE inspectors not directed to assess existing wells during inspections.
- MOE inspectors not provided with criteria for determining whether a given well was at risk. Not directed to examine daily operating sheets.
- Inadequate inspections and improperly structured and administered inspection program.
- Approval of Well 5 without attaching operating conditions or special monitoring or inspection requirements.
- No followup on inspection reports noting serious deficiencies.
- Did not inform Walkerton Medical Officer of Health about adverse test results in January to April 2000 as required to do.
- Private labs not informed about reporting guidelines.
- No certification or training requirements for grandfathered operators.
- No enforcement of continuing training requirements.
- Inadequate training of MOE personnel.

Mental Model Flaws:

- Incorrect model of state of compliance with water quality regulations and guidelines.
- Several local MOE personnel did not know *E. coli* could be fatal.

Feedback:

- Did not monitor effects of privatization on reporting of adverse test results.
- Inadequate feedback about state of water quality and water test results.

Coordination:

- Neither MOE nor MOH took responsibility for enacting notification legislation.

Ministry of Health

Safety Requirements and Constraints:

- Ensure adequate procedures exist for notification and risk abatement if water quality is compromised.

Inadequate Control Actions:

- No written protocol provided to local public health inspector on how to respond to adverse water quality or inspection reports.

Coordination:

- Neither MOE nor MOH took responsibility for enacting notification legislation.

Figure C.5
The role of the ministries in the accident.

operators (like the Koebel brothers) of the new requirement or to assess during inspections if a well required continuous monitoring.

Stan and Frank Koebel lacked the training and expertise to identify the vulnerability of Well 5 themselves and to understand the resulting need for continuous chlorine residual and turbidity monitors. After the introduction of mandatory certification in 1993, the Koebel brothers were certified on the basis of experience even though they did not meet the certification requirements. The new rules also required forty hours of training a year for each certified operator. Stan and Frank Koebel did not take the required amount of training, and the training they did take did not adequately address drinking water safety. The MOE did not enforce the training requirements and did not focus the training on drinking water safety.

The Koebel brothers and the Walkerton commissioners were not the only ones with inadequate training and knowledge of drinking water safety. Evidence at the inquiry showed that several environmental officers in the MOE's local office were unaware that *E. coli* was potentially lethal and their mental models were also incorrect with respect to other matters essential to water safety.

At the time of the privatization of the government water testing laboratories in 1996, the MOE sent a guidance document to those municipalities that requested it. The document strongly recommended that a municipality include in any contract with a private lab a clause specifying that the laboratory directly notify the MOE and the local medical officer of health about adverse test results. There is no evidence that the Walkerton PUC either requested or received this document. The MOE had no mechanism for informing private laboratories of the existing guidelines for reporting adverse results to the MOE and the MOH.

In 1997, the MOH took the unusual step of writing to the MOE requesting that legislation be amended to ensure that the proper authorities would be notified of adverse water test results. The MOE declined to propose legislation, indicating that the existing guidelines dealt with the issue. On several occasions, officials in the MOH and the MOE expressed concerns about failures to report adverse test results to local medical officers of health in accordance with the protocol. But the antiregulatory culture and the existence of the Red Tape Commission discouraged any proposals to make notification legally binding on the operators or municipal water systems and private labs.

Another important impact of the 1996 law was a reduction in the MOE water system inspection program. The cutbacks at the MOE negatively impacted the number of inspections, although the inspection program had other deficiencies as well.

The MOE inspected the Walkerton water system in 1991, 1995, and 1998. At the time of the inspections, problems existed relating to water safety. Inspectors identified some of them, but unfortunately two of the most significant problems—the vulnerability of Well 5 to surface contamination and the improper chlorination and monitoring practices of the WPUC—were not detected. Information about the vulnerability of Well 5 was available in MOE files, but inspectors were not directed to look at relevant information about the security of water sources and the archived information was not easy to find. Information about the second problem, improper chlorination and monitoring practices of the WPUC, was there to be seen in the operating records maintained by the WPUC. The Walkerton inquiry report concludes that a proper examination of the daily operating sheets would have disclosed the problem. However, the inspectors were not instructed to carry out a thorough review of operating records.

The 1998 inspection report did show there had been problems with the water supply for years: detection of *E. coli* in treated water with increasing frequency, chlorine residuals in treated water at less than the required 0.5 mg/L, noncompliance with minimum bacteriological sampling requirements, and not maintaining proper training records.

The MOE outlined improvements that should be made, but desperately short of inspection staff and faced with small water systems across the province that were not meeting standards, it never scheduled a follow-up inspection to see if the improvements were in fact being carried out. The Walkerton inquiry report suggests that the use of guidelines rather than regulations had an impact here. The report states that had the WPUC been found to be in noncompliance with a legally enforceable regulation, as opposed to a guideline, it is more likely that the MOE would have taken stronger measures to ensure compliance—such as the use of further inspections, the issuance of a director's order (which would have required the WPUC to comply with the requirements for treatment and monitoring), or enforcement proceedings. The lack of any follow-up or enforcement efforts may have led the Koebel brothers to believe the recommendations were not very important, even to the MOE.

Between January and April of 2000 (the months just prior to the May *E. coli* outbreak), the lab that tested Walkerton's water repeatedly detected coliform bacteria—an indication that surface water was getting into the water supply. The lab notified the MOE on five separate occasions. The MOE in turn phoned the WPUC, was assured the problems were being fixed, and let it go at that. The MOE failed to inform the medical officer of health, as by law it was required to do.

Looking at the role of this hierarchical level in the Ontario water quality control system provides greater understanding of the reasons for the Walkerton accident and suggests more corrective actions that might be taken to prevent future accidents. But examining the control flaws at this level is not enough to understand completely the actions or lack of actions of the MOE. A larger view of the provincial government role in the tragedy is necessary.

C.7 Provincial Government

The last component in the Ontario water quality control structure is the provincial government. Figure C.6 summarizes its role in the accident.

All of the weaknesses in the water system operations at Walkerton (and other municipalities) might have been mitigated if the source of contamination of the water had been controlled. A weakness in the basic Ontario water control structure was the lack of a government watershed and land use policy for agricultural activities that can impact drinking water sources. In fact, at a meeting of the Walkerton town council in November 1978 (when Well 5 was constructed), MOE representatives suggested land use controls for the area around Well 5, but the municipality did not have the legal means to enforce such land use regulations because the government of Ontario had not provided the legal basis for such controls.

Provincial Government

Safety Requirements and Constraints:

- Establish regulatory bodies and codes of responsibilities, authority, and accountability for the province.
- Provide adequate resources to regulatory bodies to carry out their responsibilities.
- Provide oversight and feedback loops to ensure that provincial regulatory bodies are doing their jobs adequately.
- Ensure adequate risk assessment is conducted and effective risk management plan is in place.
- Enact legislation to protect water quality.

Context in Which Decisions Made:

- Antiregulatory culture.
- Efforts to reduce red tape.

Inadequate Control Actions:

- No risk assessment or risk management plan created to determine extent of known risks, whether risks should be assumed, and if assumed, whether they could be managed.
- Privatized laboratory testing of drinking water without requiring labs to notify MOE and health authorities of adverse test results (privatizing without establishing adequate governmental oversight).
- Relied on guidelines rather than legally enforceable regulations.
- No regulatory requirements for agricultural activities that create impacts on drinking water sources.
- Spreading of manure exempted from EPA requirements for Certificates of Approval.
- Water Sewage Services Improvement Act ended provincial Drinking Water Surveillance program.
- No accreditation of water testing labs (no criteria established to govern quality of testing personnel, no provisions for licensing, inspection, or auditing by government).
- Disbanded ACES.
- Ignored warnings about deteriorating water quality.
- No law to legislate requirements for drinking water standards, reporting requirements, and infrastructure funding.
- Environmental controls systematically removed or negated.

Feedback:

- No monitoring or feedback channels established to evaluate impact of changes.

Figure C.6
The role of the provincial government in the accident.

At the same time as the increase in factory farms was overwhelming the ability of the natural filtration process to prevent the contamination of the local water systems, the spreading of manure had been granted a long-standing exemption from EPA requirements. Annual reports of the Environment Commissioner of Ontario for the four years before the Walkerton accident included recommendations that the government create a groundwater strategy. A Health Canada study stated that the cattle counties of southwestern Ontario, where Walkerton is located, are high-risk areas for *E. coli* infections. The report pointed out the direct link between cattle density and E. coli infection, and showed that 32 percent of the wells in rural Ontario showed fecal contamination. Dr. Murray McQuigge, the medical officer of health for the BGOS health unit (and the man who handled the Walkerton *E. coli* outbreak) warned in a memo to local authorities that "poor nutrient management on farms is leading to a degradation of the quality of ground water, streams, and lakes." Nothing was done in response.

With the election of a conservative provincial government in 1995, a bias against environmental regulation and red tape led to the elimination of many of the government controls over drinking water quality. A Red Tape Commission was established by the provincial government to minimize reporting and other requirements on government and private industry. At the same time, the government disbanded groups like the Advisory Committee on Environmental Standards (ACES), which reviewed ministry standards, including those related to water quality. At the time of the Walkerton contamination, there was no opportunity for stakeholder or public review of the Ontario clean water controls.

Budget and staff reductions by the conservative government took a major toll on environmental programs and agencies (although budget reductions had started before the election of the new provincial government). The MOE budget was reduced by 42 percent and 900 of the 2,400 staff responsible for monitoring, testing, inspection, and enforcement of environmental regulations were laid off. The official Walkerton inquiry report concludes that the reductions were not based on an assessment of the requirements to carry out the MOE's statutory requirements, or on any risk assessment of the potential impact on the environment or, in particular, on water quality. After the reductions, the provincial ombudsman issued a report saying that cutbacks had been so damaging that the government was no longer capable of providing the services that it was mandated to provide. The report was ignored.

In 1996, the Water Sewage Services Improvement Act was passed, which shut down the government water testing laboratories, downloaded control of provincially owned water and sewage plants to the municipalities, eliminated funding for municipal water utilities, and ended the provincial Drinking Water Surveillance Program, under which the MOE had monitored drinking water across the province.

The provincial water quality guidelines directed testing labs to report any indications of unsafe water quality to the MOE and to the local medical officer of health. The latter would then decide whether to issue a boil water advisory. When government labs conducted all of the routine drinking water tests for municipal water systems throughout the province, it was acceptable to keep the notification protocol in the form of a guideline rather than a legally enforceable law or regulation. However, the privatization of water testing and the exit of government labs from this duty in 1996 made the use of guidelines ineffective in ensuring necessary reporting would occur. At the time, the government did not regulate private environmental labs. No criteria were established to govern the quality of testing or the qualifications or experience of private lab personnel, and no provisions were made for licensing, inspection, or auditing of private labs by the government. In addition, the government did not implement any program to monitor the effect of privatization on the notification procedures followed whenever adverse test results were found.

In 1997, the MOH took the unusual step of writing to the Minister of the Environment requesting that legislation be amended to ensure that the proper authorities would be notified of adverse water test results. The Minister of the Environment declined to propose legislation, indicating that the Provincial water quality guidelines dealt with the issue. On several occasions, officials in the MOH and the MOE expressed concerns about failures to report adverse test results to local Medical Officers of Health in accordance with the protocol. But the anti-regulatory culture and the existence of the Red Tape Commission discouraged any proposals to make notification legally binding on the operators or municipal water systems and private labs.

A final important change in the safety control structure involved the drinking water surveillance program in which the MOE monitored drinking water across the province. In 1996, the provincial government dropped *E. coli* testing from its Drinking Water Surveillance Program. The next year, the program was shut down entirely. At the same time, the provincial government directed MOE staff not to enforce dozens of environmental laws and regulations still on the books. Farm operators, in particular, were to be treated with understanding if they were discovered to be in violation of livestock and wastewater regulations. By June 1998, the Walkerton town council was concerned enough about the situation to send a letter directly to Premier Mike Harris appealing for the province to resume testing of municipal water. There was no reply.

MOE officials warned the government that closing the water-testing program would endanger public health. Their concerns were dismissed. In 1997, senior MOE officials drafted another memo that the government *did* heed [55]. This memo warned that cutbacks had impaired the ministry's ability to enforce environmental regulations to the point that the MOE could be exposed to lawsuits for negligence if and when an environmental accident occurred. In response, the provincial government called a meeting of the ministry staff to discuss how to protect itself from liability, and it passed a bill (the Environmental Approvals Improvement Act) that, among other things, prohibited legal action against the government by anyone adversely affected by the Environment Minister's failure to apply environmental regulations and guidelines.

Many other groups warned senior government officials, ministers, and the Cabinet of the danger of what it was doing, such as reducing inspections and not making the notification guidelines into regulations. The warnings were ignored. Environmental groups prepared briefs. The Provincial Auditor, in his annual reports, criticized the MOE for deficient monitoring of groundwater resources and for failing to audit small water plants across the province. The International Joint Commission expressed its concerns about Ontario's neglect of water quality issues, and the Environmental Commissioner of Ontario warned that the government was compromising environmental protection, pointing specifically to the testing of drinking water as an area of concern.

In January 2000, three months before the Walkerton accident, staff at the MOE's Water Policy Branch submitted a report to the provincial government, warning, "Not monitoring drinking water quality is a serious concern for the Ministry in view of its mandate to protect public health." The report stated that a number of smaller municipalities were not up to the job of monitoring the quality of their drinking water. It further warned that because of the privatization of the testing labs, there was no longer a mechanism to ensure that the MOE and the local medical officer of health were informed if problems were detected in local water systems. The provincial government ignored the report.

The warnings were not limited to groups or individuals. Many adverse water quality reports had been received from Walkerton between 1995 and 1998. During the mid- to late 1990s, there were clear indications that the water quality was deteriorating. In 1996, for example, hundreds of people in Collingswood, a town near Walkerton, became ill after cryptosporidium (a parasite linked to animal feces) contaminated the drinking water. Nobody died, but it should have acted as a warning that the water safety control structure had degraded.

The Walkerton inquiry report notes that the decisions to remove the water safety controls in Ontario or to reduce their enforcement were taken without an assessment of the risks or the preparation of a risk-management plan. The report says there was evidence that those at the most senior levels of government who were responsible for the decisions considered the risks to be manageable, but there was no evidence that the specific risks were properly assessed or addressed.

Up to this point, the Walkerton accident has been viewed in terms of inadequate control and enforcement of safety constraints. But systems are not static. The next section describes the dynamic aspects of the accident.

C.8   The Structural Dynamics

Most hazard analysis and other safety-engineering techniques treat systems and their environments as a static design. But systems are never static: They are continually adapting and changing to achieve their ends and to react to changes within themselves, in their goals, and in their environment. The original design must not only enforce appropriate constraints on behavior to ensure safe operation, but it must continue to operate safely as changes and adaptations occur over time. Accidents in a systems-theoretic framework are viewed as the result of flawed processes and control structures that evolve over time.

The public water safety control structure in Ontario started out with some weaknesses, which were mitigated by the presence of other controls. In some cases, the control over hazards was improved over time, for example, by the introduction of operator certification requirements and by requirements added in 1994 for continuous monitoring of chlorine residuals and turbidity in wells directly influenced by surface water. While these improvements were helpful for new wells, the lack of a policy to apply them to the existing wells and existing operators left serious weaknesses in the overall public health structure.

At the same time, other actions, such as the reduction in inspections and the elimination of the surveillance program reduced the feedback to the MOE and the MOH about the state of the system components. The water-testing laboratory privatization by itself did not degrade safety; it was the way the privatization was implemented, that is, without mandatory requirements for the private testing labs to inform the government agencies about adverse test results and without informing the private labs about the guidelines for this notification. Without regulations or oversight or enforcement of safe operating conditions, and with inadequate mental models of the safety requirements, operating practices have a tendency to change over time in order to optimize a variety of goals that conflict with safety, in this case, cutting budgets, reducing government, and reducing red tape.

An example of asynchronous evolution of the control structure is the assumption by the municipal government (mayor and city council) that appropriate oversight of the public water system operations was being done by the WPUC commissioners. This assumption was true for the early operations. But the elected commissioners over time became more interested in budgets and less expert in water system operation until they were not able to provide the necessary oversight. The municipal government, not understanding the changes, did not make an appropriate response.

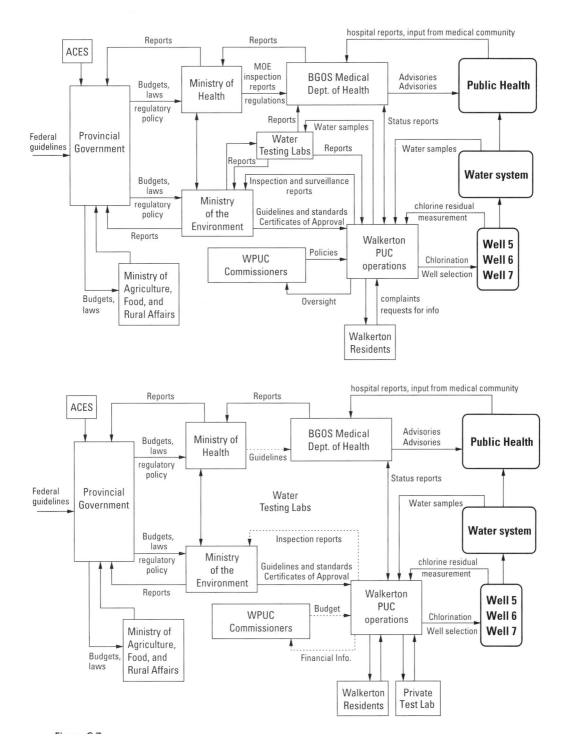

Figure C.7
The theoretical water safety control structure (top) and the structure existing at the time of the accident (bottom). Note the elimination of many feedback loops.

Changes may also involve the safety control structure environment. The lack of a provincial watershed protection plan was compensated for by the MOE ensuring that the water systems could not be affected by such runoff. The original Walkerton design satisfied this safety constraint. But factory farms and farming operations increased dramatically and the production of animal waste overwhelmed the existing design safeguards. The environment had changed, but the existing controls were not revisited to determine whether they were still adequate.

All of these changes in the Ontario water safety control structure over time led to the modified control structure shown in figure C.7. Dotted lines represent communication, control or feedback channels that still existed but had become ineffective. One thing to notice in comparing the original structure at the top and the one at the bottom is the disappearance of many of the feedback loops.

C.9 Addendum to the Walkerton Accident Analysis

Government representatives argued during the investigation that the accident cause was simply the actions of the Koebel brothers and that government actions or inactions were irrelevant. The Walkerton inquiry report rejected this viewpoint. Instead, the report included recommendations to establish regulatory requirements for agricultural activities with potential impacts on drinking water sources, updating of standards and technology, improving current practices in setting standards, establishing legally enforceable regulations rather than guidelines, requiring mandatory training for all water system operators and requiring grandfathered operators to pass certification examinations within two years, developing a curriculum for operator training and mandatory training requirements specifically emphasizing water quality and safety issues, adopting a province-wide drinking water policy and a Safe Drinking Water Act, strictly enforcing drinking water regulations, and committing sufficient resources (financial and otherwise) to enable the MOE to play their role effectively. By 2003, most of these recommendations had not been implemented, but three years after the accident, the Koebel brothers were arrested for their part in the events. Water contamination incidents continued to occur in small towns in Ontario.

D   A Brief Introduction to System Dynamics Modeling

By focusing on the events immediately preceding accidents, event chains treat a system as a static, unchanging structure. But systems and organizations continually experience change and adapt to existing conditions. Systems dynamics models are one way to illustrate and model the dynamic change in systems. They have been primarily used to examine the potential undesired consequences of organizational decision making [194].

As noted in part I of this book, a system's defenses or safety controls may degrade over time because of changes in the behavior of the components of the safety control loop. The reasons for the migration of the system toward a state of higher risk will be system-specific and can be quite complex. In contrast to the usually simple and direct relationships represented in event-chain accident models, most accidents in complex sociotechnical systems involve relationships between events and human actions that are highly nonlinear, involving multiple feedback loops. The prevention of accidents in these systems therefore requires an understanding not only of the static structure of the system (the *structural complexity*) and of the changes to this structure over time (the *structural dynamics*), but also the dynamics behind these changes (the *behavioral dynamics*). System dynamics provides a way to model and understand the dynamic processes behind the changes to the static safety control structure: how and why the safety control structure might change over time, potentially leading to ineffective controls and unsafe or hazardous states.

The field of system dynamics, created at MIT in the 1950s by Jay Forrester, is designed to help decision makers learn about the structure and dynamics of complex systems, to design high leverage policies for sustained improvement, and to catalyze successful implementation and change. System dynamics provides a framework for dealing with dynamic complexity, where cause and effect are not obviously related. It is grounded in the theory of nonlinear dynamics and feedback control, but it also draws on cognitive and social psychology, organization theory, economics, and other social sciences [194]. System dynamics models are formal and can be executed. The

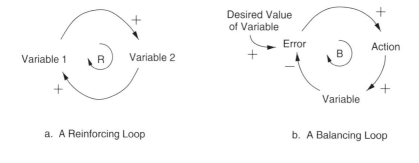

a. A Reinforcing Loop   b. A Balancing Loop

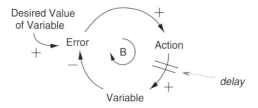

c. A Balancing Loop with a Delay

Figure D.1
The three basic components of system dynamics models.

models and simulators help to capture complex dynamics and to create an environment for organizational learning and policy design.

System dynamics is particularly relevant in safety engineering when analyzing the organizational aspects of accidents and using STPA on the higher levels of the safety control structure. The world is dynamic, evolving, and interconnected, but we tend to make decisions using mental models that are static, narrow, and reductionist. Thus decisions that might appear to have no effect on safety—or even appear to be beneficial—may in fact degrade safety and increase risk. System dynamics modeling assists in understanding and predicting instances of policy resistance or the tendency for well-intentioned interventions to be defeated by the response of the system to the intervention itself.

System behavior in system dynamics is modeled by using feedback (causal) loops, stock and flows (levels and rates), and the nonlinearities created by interactions between system components. In this view of the world, behavior over time (the dynamics of the system) can be explained by the interaction of positive and negative feedback loops [185]. The models are constructed from three basic building blocks:

positive feedback or reinforcing loops, negative feedback or balancing loops, and delays. Positive loops (called reinforcing loops) are self-reinforcing, while negative loops tend to counteract change. Delays introduce potential instability into the system.

Figure D.1a shows a *reinforcing loop*, which is a structure that feeds on itself to produce growth or decline. Reinforcing loops correspond to positive feedback loops in control theory. An increase in variable 1 leads to an increase in variable 2 (as indicated by the "+" sign), which leads to an increase in variable 1, and so on. The "+" does not mean that the values necessarily increase, only that variable 1 and variable 2 will change in the same direction. If variable 1 decreases, then variable 2 will decrease. A "−" indicates that the values change in opposite directions. In the absence of external influences, both variable 1 and variable 2 will clearly grow or decline exponentially.

Reinforcing loops generate growth, amplify deviations, and reinforce change [194].

A *balancing loop* (figure D.1b) is a structure that changes the current value of a system variable or a desired or reference variable through some action. It corresponds to a negative feedback loop in control theory. The difference between the current value and the desired value is perceived as an error. An action proportional to the error is taken to decrease the error so that, over time, the current value approaches the desired value.

The third basic element is a *delay*, which is used to model the time that elapses between cause and effect. A delay is indicated by a double line as shown in figure D.1c. Delays make it difficult to link cause and effect (dynamic complexity) and may result in unstable system behavior. For example, in steering a ship there is a delay between a change in the rudder position and a corresponding course change, often leading to overcorrection and instability.

References

1. Ackoff, Russell L. (July 1971). Towards a system of systems concepts. *Management Science* 17 (11):661–671.

2. Aeronautica Civil of the Republic of Colombia. AA965 Cali Accident Report. September 1996.

3. Air Force Space Division. *System Safety Handbook for the Acquisition Manager.* SDP 127-1, January 12, 1987.

4. Aircraft Accident Investigation Commission. Aircraft Accident Investigation Report 96–5. Ministry of Transport, Japan, 1996.

5. James G. Andrus. Aircraft Accident Investigation Board Report: U.S. Army UH-60 Black Hawk Helicopters 87-26000 and 88-26060. Department of Defense, July 13, 1994.

6. Angell, Marcia. 2005. *The Truth about the Drug Companies: How They Deceive Us and What to Do about It.* New York: Random House.

7. Anonymous. American Airlines only 75% responsible for 1995 Cali crash. Airline Industry Information, June 15, 2000.

8. Anonymous. USS Scorpion (SSN-589). Wikipedia.

9. Arnold, Richard. *A Qualitative Comparative Analysis of STAMP and SOAM in ATM Occurrence Investigation.* Master's thesis, Lund University, Sweden, June 1990.

10. Ashby, W. R. 1956. *An Introduction to Cybernetics.* London: Chapman and Hall.

11. Ashby, W. R. 1962. Principles of the self-organizing system. In *Principles of Self-Organization*, ed. H. Von Foerster and G. W. Zopf, 255–278. Pergamon.

12. Ayres, Robert U., and Pradeep K. Rohatgi. 1987. Bhopal: Lessons for technological decision-makers. *Technology in Society* 9:19–45.

13. Associated Press. Cali crash case overturned. CBS News, June 16, 1999 (http://www.csbnews.com/stories/1999/06/16/world/main51166.shtml).

14. Bainbridge, Lisanne. 1987. Ironies of automation. In *New Technology and Human Error*, ed. Jens Rasmussen, Keith Duncan, and Jacques Leplat, 271–283. New York: John Wiley & Sons.

15. Bachelder, Edward, and Nancy Leveson. Describing and probing complex system behavior: A graphical approach. *Aviation Safety Conference*, Society of Automotive Engineers, Seattle, September 2001.

16. Baciu, Alina, Kathleen R. Stratton, and Sheila P. Burke. 2007. *The Future of Drug Safety: Promoting and Protecting the Health of the Public, Institute of Medicine.* Washington, D.C.: National Academies Press.

17. Baker, James A. (Chair). The Report of the BP U.S. Refineries Independent Safety Review Panel. January 2007.

18. Barstow, David, Laura Dood, James Glanz, Stephanie Saul, and Ian Urbina. Regulators failed to address risk in oil rig fail-safe device. *New York Times*, New York Edition, June 21, 2010, Page Al.

19. Benner, Ludwig, Jr., Accident investigations: Multilinear events sequencing methods. (June 1975). *Journal of Safety Research* 7 (2):67–73.

20. Bernstein, D. A., and P. W. Nash. 2005. *Essentials of Psychology*. Boston: Houghton Mifflin.

21. Bertalanffy, Ludwig. 1969. *General Systems Theory: Foundations*. New York: Braziller.

22. Billings, Charles. 1996. *Aviation Automation: The Search for a Human-Centered Approach*. New York: CRC Press.

23. Bogard, William. 1989. *The Bhopal Tragedy*. Boulder, Colo.: Westview Press.

24. Booten, Richard C., Jr., and Simon Ramo. (July 1984). The development of systems engineering. *IEEE Transactions on Aerospace and Electronic Systems* AES-20 (4):306–309.

25. Brehmer, B. 1992. Dynamic decision making: Human control of complex systems. *Acta Psychologica* 81:211–241.

26. Brookes, Malcolm J. 1982. Human factors in the design and operation of reactor safety systems. In *Accident at Three Mile Island: The Human Dimensions*, ed. David L. Sills, C. P. Wolf, and Vivien B. Shelanski, 155–160. Boulder, Colo.: Westview Press.

27. Brown, Robbie, and Griffin Palmer. Workers on doomed rig voiced concern about safety. *New York Times*, Page Al, July 22, 2010.

28. Bundesstelle für Flugunfalluntersuchung. Investigation Report. German Federal Bureau of Aircraft Accidents Investigation, May 2004.

29. Cameron, R., and A. J. Millard. 1985. *Technology Assessment: A Historical Approach*. Dubuque, IA: Kendall/Hunt.

30. Cantrell, Rear Admiral Walt. (Ret). Personal communication.

31. Carrigan, Geoff, Dave Long, M. L. Cummings, and John Duffer. Human factors analysis of Predator B crash. *Proceedings of AUVSI: Unmanned Systems North America*, San Diego, CA 2008.

32. Carroll, J. S. 1995. Incident reviews in high-hazard industries: Sensemaking and learning under ambiguity and accountability. *Industrial and Environmental Crisis Quarterly* 9:175–197.

33. Carroll, J. S. (November 1998). Organizational learning activities in high-hazard industries: The logics underlying self-analysis. *Journal of Management Studies* 35 (6):699–717.

34. Carroll, John, and Sachi Hatakenaka. Driving organizational change in the midst of crisis. *MIT Sloan Management Review* 42:70–79.

35. Carroll, J. M., and J. R. Olson. 1988. Mental models in human-computer interaction. In *Handbook of Human-Computer Interaction*, ed. M. Helander, 45–65. Amsterdam: Elsevier Science Publishers.

36. Checkland, Peter. 1981. *Systems Thinking, Systems Practice*. New York: John Wiley & Sons.

37. Childs, Charles W. Cosmetic system safety. *Hazard Prevention*, May/June 1979.

38. Chisti, Agnees. 1986. *Dateline Bhopal*. New Delhi: Concept.

39. Conant, R. C., and W. R. Ashby. 1970. Every good regulator of a system must be a model of that system. *International Journal of Systems Science* 1:89–97.

40. Cook, Richard I. Verite, abstraction, and ordinateur systems in the evolution of complex process control. *3rd Annual Symposium on Human Interaction with Complex Systems (HICS '96)*, Dayton Ohio, August 1996.

41. Cook, R. I., S. S. Potter, D. D. Woods, and J. M. McDonald. 1991. Evaluating the human engineering of microprocessor-controlled operating room devices. *Journal of Clinical Monitoring* 7:217–226.

42. Council for Science and Society. 1977. *The Acceptability of Risks (The Logic and Social Dynamics of Fair Decisions and Effective Controls)*. Chichester, UK: Barry Rose Publishers Ltd.

43. Couturier, Matthieu. *A Case Study of Vioxx Using STAMP*. Master's thesis, Technology and Policy Program, Engineering Systems Division, MIT, June 2010.

44. Couturier, Matthieu, Nancy Leveson, Stan Finkelstein, John Thomas, John Carroll, David Weirz, Bruce Psaty, and Meghan Dierks. 2010. Analyzing the Efficacy of Regulatory Reforms after Vioxx Using System Engineering, MIT Technical Report. Engineering Systems Division.

45. Cox, Lauren, and Joseph Brownstein. Aussie civil suit uncovers fake medical journals. ABC News Medical Unit, May 14, 2009.

46. Cutcher-Gershenfeld, Joel. Personal communication.

47. Daouk, Mirna. *A Human-Centered Approach to Developing Safe Systems*. Master's thesis, Aeronautics and Astronautics, MIT, Dec. 2001.

48. Daouk, Mirna, and Nancy Leveson. An approach to human-centered design. *International Workshop on Humana Error, Safety, and System Design (HESSD '01)*, Linchoping, Sweden, June 2001.

49. Dekker, Sidney. 2004. *Ten Questions about Human Error*. New York: CRC Press.

50. Dekker, Sidney. 2006. *The Field Guide to Understanding Human Error*. London: Ashgate.

51. Dekker, Sidney. 2007. *Just Culture: Balancing Safety and Accountability*. London: Ashgate.

52. Dekker, Sidney. *Report on the Flight Crew Human Factors Investigation Conducted for the Dutch Safety Board into the Accident of TK1951, Boeing 737–800 near Amsterdam Schiphol Airport, February 25, 2009*. Lund University, Sweden 2009.

53. Department of Defense. *MIL-STD-882D: Standard Practice for System Safety*. U.S. Department of Defense, January 2000.

54. Department of Employment. 1975. *The Flixborough Disaster: Report of the Court of Inquiry*. London: Her Majesty's Stationery Office.

55. Diemer, Ulli. Contamination: The poisonous legacy of Ontario's environment cutbacks. *Canada Dimension Magazine*, July–August, 2000.

56. Dorner, D. 1987. On the difficulties people have in dealing with complexity. In *New Technology and Human Error*, ed. Jens Rasmussen, Keith Duncan, and Jacques Leplat, 97–109. New York: John Wiley & Sons.

57. Dowling, K., R. Bennett, M. Blackwell, T. Graham, S. Gatrall, R. O'Toole, and H. Schempf. A mobile robot system for ground servicing operations on the space shuttle. *Cooperative Intelligent Robots in Space*, SPIE, November, 1992.

58. Dulac, Nicolas. *Empirical Evaluation of Design Principles for Increasing Reviewability of Formal Requirements Specifications through Visualization*. Master's thesis, MIT, August 2003.

59. Dulac, Nicolas. Incorporating safety risk in early system architecture trade studies. *AIAA Journal of Spacecraft and Rockets* 46 (2) (Mar–Apr 2009).

60. Duncan, K. D. 1987. Reflections on fault diagnostic expertise. In *New Technology and Human Error*, ed. Jens Rasmussen, Keith Duncan, and Jacques Leplat, 261–269. New York: John Wiley & Sons.

61. Eddy, Paul, Elaine Potter, and Bruce Page. 1976. *Destination Disaster*. New York: Quadrangle/Times Books.

62. Edwards, M. 1981. The design of an accident investigation procedure. *Applied Ergonomics* 12:111–115.

63. Edwards, W. 1962. Dynamic decision theory and probabilistic information processing. *Human Factors* 4:59–73.

64. Ericson, Clif. Software and system safety. *5th Int. System Safety Conference*, Denver, July 1981.

65. Euler, E. E., S. D. Jolly, and H. H. Curtis. 2001. The failures of the Mars climate orbiter and Mars polar lander: A perspective from the people involved. *Guidance and Control*, American Astronautical Society, paper AAS 01-074.

66. Fielder, J. H. 2008. The Vioxx debacle revisited. *Engineering in Medicine and Biology Magazine* 27(4):106–109.

67. Finkelstein, Stan N., and Peter Temin. 2008. *Reasonable Rx: Solving the Drug Price Crisis*. New York: FT Press.

68. Fischoff, B., P. Slovic, and S. Lichtenstein. 1978. Fault trees: Sensitivity of estimated failure probabilities to problem representation. *Journal of Experimental Psychology: Human Perception and Performance* 4: 330–344.

69. Ford, Al. Personal communication.

70. Frola, F. R., and C. O. Miller. System safety in aircraft acquisition. Logistics Management Institute, Washington DC, January 1984.

71. Fujita, Y. What shapes operator performance? JAERI Human Factors Meeting, Tokyo, November 1991.

72. Fuller, J. G. (March 1984). Death by robot. *Omni* 6 (6):45–46, 97–102.

73. Government Accountability Office (GAO). 2006. *Drug Safety: Improvement Needed in FDA's Post-market Decision-making and Oversight Process.* Washington, DC: US Government Printing Office.

74. Gehman, Harold (Chair). Columbia accident investigation report. August 2003.

75. Gordon, Sallie E., and Richard T. Gill. 1997. Cognitive task analysis. In *Naturalistic Decision Making*, ed. Caroline E. Zsambok and Gary Klein, 131–140. Mahwah, NJ: Lawrence Erlbaum Associates.

76. Graham, David J. Testimony of David J. Graham, M.D. Senate 9, November 18, 2004.

77. Haddon, William, Jr. 1967. The prevention of accidents. In *Preventive Medicine*, ed. Duncan W. Clark and Brian MacMahon, 591–621. Boston: Little, Brown.

78. Haddon-Cave, Charles. (October 28, 2009). The Nimrod Review. HC 1025. London: Her Majesty's Stationery Office.

79. Hammer, Willie. 1980. *Product Safety Management and Engineering.* Englewood Cliffs, NJ: Prentice-Hall.

80. Harris, Gardiner. U.S. inaction lets look-alike tubes kill patients. *New York Times*, August 20, 2010.

81. Helicopter Accident Analysis Team. 1998. *Final Report.* NASA.

82. Hidden, Anthony. 1990. *Investigation into the Clapham Junction Railway Accident.* London: Her Majesty's Stationery Office.

83. Hill, K. P., J. S. Ross, D. S. Egilman, and H. M. Krumholz. 2009. The ADVANTAGE seeding trial: A review of internal documents. *Annals of Internal Medicine* 149:251–258.

84. Hopkins, Andrew. 1999. *Managing Major Hazards: The Lessons of the Moira Mine Disaster.* Sydney: Allen & Unwin.

85. Howard, Jeffrey. Preserving system safety across the boundary between system integrator and software contractor. *Conference of the Society of Automotive Engineers*, Paper 04AD-114, SAE, 2004.

86. Howard, Jeffrey, and Grady Lee. 2005. *SpecTRM-Tutorial.* Seattle: Safeware Engineering Corporation.

87. Ingerson, Ulf. Personal communication.

88. Ishimatsu, Takuto, Nancy Leveson, John Thomas, Masa Katahira, Yuko Miyamoto, and Haruka Nakao. Modeling and hazard analysis using STPA. *Conference of the International Association for the Advancement of Space Safety*, IAASS, Huntsville, May 2010.

89. Ito, Shuichiro Daniel. *Assuring Safety in High-Speed Magnetically Levitated (Maglev) Systems.* Master's thesis, MIT, May 2008.

90. Jaffe, M. S. *Completeness, Robustness, and Safety of Real-Time Requirements Specification.* Ph.D. Dissertation, University of California, Irvine, 1988.

91. Jaffe, M. S., N. G. Leveson, M. P. E. Heimdahl, and B. E. Melhart. (March 1991). Software requirements analysis for real-time process-control systems. *IEEE Transactions on Software Engineering* SE-17 (3):241–258.

92. Johannsen, G., J. E. Rijndorp, and H. Tamura. 1986. Matching user needs and technologies of displays and graphics. In *Analysis, Design, and Evaluation of Man–Machine Systems*, ed. G. Mancini, G. Johannsen, and L. Martensson, 51–61. New York: Pergamon Press.

93. Johnson, William G. 1980. *MORT Safety Assurance System.* New York: Marcel Dekker.

94. Joyce, Jeffrey. Personal communication.

95. JPL Special Review Board. Report on the loss of the Mars polar lander and deep space 2 missions. NASA Jet Propulsion Laboratory, 22 March 2000.

96. Juechter, J. S. Guarding: The keystone of system safety. *Proc. of the Fifth International Conference of the System Safety Society*, VB-1–VB-21, July 1981.

97. Kahneman, D., P. Slovic, and A. Tversky. 1982. *Judgment under Uncertainty: Heuristics and Biases.* New York: Cambridge University Press.

98. Kemeny, John G. 1979. *Report of the President's Commission on Three Mile Island (The Need for Change: The Legacy of TMI).* Washington, DC: U.S. Government Accounting Office.

99. Kemeny, John G. 1980. Saving American democracy: The lessons of Three Mile Island. *Technology Review* (June–July):65–75.

100. Kjellen, Urban. 1982. An evaluation of safety information systems at six medium-sized and large firms. *Journal of Occupational Accidents* 3:273–288.

101. Kjellen, Urban. 1987. Deviations and the feedback control of accidents. In *New Technology and Human Error*, ed. Jens Rasmussen, Keith Duncan, and Jacques Leplat, 143–156. New York: John Wiley & Sons.

102. Klein, Gary A., Judith Orasano, R. Calderwood, and Caroline E. Zsambok, eds. 1993. *Decision Making in Action: Models and Methods*. New York: Ablex Publishers.

103. Kletz, Trevor. Human problems with computer control. *Plant/Operations Progress* 1 (4), October 1982.

104. Koppel, Ross, Joshua Metlay, Abigail Cohen, Brian Abaluck, Russell Localio, Stephen Kimmel, and Brian Strom. (March 9, 2003). The role of computerized physical order entry systems in facilitating medication errors. *Journal of the American Medical Association* 293 (10):1197–1203.

105. Kraft, Christopher. Report of the Space Shuttle Management Independent Review. NASA, February 1995.

106. Ladd, John. Bhopal: An essay on moral responsibility and civic virtue. Department of Philosophy, Brown University, Rhode Island, January 1987.

107. La Porte, Todd R., and Paula Consolini. 1991. Working in practice but not in theory: Theoretical challenges of high-reliability organizations. *Journal of Public Administration: Research and Theory* 1:19–47.

108. Laracy, Joseph R. *A Systems-Theoretic Security Model for Large Scale, Complex Systems Applied to the U.S. Air Transportation System*. Master's thesis, Engineering Systems Division, MIT, 2007.

109. Lederer, Jerome. 1986. How far have we come? A look back at the leading edge of system safety eighteen years ago. *Hazard Prevention* (May/June):8–10.

110. Lees, Frank P. 1980. *Loss Prevention in the Process Industries, Vol. 1 and 2*. London: Butterworth.

111. Leplat, Jacques. 1987. Accidents and incidents production: Methods of analysis. In *New Technology and Human Error*, ed. Jens Rasmussen, Keith Duncan, and Jacques Leplat, 133–142. New York: John Wiley & Sons.

112. Leplat, Jacques. 1987. Occupational accident research and systems approach. In *New Technology and Human Error*, ed. Jens Rasmussen, Keith Duncan, and Jacques Leplat, 181–191. New York: John Wiley & Sons.

113. Leplat, Jacques. 1987. Some observations on error analysis. In *New Technology and Human Error*, ed. Jens Rasmussen, Keith Duncan, and Jacques Leplat, 311–316. New York: John Wiley & Sons.

114. Leveson, Nancy G. High-pressure steam engines and computer software. *IEEE Computer*, October 1994 (Keynote Address from IEEE/ACM International Conference on Software Engineering, 1992, Melbourne, Australia).

115. Leveson, Nancy G. 1995. *Safeware: System Safety and Computers*. Boston: Addison Wesley.

116. Leveson, Nancy G. The role of software in spacecraft accidents. *AIAA Journal of Spacecraft and Rockets* 41 (4) (July 2004).

117. Leveson, Nancy G. 2007. Technical and managerial factors in the NASA Challenger and Columbia losses: Looking forward to the future. In *Controversies in Science and Technology, Vol. 2: From Chromosomes to the Cosmos*, ed. D. L. Kleinman, K. Hansen, C. Matta, and J. Handelsman, 237–261. New Rochelle, NY: Mary Ann Liebert, Inc.

118. Leveson, Nancy G., Margaret Stringfellow, and John Thomas. Systems Approach to Accident Analysis. IT Technical Report, 2009.

119. Leveson, Nancy, and Kathryn Weiss. Making embedded software reuse practical and safe. *Foundations of Software Engineering*, Newport Beach, Nov. 2004.

120. Leveson, Nancy G. (January 2000). Leveson intent specifications: An approach to building human-centered specifications. *IEEE Transactions on Software Engineering* SE-26 (1):15–35.

121. Leveson, Nancy, and Jon Reese. TCAS intent specification. http://sunnyday.mit.edu/papers/tcas-intent.pdf.

122. Leveson, Nancy, Maxime de Villepin, Mirna Daouk, John Bellingham, Jayakanth Srinivasan, Natasha Neogi, Ed Bacheldor, Nadine Pilon, and Geraldine Flynn. A safety and human-centered approach to developing new air traffic management tools. *4th International Seminar or Air Traffic Management Research and Development*, Santa Fe, New Mexico, December 2001.

123. Leveson, N.G., M. P.E. Heimdahl, H. Hildreth, and J.D. Reese. Requirements specification for process-control systems. *Trans. on Software Engineering*, SE-20(9), September 1994.

124. Leveson, Nancy G., Nicolas Dulac, Karen Marais, and John Carroll. (February/March 2009). Moving beyond normal accidents and high reliability organizations: A systems approach to safety in complex systems. *Organization Studies* 30:227–249.

125. Leveson, Nancy, Nicolas Dulac, Betty Barrett, John Carroll, Joel Cutcher-Gershenfield, and Stephen Friedenthal. 2005. *Risk Analysis of NASA Independent Technical Authority. ESD Technical Report Series, Engineering Systems Division*. Cambridge, MA: MIT.

126. Levitt, R. E., and H. W. Parker. 1976. Reducing construction accidents—Top management's role. *Journal of the Construction Division* 102 (CO3):465–478.

127. Lihou, David A. 1990. Management styles—The effects of loss prevention. In *Safety and Loss Prevention in the Chemical and Oil Processing Industries*, ed. C. B. Ching, 147–156. Rugby, UK: Institution of Chemical Engineers.

128. London, E. S. 1982. Operational safety. In *High Risk Safety Technology*, ed. A. E. Green, 111–127. New York: John Wiley & Sons.

129. Lucas, D. A. 1987. Mental models and new technology. In *New Technology and Human Error*, ed. Jens Rasmussen, Keith Duncan, and Jacques Leplat, 321–325. New York: John Wiley & Sons.

130. Lutz, Robyn R. Analyzing software requirements errors in safety-critical, embedded systems. *Proceedings of the International Conference on Software Requirements*, IEEE, January 1992.

131. Machol, Robert E. (May 1975). The Titanic coincidence. *Interfaces* 5 (5):53–54.

132. Mackall, Dale A. Development and Flight Test Experiences with a Flight-Critical Digital Control System. NASA Technical Paper 2857, National Aeronautics and Space Administration, Dryden Flight Research Facility, November 1988.

133. Main Commission Aircraft Accident Investigation Warsaw. Report on the Accident to Airbus A320-211 Aircraft in Warsaw, September 1993.

134. Martin, John S. 2006. Report of the Honorable John S. Martin to the Special Committee of the Board of Directors of Merck & Company, Inc, Concerning the Conduct of Senior Management in the Development and Marketing of Vioxx., Debevoise & Plimpton LLP, September 2006.

135. Martin, Mike W., and Roland Schinzinger. 1989. *Ethics in Engineering*. New York: McGraw-Hill.

136. McCurdy, H. 1994. *Inside NASA: High Technology and Organizational Change in the U.S. Space Program*. Baltimore: Johns Hopkins University Press.

137. Miles, Ralph F., Jr. 1973. Introduction. In *Systems Concepts: Lectures on Contemporary Approaches to Systems*, ed. Ralph F. Miles, Jr., 1–12. New York: John F. Wiley & Sons.

138. Miller, C. O. 1985. A comparison of military and civil approaches to aviation system safety. *Hazard Prevention* (May/June):29–34.

139. Morgan, Gareth. 1986. *Images of Organizations*. New York: Sage Publications.

140. Mostrous, Alexi. Electronic medical records not seen as a cure-all: As White House pushes expansion, critics cite errors, drop-off in care. *Washington Post*, Sunday Oct. 25, 2009.

141. NASA Aviation Safety Reporting System Staff. Human factors associated with altitude alert systems. NASA ASRS Sixth Quarterly Report, NASA TM-78511, July 1978.

142. Nelson, Paul S. *A STAMP Analysis of the LEX Comair 5191 Accident*. Master's thesis, Lund University, Sweden, June 2008.

143. Norman, Donald A. 1990. The "problem" with automation: Inappropriate feedback and interaction, not "over-automation." In *Human Factors in Hazardous Situations*, ed. D. E. Broadbent, J. Reason, and A. Baddeley, 137–145. Oxford: Clarendon Press.

144. Norman, Donald A. (January 1981). Categorization of action slips. *Psychological Review* 88 (1):1–15.

145. Norman, Donald A. (April 1983). Design rules based on analyses of human error. *Communications of the ACM* 26 (4):254–258.

146. Norman, D. A. 1993. *Things That Make Us Smart*. New York: Addison-Wesley.

147. O'Connor, Dennis R. 2002. *Report of the Walkerton Inquiry*. Toronto: Ontario Ministry of the Attorney General.

148. Okie, Susan. 2005. What ails the FDA? *New England Journal of Medicine* 352 (11):1063–1066.

149. Orisanu, J., J. Martin, and J. Davison. 2007. Cognitive and contextual factors in aviation accidents: Decision errors. In *Applications of Naturalistic Decision Making*, ed. E. Salas and G. Klein, 209–225. Mahwah, NJ: Lawrence Erlbaum Associates.

150. Ota, Daniel Shuichiro. *Assuring Safety in High-Speed Magnetically Levitated (Maglev) Systems: The Need for a System Safety Approach*. Master's thesis, MIT, May 2008.

151. Owens, Brandon, Margaret Stringfellow, Nicolas Dulac, Nancy Leveson, Michel Ingham, and Kathryn Weiss. Application of a safety-driven design methodology to an outer planet exploration mission. *2008 IEEE Aerospace Conference*, Big Sky, Montana, March 2008.

152. Pate-Cornell, Elisabeth. (November 30, 1990). Organizational aspects of engineering system safety: The case of offshore platforms. *Science* 250:1210–1217.

153. Pavlovich, J. G. 1999. *Formal Report of the Investigation of the 30 April 1999 Titan IV B/Centaur TC-14/Milstar-3 (B32)*. U.S. Air Force.

154. Pereira, Steven J., Grady Lee, and Jeffrey Howard. A system-theoretic hazard analysis methodology for a non-advocate safety assessment of the ballistic missile defense system. *AIAA Missile Sciences Conference*, Monterey, CA, Nov. 2006.

155. Perrow, Charles. 1999. *Normal Accidents: Living with High-Risk Technology*. Princeton, NJ: Princeton University Press.

156. Perrow, Charles. 1986. The habit of courting disaster. *The Nation* (October):346–356.

157. Petersen, Dan. 1971. *Techniques of Safety Management*. New York: McGraw-Hill.

158. Pickering, William H. 1973. Systems engineering at the Jet Propulsion Laboratory. In *Systems Concepts: Lectures on Contemporary Approaches to Systems*, ed. Ralph F. Miles, Jr., 125–150. New York: John F. Wiley & Sons.

159. Piper, Joan L. 2001. *Chain of Events: The Government Cover-Up of the Black Hawk Incident and the Friendly Fire Death of Lt. Laura Piper*. London: Brasseys.

160. Psaty, Bruce, and Richard A. Kronmal. (April 16, 2008). Reporting mortality findings in trials of rofecoxib for Alzheimer disease or cognitive impairment: A case study based on documents from rofecoxib litigation. *Journal of the American Medical Association* 299 (15):1813.

161. Ramo, Simon. 1973. The systems approach. In *Systems Concepts: Lectures on Contemporary Approaches to Systems*, ed. Ralph F. Miles, Jr., 13–32. New York: John Wiley & Sons.

162. Rasmussen, Jens. Approaches to the control of the effects of human error on chemical plant safety. In *International Symposium on Preventing Major Chemical Accidents*, American Inst. of Chemical Engineers, February 1987.

163. Rasmussen, J. (March/April 1985). The role of hierarchical knowledge representation in decision making and system management. *IEEE Transactions on Systems, Man, and Cybernetics* SMC-15 (2):234–243.

164. Rasmussen, J. 1986. *Information Processing and Human–Machine Interaction: An Approach to Cognitive Engineering*. Amsterdam: North Holland.

165. Rasmussen, J. 1990. Mental models and the control of action in complex environments. In *Mental Models and Human–Computer Interaction*, ed. D. Ackermann and M. J. Tauber, 41–69. Amsterdam: North-Holland.

166. Rasmussen, Jens. 1990. Human error and the problem of causality in analysis of accidents. In *Human Factors in Hazardous Situations*, ed. D. E. Broadbent, J. Reason, and A. Baddeley, 1–12. Oxford: Clarendon Press.

167. Rasmussen, Jens. Risk management in a dynamic society: A modelling problem. *Safety Science* 27 (2/3) (1997):183–213.

168. Rasmussen, Jens, Keith Duncan, and Jacques Leplat. 1987. *New Technology and Human Error*. New York: John Wiley & Sons.

169. Rasmussen, Jens, Annelise Mark Pejtersen, and L. P. Goodstein. 1994. *Cognitive System Engineering*. New York: John Wiley & Sons.

170. Rasmussen, Jens, and Annelise Mark Pejtersen. 1995. Virtual ecology of work. In *An Ecological Approach to Human Machine Systems I: A Global Perspective*, ed. J. M. Flach, P. A. Hancock, K. Caird, and K. J. Vicente, 121–156. Hillsdale, NJ: Erlbaum.

171. Rasmussen, Jens, and Inge Svedung. 2000. *Proactive Risk Management in a Dynamic Society*. Stockholm: Swedish Rescue Services Agency.

172. Reason, James. 1990. *Human Error*. New York: Cambridge University Press.

173. Reason, James. 1997. *Managing the Risks of Organizational Accidents*. London: Ashgate.

174. Risk Management Pro. Citichem Syndicate: Introduction to the Transcription of the Accident Scenario, ABC Circle Films, shown on ABC television, March 2, 1986.

175. Roberts, Karlene. 1990. Managing high reliability organizations. *California Management Review* 32 (4):101–114.

176. Rochlin, Gene, Todd LaPorte, and Karlene Roberts. The self-designing high reliability organization. *Naval War College Review* 40 (4):76–91, 1987.

177. Rodriguez, M., M. Katahira, M. de Villepin, and N. G. Leveson. Identifying mode confusion potential in software design. *Digital Aviation Systems Conference*, Philadelphia, October 2000.

178. Rubin, Rita. How did the Vioxx debacle happen? *USA Today*, October 12, 2004.

179. Russell, Bertrand. 1985. *Authority and the Individual*. 2nd ed. London: Routledge.

180. Sagan, Scott. 1995. *The Limits of Safety*. Princeton, NJ: Princeton University Press.

181. Sarter, Nadine, and David Woods. (November 1995). How in the world did I ever get into that mode? Mode error and awareness in supervisory control. *Human Factors* 37 (1):5–19.

182. Sarter, Nadine N., and David Woods. Strong, silent, and out-of-the-loop. CSEL Report 95-TR-01, Ohio State University, February 1995.

183. Sarter, Nadine, David D. Woods, and Charles E. Billings. 1997. Automation surprises. In *Handbook of Human Factors and Ergonomics*, 2nd ed., ed. G. Salvendy, 1926–1943. New York: Wiley.

184. Schein, Edgar. 1986. *Organizational Culture and Leadership*. 2nd ed. New York: Sage Publications.

185. Senge, Peter M. 1990. *The Fifth Discipline: The Art and Practice of Learning Organizations*. New York: Doubleday Currency.

186. Shappell, S., and D. Wiegmann. The Human Factors Analysis and Classification System—HFACS. Civil Aeromedical Medical Institute, Oklahoma City, OK, Office of Aviation Medicine Technical Report COT/FAA/AN-00/7, 2000.

187. Sheen, Barry. 1987. *Herald of Free Enterprise Report Marine Accident Investigation Branch, Department of Transport (originally Report of Court No 8074 Formal Investigation)*. London: HMSO.

188. Shein, Edgar. 2004. *Organizational Culture and Leadership*. San Francisco: Jossey-Bass.

189. Shockley-Zabalek, P. 2002. *Fundamentals of Organizational Communication*. Boston: Allyn & Bacon.

190. Smith, Sheila Weiss. 2007. Sidelining safety—The FDA's inadequate response to the IOM. *New England Journal of Medicine* 357 (10):960–963.

191. Snook, Scott A. 2002. *Friendly Fire: The Accidental Shootdown of U.S. Black Hawks Over Northern Iraq*. Princeton, NJ: Princeton University Press.

192. Staff, Spectrum. 1987. Too much, too soon. *IEEE Spectrum* (June):51–55.

193. Stephenson, A. Mars Climate Orbiter: Mishap Investigation Board Report. NASA, November 10, 1999.

194. Sterman, John D. 2000. *Business Dynamics*. New York: McGraw-Hill.

195. Stringfellow, Margaret. *Human and Organizational Factors in Accidents.* Ph.D. Dissertation, Aeronautics and Astronautics, MIT, 2010.

196. Swaanenburg, H. A. C., H. J. Swaga, and F. Duijnhouwer. 1989. The evaluation of VDU-based man-machine interfaces in process industry. In *Analysis, Design, and Evaluation of Man-Machine Systems,* ed. J. Ranta, 71–76. New York: Pergamon Press.

197. Taylor, Donald H. 1987. The role of human action in man-machine system errors. In *New Technology and Human Error,* ed. Jens Rasmussen, Keith Duncan, and Jacques Leplat, 287–292. New York: John Wiley & Sons.

198. Taylor, J. R. 1982. An integrated approach to the treatment of design and specification errors in electronic systems and software. In *Electronic Components and Systems,* ed. E. Lauger and J. Moltoft, 87–93. Amsterdam: North Holland.

199. Thomas, John, and Nancy Leveson. 2010. *Analyzing Human Behavior in Accidents.* MIT Research Report, Engineering Systems Division.

199a. Thomas, John and Nancy Leveson. 2011. Performing hazard analysis on complex software and human-intensive systems. In *Proceedings of the International System Safety Society Conference,* Las Vegas.

200. U.S. Government Accounting Office, Office of Special Investigations. 1997. *Operation Provide Comfort: Review of Air Force Investigation of Black Hawk Fratricide Incident (GAO/T-OSI-98–13).* Washington, DC: U.S. Government Printing Office.

201. Vicente, Kim J. 1995. *A Field Study of Operator Cognitive Monitoring at Pickering Nuclear Generating Station. Technical Report CEL 9504, Cognitive Engineering Laboratory.* University of Toronto.

202. Vicente, Kim J. 1999. *Cognitive Work Analysis: Toward Safe, Productive, and Healthy Computer-Based Work.* Mahwah, NJ: Lawrence Erlbaum Associates.

203. Vicente, Kim J., and J. Rasmussen. Ecological interface design: Theoretical foundations. *IEEE Trams. on Systems, Man, and Cybernetics* 22 (4) (July/August 1992).

204. Watt, Kenneth E.F. 1974. *The Titanic Effect.* Stamford, CT: Sinauer Associates.

205. Weick, Karl E. 1987. Organizational culture as a source of high reliability. *California Management Review* 29 (2):112–127.

206. Weick, Karl E. 1999. K. Sutcliffe, and D. Obstfeld. Organizing for high reliability. *Research in Organizational Behavior* 21:81–123.

207. Weinberg, Gerald. 1975. *An Introduction to General Systems Thinking.* New York: John Wiley & Sons.

208. Weiner, E.L. *Human Factors of Advanced Technology ("Glass Cockpit") Transport Aircraft.* NASA Contractor Report 177528, NASA Ames Research Center, June 1989.

209. Weiner, Earl L., and Renwick E. Curry. 1980. Flight-deck automation: Promises and problems. *Ergonomics* 23 (10):995–1011.

210. Wiener, Norbert. 1965. *Cybernetics: or the Control and Communication in the Animal and the Machine.* 2nd ed. Cambridge, MA: MIT Press.

211. Weiss, Kathryn A. *Building a Reusable Spacecraft Architecture Using Component-Based System Engineering.* Master's thesis, MIT, August 2003.

212. Wong, Brian. *A STAMP Model of the Überlingen Aircraft Collision Accident.* S.M. thesis, Aeronautics and Astronautics, MIT, 2004.

213. Woods, David D. Some results on operator performance in emergency events. In *Ergonomic Problems in Process Operations,* ed. D. Whitfield, Institute of Chemical Engineering Symposium, Ser. 90, 1984.

214. Woods, David D. Lessons from beyond human error: Designing for resilience in the face of change and surprise. Design for Safety Workshop, NASA Ames Research Center, October 8–10, 2000.

215. Young, Thomas (Chairman). Mars Program Independent Assessment Team Report. NASA, March 2000.

216. Young, T. Cuyler. 1975. Pollution begins in prehistory: The problem is people. *Man in Nature: Historical Perspectives on Mara in His Environment,* ed. Louis D. Levine. Toronto: Royal Ontario Museum.

217. Zsambok, Caroline E., and Gary Klein, eds. 1997. *Naturalistic Decision Making.* Mahwah, NJ: Lawrence Erlbaum Associates.

Index